The UMTS Air-Interface in RF Engineering

ABOUT THE AUTHOR

Shing-Fong Su, Ph.D., is a distinguished member of the technical staff at Alcatel-Lucent (formerly Lucent Technologies, Inc.). He has been involved in the network design, deployment, and optimization of CDMA, GSM, and UMTS systems in North America, Taiwan, and China. Previously, he was an electrical engineering professor at the University of South Florida. He was also a senior member of the technical staff at GTE Labs. Dr. Su is the holder of seven U.S. patents and the author of over 70 papers in communications and lightwave technology. He also served as a reviewer for many industry publications, including *IEEE Transactions on Communications, Journal of Lightwave Technology, Applied Optics,* and *Electronics Letters.*

The UMTS Air-Interface in RF Engineering: Design and Operation of UMTS Networks

Shing-Fong Su, Ph.D.

Distinguished Member of Technical Staff, Alcatel-Lucent

New York   Chicago   San Francisco   Lisbon   London   Madrid
Mexico City   Milan   New Delhi   San Juan   Seoul
Singapore   Sydney   Toronto

The McGraw·Hill Companies

Library of Congress Cataloging-in-Publication Data

Su, Shing-Fong.
  The UMTS air-interface in RF engineering / Shing-Fong Su.
    p. cm.
  ISBN 0-07-148866-9 (alk. paper)
  1. Radio circuits--Design and construction.   2. Radio frequency.   I. Title.
TK6560.S87 2007
621.384'12--dc22

2007008339

McGraw-Hill books are available at special quantity discounts to use as premiums and sales promotions, or for use in corporate training programs. For more information, please write to the Director of Special Sales, Professional Publishing, McGraw-Hill, Two Penn Plaza, New York, NY 10121-2298. Or contact your local bookstore.

The UMTS Air-Interface in RF Engineering:
Design and Operation of UMTS Networks

Copyright © 2007 by The McGraw-Hill Companies. All rights reserved. Printed in the United States of America. Except as permitted under the Copyright Act of 1976, no part of this publication may be reproduced or distributed in any form or by any means, or stored in a database or retrieval system, without the prior written permission of publisher.

1234567890 DOC/DOC 01987

ISBN-13: 978-0-07-148866-2
ISBN-10:   0-07-148866-9

Sponsoring Editor Steve Chapman	**Copy Editor** Julie M. Smith	**Composition** International Typesetting and Composition
Editorial Supervisor Jody McKenzie	**Proofreader** Bev Weiler	**Illustration** International Typesetting and Composition
Project Manager Vastavikta Sharma	**Indexer** Shing-Fong Su	
Acquisitions Coordinator Katie Andersen	**Production Supervisor** George Anderson	

Information has been obtained by McGraw-Hill from sources believed to be reliable. However, because of the possibility of human or mechanical error by our sources, McGraw-Hill, or others, McGraw-Hill does not guarantee the accuracy, adequacy, or completeness of any information and is not responsible for any errors or omissions or the results obtained from the use of such information.

Contents

Preface xv
Acknowledgments xix
Nomenclature xxi

Chapter 1. Introduction to UMTS 1

1.1 What Is UMTS? 1
1.2 Evolution of WCDMA 2
1.3 UMTS Services 3
1.4 UMTS Network 4
    1.4.1 Core network 4
    1.4.2 UTRAN 5
    1.4.3 User equipment 6
1.5 UMTS Frequency Band and Channel Arrangement 6
    1.5.1 Frequency Bands 6
    1.5.2 UTRA/FDD Transmit-To-Receive (Tx-Rx) Frequency Separation 7
    1.5.3 Channel Arrangement 7
    1.5.4 Carrier Frequency 7
    1.5.5 UARFCN 8
1.6 Organization of the Book 8
References 10

Chapter 2. UMTS Fundamentals 11

2.1 UMTS Network Topology 11
2.2 UMTS Signaling Protocol Stack 12
    2.2.1 Circuit-Switched Control Plane Protocol Stack 13
    2.2.2 Packet-Switched Control Plane Protocol Stack 14
    2.2.3 Circuit-Switched User Plane Protocol Stack 14
    2.2.4 Packet-Switched User Plane Protocol Stack 15
2.3 Access Stratum Data Flow 15
    2.3.1 RRC Functions 15
    2.3.2 RLC Functions 16
    2.3.3 MAC Functions 16
    2.3.4 Physical Layer Functions 16
2.4 UMTS Channels 17

Contents

2.5	Channel Mappings	18
	2.5.1 Channel Mapping of BCCH	19
	2.5.2 Channel Mapping of PCCH	19
	2.5.3 Channel Mapping of CCCH to RACH/FACH Common Channels in Idle Mode	20
	2.5.4 Channel Mapping of DCCH and DTCH to RACH/FACH Common Channels in Connected Mode	20
	2.5.5 Channel Mapping of Dedicated Channels	21
	2.5.6 Channel Mappings of CPCH, DSCH, and HS-DSCH	22
2.6	Protocol States	24
	2.6.1 Idle Mode	24
	2.6.2 Connected Mode	25
2.7	UE and Subscriber Identifiers	29
	2.7.1 International Mobile Subscriber Identity	29
	2.7.2 Temporary Mobile Subscriber Identity	30
	2.7.3 Radio Network Temporary Identity	30
	2.7.4 International Mobile Station Equipment Identity	31
2.8	System Frame Timing	31
	2.8.1 System Frame Number	31
	2.8.2 Connection Frame Number	32
2.9	Summary	32
	References	34

Chapter 3. Overview of 3GPP and UMTS Standards — 35

3.1	Technical Specification Groups	36
	3.1.1 Services and System Aspects Group	36
	3.1.2 Core Network and Terminals Group	36
	3.1.3 GSM EDGE Radio Access Network Group	36
	3.1.4 Radio Access Network Group	37
3.2	3GPP Specification Releases	37
3.3	3GPP Specification Numbering Scheme	37
3.4	3GPP Specification Series	37
3.5	Summary	38
	Reference	38

Chapter 4. Radio Resource Control — 39

4.1	RRC Message Specifications	39
	4.1.1 Protocol Extensions	40
4.2	System Information	40
	4.2.1 System Information Blocks	42
	4.2.2 System Information Block Segmentation and Concatenation	45
	4.2.3 Example of a System Information Message	47
	4.2.4 Contents of System Information Blocks	47
4.3	Paging and Notification	55
	4.3.1 Paging Type 1	55
	4.3.2 Paging Type 2	58
4.4	RRC Connection Management	58
	4.4.1 RRC Connection Request	59
	4.4.2 RRC Connection Setup	60
	4.4.3 RRC Connection Setup Complete	61
	4.4.4 RRC Connection Release	63

4.5 Ciphering and Integrity Protection Control 63
    4.5.1 Security Mode Command Message 64
4.6 Radio Bearer Control 65
    4.6.1 Radio Bearer Establishment 65
    4.6.2 Radio Bearer Reconfiguration 66
    4.6.3 Radio Bearer Release 66
4.7 RRC Management of UE Mobility 67
    4.7.1 Cell Reselection 67
    4.7.2 Cell Update and URA Update Procedures 68
    4.7.3 Active Set Update Procedure 69
    4.7.4 Inter-RAT Mobility 69
4.8 Measurements and Reporting 71
    4.8.1 Measurement Control Message 71
    4.8.2 Quality Measurement 72
    4.8.3 UE Internal Measurements 72
4.9 NAS Message Routing 73
4.10 Summary 74
    References 76

Chapter 5. Radio Link Control   77

5.1 Layer 2 Fundamentals 77
5.2 RLC Functions 79
5.3 RLC Entities 79
5.4 RLC Data Transfer Modes 80
    5.4.1 RLC Transparent Mode 80
    5.4.2 RLC Unacknowledged Mode 83
    5.4.3 RLC Acknowledged Mode 85
5.5 RLC Ciphering 89
5.6 RLC Configurable Parameters 90
    5.6.1 SDU Discard 90
    5.6.2 AM Configurable Parameters 91
5.7 Summary 92
    References 93

Chapter 6. Medium Access Control   95

6.1 MAC Architecture 95
6.2 Logical to Transport Channel Mappings 96
6.3 MAC Header 97
    6.3.1 MAC Header for Dedicated Logical Channels 98
    6.3.2 MAC Header for Common Transport Channels 100
6.4 Transport Format Combination Selection 102
6.5 Traffic Volume Measurement 102
6.6 MAC Ciphering 103
6.7 MAC RACH Functions 103
    6.7.1 MAC RACH Procedure 104
    6.7.2 Access Class and Access Service Class 105
    6.7.3 Persistency Check 106
6.8 MAC Configurable Parameters 107
6.9 Summary 108
    References 109

Chapter 7. Physical Layer — 111

- 7.1 Orthogonal Spreading Codes — 112
  - 7.1.1 Orthogonal Sequences — 114
  - 7.1.2 Spreading and De-spreading — 114
- 7.2 Scrambling Codes — 118
  - 7.2.1 Maximum Length Pseudorandom Binary Sequences — 118
  - 7.2.2 Gold Codes — 118
  - 7.2.3 Scrambling Code Generation — 119
- 7.3 Synchronization Codes — 119
- 7.4 Physical Layer Timing — 120
- 7.5 Downlink Procedure — 121
  - 7.5.1 Transport Channel Data Delivery to Physical Channels — 122
  - 7.5.2 CRC Attachment — 126
  - 7.5.3 Transport Block Concatenation and Code Block Segmentation — 127
  - 7.5.4 Channel Coding — 127
  - 7.5.5 Rate Matching — 129
  - 7.5.6 First DTX Insertion — 130
  - 7.5.7 First Interleaving and Radio Frame Segmentation — 131
  - 7.5.8 Transport Channel Multiplexing and Second DTX Insertion — 132
  - 7.5.9 Second Interleaving — 132
  - 7.5.10 Mapping to physical channel — 134
  - 7.5.11 Spreading and Scrambling — 134
  - 7.5.12 Modulation — 136
- 7.6 Uplink Procedure — 136
  - 7.6.1 Radio Frame Equalization — 137
  - 7.6.2 Rate Matching — 137
  - 7.6.3 Mapping to Physical Channels — 138
  - 7.6.4 Spreading and Scrambling — 138
- 7.7 Physical Channel Structures and Channel Timing — 139
  - 7.7.1 PCCPCH — 139
  - 7.7.2 SCCPCH — 140
  - 7.7.3 SCH — 143
  - 7.7.4 CPICH — 148
  - 7.7.5 PICH — 149
  - 7.7.6 AICH — 150
  - 7.7.7 PRACH — 151
  - 7.7.8 Downlink DPCH — 154
  - 7.7.9 Uplink DPCH — 156
  - 7.7.10 Physical Channel Timing — 158
- 7.8 Physical Layer Procedures — 159
  - 7.8.1 Initial Acquisition Procedure — 159
  - 7.8.2 Physical Random Access Procedure — 160
  - 7.8.3 Page Procedure — 163
  - 7.8.4 DPDCH/DPCCH Synchronization — 164
  - 7.8.5 Radio Link Establishment and Radio Link Failure — 165
  - 7.8.6 Measurements — 168
- 7.9 Summary — 172
- References — 175

Chapter 8. Cell Reselection — 177

- 8.1 Types of Cell Reselection — 177
- 8.2 Cell Reselection Fundamentals — 179
    - 8.2.1 Cell Reselection Criteria — 179
    - 8.2.2 Cell Reselection Ranking Process — 181
    - 8.2.3 Inter-Frequency Cell Reselection — 183
    - 8.2.4 Inter-RAT Cell Reselection — 183
- 8.3 Summary — 184
- References — 184

Chapter 9. Handover — 185

- 9.1 UE Measurements and Reporting — 186
    - 9.1.1 UE Measurements for Handover — 187
- 9.2 Categories of Cells — 188
- 9.3 Soft and Softer Handover — 189
    - 9.3.1 Soft and Softer Handovers at Call Setup — 189
    - 9.3.2 Soft and Softer Handovers in Cell_DCH State — 190
- 9.4 Inter-Frequency Handover — 198
    - 9.4.1 Virtual Active Set — 198
    - 9.4.2 Inter-Frequency Handover Procedure — 198
    - 9.4.3 Inter-Frequency Reporting Events — 199
- 9.5 Inter-RAT Handover — 204
    - 9.5.1 Inter-RAT Handover Triggering Events — 204
- 9.6 Compressed Mode — 206
    - 9.6.1 Compressed Mode Basics — 206
    - 9.6.2 Transmission Gap Pattern Sequence — 208
    - 9.6.3 Transmission Gap Pattern — 209
- 9.7 Summary — 212
- References — 214

Chapter 10. Power Control — 215

- 10.1 Power Control for Downlink Dedicated Channels — 216
    - 10.1.1 Open Loop Power Control — 216
    - 10.1.2 Closed Loop Power Control — 217
- 10.2 Power Control for Downlink Common Channels — 225
    - 10.2.1 Common Pilot Channel and Synchronization Channel Power Levels — 226
    - 10.2.2 Primary Common Control Physical Channel Power level — 227
    - 10.2.3 Secondary Common Control Physical Channel Power Level — 227
    - 10.2.4 Paging Indicator Channel and Acquisition Indicator Channel Power Levels — 228
- 10.3 Power Control for Uplink Dedicated Channels — 229
    - 10.3.1 Initial Transmit Power for Uplink Dedicated Channels — 229
    - 10.3.2 Outer Loop Power Control for Uplink Dedicated Channels — 230
    - 10.3.3 Inner Loop Power Control for Uplink Dedicated Channels — 232
- 10.4 Power Control for Uplink Common Channels — 239
    - 10.4.1 Initial Preamble Power — 239
    - 10.4.2 Successive Preamble Power — 239
    - 10.4.3 RACH Message Part Power — 240

10.5	Power Control in Compressed Mode	240
	10.5.1 Downlink Power Control in Compressed Mode	241
	10.5.2 Uplink Power Control in Compressed Mode	242
10.6	Summary	244
	References	246

Chapter 11. HSDPA Overview — 247

11.1	HSDPA Key Features	248
	11.1.1 Adaptive Modulation and Coding	248
	11.1.2 Transmission Time Interval	250
	11.1.3 Scheduling	250
	11.1.4 Retransmission	251
	11.1.5 Code Allocation and Code Multiplexing of Packet Transmissions	253
	11.1.6 Power Allocation	254
	11.1.7 No Downlink Soft Handover	254
	11.1.8 HSDPA UE Capability	255
11.2	HSDPA Channels	256
	11.2.1 HS-DSCH	256
	11.2.2 HS-PDSCH	256
	11.2.3 HS-SCCH	257
	11.2.4 HS-DPCCH	258
11.3	HSDPA Physical Layer Procedure	260
	11.3.1 CQI Reporting	261
11.4	HSDPA Related Parameters	261
	11.4.1 Cell Parameters	262
	11.4.2 UE Parameters	262
	11.4.3 Fixed Parameters	265
11.5	General Considerations for HSDPA Deployment	265
	11.5.1 Deployment Strategies	265
	11.5.2 Deployment Options	265
	11.5.3 User Mobility	265
	11.5.4 Impact of HSDPA on Release 99 Networks	266
11.6	Summary	267
	References	269

Chapter 12 WCDMA RF Network Planning — 271

12.1	Capacity and Coverage	271
12.2	Uplink Analysis	272
	12.2.1 Bit Rate	272
	12.2.2 Traffic Load	272
	12.2.3 Processing Gain	273
	12.2.4 Required Eb/Io	273
	12.2.5 Propagation Environment	273
	12.2.6 Node B Receiver Noise Figure	274
	12.2.7 Receiver Sensitivity, Pole Capacity and Loading	274
	12.2.8 Noise Rise Due to Interference	276
	12.2.9 Antenna Gain, Feeder Cable Loss, and Body Loss	277
	12.2.10 Shadow Fading, Coverage Probability, and Shadow Fade Margin	278

12.2.11	Fast Fade Margin	280
12.2.12	Soft Handover Gain	281
12.2.13	UE Transmit Power	282
12.2.14	Penetration Loss	282
12.2.15	Uplink Budget	282
12.3	Propagation Models	283
12.3.1	Okumura Model	283
12.3.2	Hata Model	285
12.3.3	COST 231-Hata Model	285
12.3.4	Lee Model	286
12.4	Downlink Analysis	286
12.5	Overhead Channel Power Allocation	287
12.6	Scrambling Code Planning	289
12.6.1	Scrambling Code Planning Example	290
12.7	Base Station Antennas	291
12.7.1	Antenna Gain and Beam Width	291
12.7.2	Antenna Down Tilt	292
12.7.3	Side Lobe Suppression and Null Fill	292
12.7.4	Dual Polarization Antenna	293
12.7.5	Voltage Standing Wave Ratio and Front-to-Back Ratio	293
12.7.6	Mechanical Specifications of Antennas	294
12.8	WCDMA RF Planning Process	294
12.9	Summary	295
	References	297

Chapter 13. WCDMA RF Network Optimization   299

13.1	RF Optimization Overview	300
13.2	Issues in RF Optimization	300
13.2.1	Cell Breathing	301
13.2.2	Pilot Pollution	301
13.2.3	Near-Far Problem	301
13.2.4	Around-the-Corner Problem	302
13.2.5	Handover Problem	302
13.2.6	Incomplete Neighbor List	302
13.3	Pre-Optimization Preparation	303
13.3.1	Hardware Check	303
13.3.2	Antenna Audit	303
13.4	RF Drive-Test-Based Optimization	303
13.4.1	RF Optimization Planning	303
13.4.2	Sector Verification	305
13.4.3	Cluster Optimization	306
13.4.4	System Verification	309
13.4.5	RF Optimization Tools	310
13.4.6	WCDMA Performance Metrics	310
13.5	Traffic-Statistics-Based Optimization	312
13.5.1	Traffic Statistics Data Collection and Processing	312
13.5.2	Key Traffic Statistics Metrics for WCDMA Networks	312
13.6	Summary	313

Chapter 14. Applications of Repeaters and Tower Mounted Amplifiers in WCDMA Networks — 315

- 14.1 Repeater Engineering Considerations — 316
  - 14.1.1 Repeater Coverage Objectives — 316
  - 14.1.2 Base Station Desensitization — 317
  - 14.1.3 Repeater Noise Figure Rise — 318
  - 14.1.4 Donor Link Characteristics — 318
  - 14.1.5 Pilot Discrimination — 319
  - 14.1.6 Antenna Isolation and Gain Setting — 319
  - 14.1.7 Handover Issue — 319
  - 14.1.8 Donor Cell Traffic Overload — 320
  - 14.1.9 Narrow-Band Interference Amplified by Repeaters — 320
- 14.2 Major Repeater-Related Problems in WCDMA Networks — 320
  - 14.2.1 Unable to Make a Call in the Repeater Coverage Area — 321
  - 14.2.2 High Dropped-Call Rate in the Repeater Coverage Area — 321
  - 14.2.3 High Mobile Station Transmit Power in the Repeater Coverage Area — 321
  - 14.2.4 Repeater Coverage Area Smaller Than Expected — 322
  - 14.2.5 Frequent Handovers Within the Repeater Coverage Area — 322
  - 14.2.6 Unable to Make Handovers with Other Cells — 322
  - 14.2.7 Long Access Time in the Repeater Coverage Area — 322
  - 14.2.8 Donor Cell Coverage Area Shrinkage — 323
  - 14.2.9 High Dropped-Call Rate and High Mobile Station Transmit Power in the Donor Cell Coverage Area — 323
  - 14.2.10 Pilot Pollution — 323
- 14.3 Repeater Deployment Guidelines — 323
  - 14.3.1 Selection of Repeater Locations — 324
  - 14.3.2 Repeater Installation — 324
  - 14.3.3 General Adjustments of Repeaters After Installation — 324
- 14.4 Tower Mounted Amplifiers — 325
  - 14.4.1 Analysis on the Improvement of Base Station Noise Figure and Link Budget — 326
  - 14.4.2 Applications of TMAs — 326
- 14.5 Summary — 327

Chapter 15. Intersystem Interferences — 329

- 15.1 Adjacent Channel Performance — 329
  - 15.1.1 Adjacent Channel Interference Power Ratio — 329
  - 15.1.2 Adjacent Channel Leakage Power Ratio — 330
  - 15.1.3 Adjacent Channel Selectivity — 331
  - 15.1.4 Relation Between ACIR, ACLR, and ACS — 332
- 15.2 Interferences Between UMTS and CDMA2000 Systems — 332
- 15.3 Interferences Between UMTS and PHS Systems — 335
  - 15.3.1 Analysis of the Interference Between PHS Base Station Transmitter and UMTS Base Station Receiver — 335
  - 15.3.2 Simulation of the Interferences Between UMTS and PHS Systems — 340
- 15.4 Interferences Between UMTS and GSM systems — 341
  - 15.4.1 Isolation Requirement for Spurious Emissions — 341
  - 15.4.2 Isolation Requirement for Third-Order Inter-Modulation — 342
  - 15.4.3 Isolation Requirement for Carrier Overload — 342
  - 15.4.4 Overall Isolation Requirement — 342
- 15.5 Summary — 343
- References — 343

Chapter 16. Comparison of WCDMA and CDMA2000 — 345

16.1 Similarities Between WCDMA and CDMA2000 — 345
    16.1.1 Physical Layer Concepts — 345
    16.1.2 Physical Layer Procedures — 345
    16.1.3 Channelization and Spreading Concepts — 346
    16.1.4 Power Control — 346
    16.1.5 Physical Channels with Similar Functions — 346
    16.1.6 Different Terminologies for Similar Functions — 346

16.2. Differences Between WCDMA and CDMA2000 — 347
    16.2.1 Network Synchronization — 347
    16.2.2 RF Characteristics — 347
    16.2.3 Channel Structure — 348
    16.2.4 Overhead — 348
    16.2.5 Paging Operation — 348
    16.2.6 Inter-Frequency and Inter-RAT Hard Handover — 348

16.3 Summary — 349

Index — 351

Preface

UMTS technology is no doubt one of the top contenders in the arena of third-generation mobile communications. The 3GPP organization has developed and published numerous specifications on UMTS for the telecommunication communities to follow. The 3GPP specifications start with Release 99 and continue on to Release 4, Release 5, Release 6, and beyond. Network equipment and mobile vendors have developed or are developing UMTS equipment and devices complying with the 3GPP specifications in order to ensure the interoperability of different equipment vendors' systems. Currently, many UMTS systems have been deployed in a number of markets worldwide and more are to come in the near future.

The main objective of this book is to provide wireless telecommunication professionals and students with a basic understanding of UMTS air interface and RF engineering. Understanding the UMTS air interface and RF engineering is indispensable for RF system and field engineers to do a good job in planning, designing, deploying, optimizing, operating, and maintaining UMTS networks. Although the UMTS air interface has been described in detail in 3GPP specifications, many engineers and managers find that reading through all the specifications can be impractical. Faced with an explosive amount of detailed information, engineers may not have the time and luxury to absorb all air interface specifications in a comprehensive way and keep up with their daily work. They need a book that explains the key parts of the UMTS air interface in an organized fashion so that they can grasp the air interface concepts easily and quickly. In addition to air interface, many engineers and managers also desire to have a book that can provide them with knowledge in all aspects of RF engineering so that they can do their jobs more efficiently. These are the two main reasons that prompted me to write this book.

This book presents not only theoretical discussions but also engineering practices. This is especially true in the area of RF engineering, where the material presented is essentially based on my own engineering experience. The integration of both theory and practice provides one-stop shopping for the required knowledge that engineers and managers need for understanding the UMTS air interface and for designing, deploying, optimizing, operating, and maintaining UMTS networks. Aside from being used as a handbook for engineers and managers, this book may be utilized as a reference book for engineering students as well.

As I kept readability and succinctness in mind when writing this book, I omitted many details of the UMTS air interface described in the 3GPP specifications. However, I adopted the same symbols and terminologies defined in the 3GPP specifications to avoid confusion wherever applicable. In addition, some figures, tables, and sentences are extracted from 3GPP TSs and TRs with ETSI's permission and are indicated by "courtesy of ETSI." The following statement applies to all material extracted from 3GPP TSs and TRs: "3GPP™ TSs and TRs are the property of ARIB, ATIS, ETSI, CCSA, TTA, and TTC who jointly own the copyright in them. They are subject to further modification and are therefore provided to you 'as is' for information purposes only. Further use is strictly prohibited."

UMTS may encompass different access technologies. This book only focuses on the UMTS system with a WCDMA access network operating in FDD mode. There are many concepts that apply not only to UMTS technology but also to other technologies. Therefore, UMTS specific terminologies and general terminologies are used interchangeably throughout this book. For example, Node B and base station are used synonymously, as are UE and mobile.

This book contains 16 chapters. Following is a summary of the topics covered in each one:

- Chapter 1 presents a brief introduction of UMTS, including definitions of UMTS, WCDMA evolution, UMTS services, and UMTS network architecture.

- Chapter 2 describes in detail the UMTS basics, including the protocol stack, UMTS signaling, control planes, user planes, UMTS channels and channel mapping, frame timing, system frame number, connection frame number, UE call states, and UE/subscriber identifiers. Chapter 3 briefly introduces the 3GPP organization and UMTS standards, including standard evolution and standard Releases, with emphasis on radio access specifications.

- Chapters 4 to 7 deal with UMTS access stratum, including Radio Resource Control (RRC), Radio Link Control (RLC), Medium Access Control (MAC), and Physical Layer concepts.

- Chapter 8 depicts cell reselection in detail.

- Chapter 9 discusses handover mechanisms, including soft handover, softer handover, and hard handover. Also discussed in this chapter are measurement control and reporting, and the compressed mode.
- Chapter 10 talks about UMTS power control.
- Chapter 11 offers a brief overview of High-Speed Downlink Packet Access (HSDPA).
- Chapter 12 presents WCDMA radio network planning.
- Chapter 13 gives general concepts on WCDMA RF network optimization, including RF network optimization procedures, drive test, and data analysis.
- Chapter 14 describes the application of repeaters and tower-mounted amplifiers in WCDMA systems. It addresses the engineering considerations for the usage of repeaters and tower top amplifiers in WCDMA networks from a practical point of view and provides some guidelines for repeater deployment.
- Chapter 15 discusses intersystem interference, including interference between WCDMA and CDMA2000, between WCDMA and GSM, and between WCDMA and PHS systems.
- Finally, Chapter 16 gives a comparison of WCDMA and CDMA2000. Similarities and differences of the two systems are examined.

—Shing-Fong Su

Acknowledgments

Many people have helped me in the process of writing this book. Firstly, I would like to thank Sarah Chan. Two years ago, Sarah inspired me to write this book with her suggestion that I should utilize my expertise in the RF engineering field to write a book on UMTS. As my manager, she provided continuous support and encouragement during the past two years. This book would not have been possible without her assistance and the simultaneous support of all the members of the RF engineering group headed by her.

Secondly, I would like to thank Chuck Adelman, Rick Shaw, Gopal Jaisingh, and Talmage Bursh for shepherding this book through the approval process, and my thanks also goes to Paul Mankiewich for granting the final approval. I would also like to thank the European Telecommunications Standards Institute (ETSI) for permitting the usage of many of the figures and tables in the 3GPP specifications.

My colleague, Li Meng, took the time to patiently read the manuscript and to respond with numerous suggestions and corrections. I would like to express my thanks for her contributions. I also thank the anonymous reviewers who provided me with feedback.

I thank Kang Lee and Kelvin Ho for providing me with the opportunity to work with the Lucent China RF engineering and system engineering teams in China. The numerous technical questions that were posed by the teams for me to answer generated an excellent environment for me to enhance my knowledge.

The McGraw-Hill Companies and International Typesetting and Composition provided great support for the publication of this book. I would like to thank both companies, especially Steve Chapman, Jody McKenzie, Vastavikta Sharma, and Julie M. Smith for their assistance during the publication and production processes of this book.

Last but not least, I am indebted to my family. I am profoundly grateful to my wife, Bih-Hwa, for her love and support through the years. Without her encouragement and support of my professional endeavors, I would not have had the capacity and concentration to complete this book in a timely manner.

Nomenclature

16-QAM	16-Quadrature Amplitude Modulation
1x-EVDO	1x Evolution Data Optimized
3G	3rd Generation
3GPP	3rd Generation Partnership Project
AC	access class
ACIF	Australian Communications Industry Forum
ACIR	adjacent channel interference power ratio
ACK	acknowledgment
ACLR	adjacent channel leakage power ratio
ACS	adjacent channel selectivity
AI	acquisition indicator
AICH	acquisition indicator channel (physical channel)
AM	acknowledged mode
AMD	acknowledged mode data
AMR	adaptive multi-rate
AN	access network
ARFCN	absolute radio frequency channel number
ARIB	Association of Radio Industries and Business, Japan
ARQ	automatic repeat request
AS	access stratum
ASC	access service class
ASN.1	abstract syntax notation one
ATIS	Alliance for Telecommunications Industry Solutions, North America
AuC	authentication center
AWGN	additive white Gaussian noise
BCCH	broadcast control channel (logical channel)
BCH	broadcast channel (transport channel)
BER	bit error rate
BLER	block error rate
BMC	broadcast/multicast control
BPSK	binary phase shift keying
BS	base station
BSC	base station controller
BSIC	base station identification code
BTS	base transceiver subsystem
BTSD	base station desensitization
BWAF	bandwidth adjustment factor

Nomenclature

C-RNTI	cell radio network temporary identity
CBS	cell broadcast service
CC	call control
CCCH	common control channel (logical channel)
CCITT	Consultative Committee on International Telephony and Telegraphy
CCPCH	common control physical channel
CCSA	China Communications Standards Association, China
CCTrCh	coded composite transport channel
CDF	cumulative distribution function
CDMA	code division multiple access
CFN	connection frame number
C/I	carrier-to-interference ratio
CIO	cell individual offset
CK	cipher key
CM	connection management; compressed mode
CN	core network
CPCH	common packet channel (transport channel)
CPICH	common pilot channel (physical channel)
CQI	channel quality indicator
CRC	cyclic redundancy check
CRCI	cyclic redundancy check indicator
CS	circuit switched; PHS base station
CTCH	common traffic channel (logical channel)
Cu	interface for mobile equipment and USIM
CWTS	China Wireless Telecommunications Standard Group
dB	Decibel
D/C	data/control
DCCH	dedicated control channel (logical channel)
DCH	dedicated channel (transport channel)
DGPS	differential global positioning system
DL	downlink
DOFF	DPCH offset
DPCCH	dedicated physical control channel (physical channel)
DPCH	dedicated physical channel
DPDCH	dedicated physical data channel (physical channel)
DRAC	dynamic resource allocation control
DRX	discontinuous reception
DSCH	downlink shared channel (transport channel)
DS-CDMA	direct spread CDMA
DTCH	dedicated traffic channel (logical channel)
DTX	discontinuous transmission
E	extension bit
EDGE	enhanced data rates for GSM evolution
EFR	enhanced full rate
EIRP	equivalent isotropic radiated power
EM	electromagnetic
EPC	estimated PDU counter
ETSI	European Telecommunications Standards Institute
EVDO	evolution data optimized
FAC	final assembly code

FACH	forward access channel (transport channel)
FBI	feedback indicator bits
FDD	frequency division duplex
FEC	forward error correction
FER	frame error rate
FP	frame protocol
F-FCH	forward fundamental channel
F-PICH	forward pilot channel
F-QPCH	forward quick paging channel
F-SCH	forward supplemental channel
F-SYNCH	forward synchronization channel
FSPL	free space path loss
FTP	file transfer protocol
GERAN	GSM/EDGE radio access network
GGSN	GPRS gateway support node
GMM	GPRS mobility management
GMSC	gateway mobile switching center
GPRS	general packet radio service
GPS	global positioning system
GSA	Global Mobile Suppliers Association, UK
GSM	Global System for Mobile communications
HARQ	hybrid automatic repeat request
HCS	hierarchical cell structure
HE	header extension
HFN	hyper-frame number
HFNI	hyper-frame number indicator
HLR	home location register
HSDPA	high speed downlink packet access
HS-DPCCH	high speed dedicated physical control channel
HS-DSCH	high speed downlink shared channel (transport channel)
HS-PDSCH	high speed physical downlink shared channel (physical channel)
HS-SCCH	high speed shared control channel (physical channel)
HTTP	hyper text transfer protocol
Hz	Hertz
IE	information element
IK	integrity key
IMEI	international mobile equipment identity
IMP	inter-modulation product
IMSI	international mobile subscriber identity
IMT-2000	International Mobile Telephony
IP	internet protocol
IR	incremental redundancy
ISACC	ICT Standards Advisory Council, Canada
ISCP	interference on signal code power
ITU	International Telecommunications Union
Iu	interface for RNC-to-CN communication
Iub	interface for RNC and Node B communication
Iur	interface for RNC-to-RNC communication
kbps	kilobits per second
km	kilometer

ksps	kilosymbols per second
L1	Layer 1
L2	Layer 2
L3	Layer 3
LA	location area
LAC	location area code
LAI	location area identity
LCR TDD	low chip rate time division duplex
LNA	low noise amplifier
MAC	medium access control (protocol layering context); message authentication code (security context)
MAC-b	MAC for broadcast channels
MAC-c/sh	MAC for common and shared channels
MAC-d	MAC for dedicated channels
MAC-hs	HSDPA medium access control
MAC-I	message authentication code for integrity protection
MAP	mobile application part
Mbps	megabits per second
MCC	mobile country code
Mcps	megachips per second
ME	mobile equipment
MHz	megahertz
MIB	master information block
MIMO	multiple input multiple output
MM	mobility management
MNC	mobile network code
MRP	market representation partners
MRW	move receiving window
MS	mobile station
ms or msec	millisecond
MSC	mobile switching center
MSIN	mobile subscriber identification number
MUX	multiplex
NACK	negative acknowledgment
NAS	non-access stratum
NBAP	Node B application part
NIM	noise injection margin
NMSI	national mobile subscriber identity
NMT	Nordic Mobile Telephone
Node B	UMTS base station
OA&M	operations, administration and maintenance
OMC	operation and maintenance center
OTDOA	observed time difference of arrival
OVSF	orthogonal variable spreading factor
P	polling bit
PAD	padding
P-CPICH	primary common pilot channel (physical channel)
P-SCH	primary synchronization channel (physical channel)
P-TMSI	packet temporary mobile subscriber identity
PCA	power control algorithm
PCCC	parallel concatenated convolutional code

PCCH	paging control channel (logical channel)
PCCPCH	primary common control physical channel (physical channel)
PCH	paging channel (transport channel)
PCP	power control preamble
PCPCH	physical common packet channel (physical channel)
PCS	personal communication systems
PDCP	packet data convergence protocol
PDSCH	physical downlink shared channel (physical channel)
PDSN	packet data switching network
PDU	protocol data unit
PHS	personal handy-phone system
PHY	physical layer
PI	page indicator
PICH	paging indicator channel (physical channel)
PLMN	public land mobile network
PN	pseudorandom noise
PO	power offset
PPP	point-to-point protocol
PRACH	physical random access channel (physical channel)
PS	packet switched; PHS mobile
PSC	primary scrambling code
P-SCH	primary synchronization channel
PSTN	public switched telephone network
P-TMSI	packet temporary mobile subscriber identity
QE	quality estimate
QoS	quality of service
QPSK	quadrature phase shift keying
R1	reserved field
RA	routing area
RAB	radio access bearer
RAC	routing area code
RACH	random access channel (transport channel)
RAI	routing area identity
RAN	radio access network
RAND	random number
RAT	radio access technology
RB	radio bearer
R-DCCH	reverse dedicated control channel
R-EACH	reverse enhanced access channel
R-FCH	reverse fundamental channel
R-SCH	reverse supplemental channel
RF	radio frequency
RLC	radio link control
RNC	radio network controller
RNSAP	radio network subsystem application part
RNTI	radio network temporary identity
RRC	radio resource control; root-raised-cosine
RRM	radio resource management
RSCP	received signal code power
RSN	reset sequence number

RSSI	received signal strength indicator
RTP	real time protocol
RTT	round trip time
Rx	receive
SAW	stop and wait
S-CPICH	secondary common pilot channel (physical channel)
S-SCH	secondary synchronization channel (physical channel)
SCCPCH	secondary common control physical channel (physical channel)
SCH	synchronization channel (physical channel)
SDU	service data unit; segmented data unit
SF	spreading factor
SFN	system frame number
SGSN	serving GPRS support node
SIB	system information block
SID	size index
SIR	signal-to-interference ratio
SM	session management
SMS	short message service
SN	sequence number
SRB	signal radio bearer
SS	supplementary services
SSC	secondary scrambling code
SSDT	site selection diversity transmit
STTD	space time transmit diversity
SUFI	superfield
TAC	type approval code
TCP	transmission control protocol
TCTF	target channel type field
TD-SCDMA	time division synchronous code division multiple access
TDD	time division duplex
TF	transport format
TFC	transport format combination
TFCI	transport format combination indicator
TFCS	transport format combination set
TFRC	transport format resource combination
TFRI	transport format resource indicator
TFS	transport format set
TG	transmission gap
TGCFN	transmission gap connection frame number
TGD	transmission gap start distance
TGL	transmission gap length
TGP	transmission gap pattern
TGPL	transmission gap pattern length
TGPRC	transmission gap pattern repetition count
TGPS	transmission gap pattern sequence
TGPSI	transmission gap pattern sequence identifier
TIA	Telecommunications Industries Association, US
TM	transparent mode
TMSI	temporary mobile subscriber identity
TOI	third order intercept

TPC	transmission power control
TRX	transmitter/receiver, also transceiver
TTI	transmission time interval
TTA	Telecommunications Technology Association, Korea
TTC	Telecommunications Technology Committee, Japan.
Tx	transmit
U-RNTI	UTRAN radio network temporary identity
UARFCN	UTRA absolute radio frequency channel number
UDP	user datagram protocol
UE	user equipment
UEA1	UMTS encryption algorithm 1
UIA1	UMTS integrity algorithm 1
UL	uplink
UM	unacknowledged mode
UMTS	Universal Mobile Telecommunications System
URA	UTRAN registration area
USIM	universal subscriber identity module
UTC	coordinated universal time
UTRA	universal terrestrial radio access
UTRAN	UMTS (universal) terrestrial radio access network
Uu	over-the-air interface between the UE and the Node Bs
VLR	visitor location register
VoIP	voice over IP
VSWR	voltage standing wave ratio
WCDMA	wideband code division multiple access

Chapter 1

Introduction to UMTS

Third generation (3G) mobile communication systems are showing their strength around the world. Many service providers have already deployed 3G systems, while many more are planning to do so in the near future. 3G systems arguably can provide a global mobility with a wide range of services including both *teleservices* and *bearer services*. Teleservices include speech and short message services, whereas bearer services enable information transfers between access points. The characteristics of a bearer service can be negotiated at connection or session establishment. Typical examples of bearer services include video, video telephony, video-on-demand, multimedia, movies, and internet applications such as web browsing and web casts.

UMTS and CDMA2000 are the two major 3G technologies currently being considered as the leading contenders for deployment worldwide. Another 3G technology is TD-SCDMA, which is just maturing now and is likely to be deployed in many countries including China. This book covers the UMTS system in great detail. CDMA2000 and TD-SCDMA are not discussed.

1.1 What Is UMTS?

UMTS stands for universal mobile telecommunications system. It supports multiple services and multiple qualities of service (QoS) at data rates up to 2 Mbps for packet-switched data, and 384 kbps for circuit-switched data without high speed downlink packet access (HSDPA). With HSDPA, the data rate could go beyond 10 Mbps.

UMTS provides high voice quality, improved spectral efficiency, and high system capacity. The high voice quality can be attributed to the usage of the adaptive multi-rate (AMR) vocoder, which provides an improvement over the enhanced full-rate (EFR) vocoder used in global

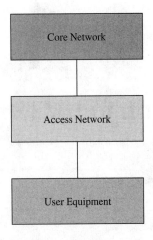

Figure 1.1 UMTS system constituents.

systems for mobile communications (GSM). The improvement in system capacity is achieved through fast power control. Power control in UMTS runs at 1500 Hz on both uplink and downlink, as opposed to 2 Hz in GSM systems.

Figure 1.1 shows a high level diagram of a UMTS system, which encompasses three parts: core network (CN), radio access network (RAN), and user equipment (UE). The core network consists of a circuit-switched (CS) core network and a packet-switched (PS) core network. The user equipment could be a mobile or a data terminal. The radio access network is UMTS terrestrial radio access network (UTRAN), which may operate in frequency division duplex (FDD) mode or time division duplex (TDD) mode. The UTRAN-FDD access network uses wideband code division multiple access (WCDMA) protocols and operates at 3.84 Mcps, and is conventionally called WCDMA access network. A UTRAN-TDD access network may operate either at 3.84 Mcps or at 1.28 Mcps. The radio access operating at 1.28 Mcps is also called low chip rate TDD (LCR TDD) or time-division synchronous code division multiple access (TD-SCDMA). However, this book only focuses on the UMTS system with a WCDMA access network.

1.2 Evolution of WCDMA

UMTS is deployed under the GSM or general packet radio service (GPRS) mobile application part (MAP) network. In the GSM MAP core network family, GSM systems provide voice and basic data services, while GPRS and/or enhanced data rates for GSM evolution (EDGE) provide higher speed data services. From an evolution standpoint, the next step is WCDMA, which uses the GSM MAP core network architecture but with

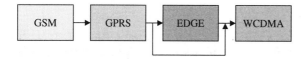

Figure 1.2 Evolution paths from GSM to WCDMA.

a different access network. Figure 1.2 shows the evolution paths from GSM to WCDMA.

Since WCDMA is evolved from GSM, the higher layers of the two systems are essentially the same. These include mobility management (MM), GPRS mobility management (GMM), connection management (CM), and session management (SM). Mobility management and connection management are in a circuit-switched domain while GPRS mobility management and session management are in a packet-switched domain. In connection management, there are call control (CC) services, short message services (SMS), and supplementary services (SS). In addition to the higher layers being the same, WCDMA shares another similarity with GSM. The universal subscriber identity module (USIM) for UMTS is based on the subscriber identity module (SIM) for GSM.

1.3 UMTS Services

Network services should be considered end-to-end. This means that network services are from a user to another user. An end-to-end service may have a certain quality of service (QoS) provided to the user. The user should decide whether the provided QoS is acceptable or not.

UMTS offers different bit rates under different environments. Excluding HSDPA, the typical bit rates that UMTS offers are 144 kbps for rural outdoor services, 384 kbps for urban outdoor services, and about 2 Mbps for indoor and short range outdoor services. If HSDPA is included, the offered bit rate could be higher than 10 Mbps.

Based on the types of traffic, UMTS network can provide services with different QoS classes. Normally, the user traffic can be categorized into four classes: conversational, streaming, interactive, and background. Conversational class includes voice, videophone, and video games. This class has a time-delay tolerance of much less than one second. Streaming class, such as multimedia, video-on-demand, and web cast, has a time-delay tolerance of about one second. Interactive class, such as web browsing, network gaming, and database access, has a time-delay tolerance of less than ten seconds. Background class, such as e-mail, SMS, and downloading, may allow a time delay of longer than ten seconds.

1.4 UMTS Network

A UMTS network includes a variety of network elements with specific functions. Firstly, those network elements that switch and route calls and data connections to external networks are called core network elements. Secondly, those that handle all radio link functions are called radio access network elements. Thirdly, those that serve as an interface between humans and the network are called user equipment. Iu interface is defined between core network and radio access network. Meanwhile, Uu interface is defined between radio access network and user equipment. These open interfaces specify the standard connections between the network elements, allowing inter-operability between network equipment from different manufacturers.

1.4.1 Core network

The basic core network architecture for UMTS is fundamentally based on a GSM network with GPRS. The structure of the core network is shown in Figure 1.3. It consists of a circuited-switched domain and a packet-switched domain. In the circuit-switched domain, the network elements include mobile switching center (MSC), visitor location register (VLR), and gateway mobile switching center (GMSC). In the packet-switched domain, the network elements include serving GPRS support node (SGSN) and gateway GPRS support node (GGSN).

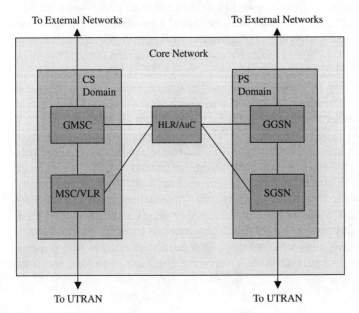

Figure 1.3 Core network.

Other network elements, such as home location register (HLR) and authentication center (AUC), are shared by both domains. The circuit-switched domain network elements handle circuit-switched traffic such as voice and low rate data, while the packet-switched network elements handle packet-switched traffic such as multimedia.

1.4.2 UTRAN

The radio access network called UTRAN provides the air interface access for the UE. Figure 1.4 shows the UTRAN architecture. Base stations in UMTS are called Node B. The equipment controlling Node Bs is called a radio network controller (RNC). Within UTRAN, two interfaces are defined: Iub and Iur. The Iub interface connects the Node B and the RNC, allowing inter-operability between Node B and the RNC from different equipment vendors. The Iur interface connects two RNCs supporting soft handover between RNCs from different vendors.

UTRAN uses the direct sequence CDMA (DS-CDMA) technique, in which user data is multiplied with orthogonal variable spreading factor (OVSF) codes. In addition, scrambling codes and synchronization codes are respectively employed for the scrambling and synchronization of the user data.

Node B is responsible for air interface transmission and reception, modulation and demodulation, physical channel coding, micro-diversity, error correction, and closed loop power control. The RNC is responsible for radio resource control, admission control, channel allocation, power control settings, handover control, macro-diversity, ciphering, segmentation and reassembly, broadcast signaling, and open loop power control.

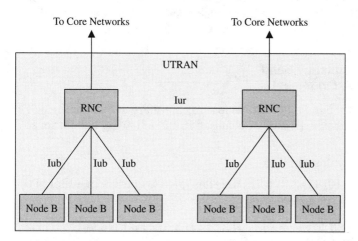

Figure 1.4 UTRAN.

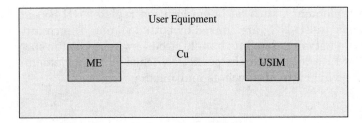

Figure 1.5 User equipment.

1.4.3 User equipment

User equipment (UE) is the air interface counterpart of Node B. The basic structure of a user equipment is shown in Figure 1.5. Generically, it contains a mobile equipment (ME) and a UMTS subscriber identity module (USIM). The Cu interface is introduced between the ME and the USIM to provide electrical connections. A UE has many different types of identities and functionalities. The key UE identities include international mobile subscriber identity (IMSI), temporary mobile subscriber identity (TMSI), packet temporary mobile subscriber identity (P-TMSI), and international mobile station equipment identity (IMEI).

A UE can operate in circuit-switched (CS) mode, packet-switched (PS) mode, or hybrid mode. When operating in circuit-switched mode, the UE attaches to the circuit-switched domain and offers circuit-switched services only. If the UE operates in packet-switched mode, it attaches to the packet-switched domain and offers packet-switched services only. However, some special circuit-switched services, such as voice over IP (VoIP), may also be offered over the packet-switched domain. When the UE operates in hybrid mode, it attaches to both the circuit-switched and the packet-switched domains. Both circuit-switched and packet-switched services are offered simultaneously. This mode is also known as CS/PS mode.

1.5 UMTS Frequency Band and Channel Arrangement

UMTS frequency band and channel arrangement are defined in the 3GPP specifications. They are briefly discussed in the following subsections.

1.5.1 Frequency Bands

Based on TS 25.101 [1], UTRA/FDD is designed to operate in either of the following paired frequency bands:

- (I) 1920 – 1980 MHz: uplink (UE transmits and Node B receives)
  2110 – 2170 MHz: downlink (Node B transmits and UE receives)

- (II) 1850 – 1910 MHz: uplink (UE transmits and Node B receives)
  1930 – 1990 MHz: downlink (Node B transmits and UE receives)

Deployment in other frequency bands is not precluded. UTRA/TDD operates in the 1900–1920 and 2010–2025 MHz bands. The TDD channels are unpaired with a channel spacing of 5 MHz and a raster of 200 kHz. The transmission and reception are not separated in frequency.

1.5.2 UTRA/FDD Transmit-To-Receive (Tx-Rx) Frequency Separation

TS25.101 also specifies that UTRA/FDD is designed to operate with the transmit-to-receive frequency separation listed in Table 1.1. However, it can also support both fixed and variable Tx-Rx frequency separations. Furthermore, the use of other transmit-to-receive frequency separations in existing or other frequency bands is not precluded.

1.5.3 Channel Arrangement

The nominal channel spacing is 5 MHz, but this can be adjusted to optimize performance in a particular deployment scenario. The channel raster is 200 kHz, which means that the carrier center frequency must be an integer multiple of 200 kHz.

1.5.4 Carrier Frequency

The carrier frequency is represented using the UTRA absolute radio frequency channel number (UARFCN). The values are defined in Table 1.2.

For the frequency band (Band II) described in 1.5.1 (II), the carrier frequencies of additional channels are defined in Table 1.3.

TABLE 1.1 Tx-Rx Frequency Separation (Courtesy of ETSI)

Frequency Band	Tx-Rx Frequency Separation
For operation in frequency band defined in 1.5.1 (I)	190 MHz
For operation in frequency band defined in 1.5.1 (II)	80 MHz

TABLE 1.2 UARFCN Definition of Carrier Frequency (Courtesy of ETSI)

	UARFCN	Carrier Frequency (MHz)
Uplink	$N_u = 5 * F_{uplink}$	$0.0 \text{ MHz} \leq F_{uplink} \leq 3276.6 \text{ MHz}$ where F_{uplink} is the uplink frequency in MHz
Downlink	$N_d = 5 * F_{downlink}$	$0.0 \text{ MHz} \leq F_{downlink} \leq 3276.6 \text{ MHz}$ where $F_{downlink}$ is the downlink frequency in MHz

TABLE 1.3 UARFCN Definition of Additional Band II Carrier Frequency (Courtesy of ETSI)

	UARFCN	Carrier Frequency (MHz)
Uplink	$N_u = 5 * [(F_{uplink} - 100 \text{ kHz}) - 1850 \text{ MHz}]$	1852.5, 1857.5, 1862.5, 1867.5, 1872.5, 1877.5, 1882.5, 1887.5, 1892.5, 1897.5, 1902.5, 1907.5
Downlink	$N_d = 5 * [(F_{downlink} - 100 \text{ kHz}) - 1850 \text{ MHz}]$	1932.5, 1937.5, 1942.5, 1947.5, 1952.5, 1957.5, 1962.5, 1967.5, 1972.5, 1977.5, 1982.5, 1987.5

TABLE 1.4 UTRA Absolute Radio Frequency Channel Number (Courtesy of ETSI)

Frequency Band	Uplink UE Transmits, Node B Receives	Downlink UE Receives, Node B Transmits
For operation in frequency band as defined in 1.5.1 (I)	9612 to 9888	10562 to 10838
For operation in frequency band as defined in 1.5.1 (II)	9262 to 9538, And for additional channels: 12, 37, 62, 87, 112, 137, 162, 187, 212, 237, 262, 287	9662 to 9938 And for additional channels: 412, 437, 462, 487, 512, 537, 562, 587, 612, 637, 662, 687

1.5.5 UARFCN

The UARFCN ranges defined in Table 1.4 [1] are supported for each paired frequency band.

1.6 Organization of the Book

The main objective of this book is to provide wireless communication professionals in general, and RF engineers in particular, with a basic knowledge of UMTS air interface and RF engineering. The author does not try to cover everything in this book, but instead focuses on the topics considered to be important or helpful for RF engineers in deploying, optimizing, operating, and maintaining UMTS networks. This book is organized as follows:

Chapter 1 presents a brief introduction of UMTS including the definition of UMTS, WCDMA evolution, UMTS services, UMTS network architecture, UMTS frequency band, and channel arrangement.

Chapter 2 discusses the UMTS fundamentals in detail, including UMTS network topology, UMTS signaling protocol stack, access stratum data flow, UMTS channels and channel mapping, protocol states, UE and subscriber identifiers, and system frame timing.

Chapter 3 briefly describes the 3GPP organization and UMTS standards. It mainly introduces the subjects, technical specification groups,

3GPP specification releases, 3GPP specification-numbering scheme, and 3GPP specification series.

Chapter 4 explains radio resource control (RRC). Major topics discussed include RRC message specifications, system information, paging and notification, RRC connection management, ciphering and integrity protection control, radio bearer control, RRC management of UE mobility, measurements and reporting, and NAS message routing.

Chapter 5 concerns radio link control (RLC). It starts with Layer 2 fundamentals, continues on to RLC functions, RLC entities, RLC data transfer modes, RLC ciphering, and ends with RLC configurable parameters.

Chapter 6 deals with medium access control (MAC). It explains MAC architecture, logical to transport channel mappings, MAC header, transport format combination selection, traffic volume measurement, MAC ciphering, MAC RACH functions, and MAC configurable parameters.

Chapter 7 presents the physical layer concepts. Topics covered in this chapter include orthogonal spreading codes, scrambling codes, synchronization codes, physical layer timing, downlink procedure, uplink procedure, physical channel structures and channel timing, and physical layer procedures.

Chapter 8 explains cell reselection in detail. Major subjects in this chapter are cell reselection criteria, cell reselection ranking, inter-frequency cell reselection, and inter-RAT cell reselection.

Chapter 9 discusses handover mechanisms, including soft handover, softer handover, and hard handover. Also in this chapter are measurement control and reporting, such as intra-frequency measurements and reporting, inter-frequency measurements and reporting, inter-RAT measurements and reporting, and compressed mode.

Chapter 10 covers power control. This chapter begins with power control basics, continues on to downlink power control, uplink power control, open loop power control, closed loop power control, and ends with power control during compressed mode.

Chapter 11 is a brief overview of high-speed downlink packet access (HSDPA) and reveals the features unique to HSDPA. It also addresses HSDPA channels, HSDPA physical layer procedure, and HSDPA configuration parameters.

Chapter 12 is about WCDMA radio network planning. Its scope includes capacity and coverage, uplink analysis, propagation models, downlink analysis, overhead channel power allocation, scrambling code planning, base station antennas, and WCDMA RF planning processes.

Chapter 13 elaborates on the general concepts relating to RF network optimization. It gives an RF optimization overview, addresses a variety of RF optimization issues, and describes the pre-optimization preparation, the RF drive-test-based optimization, and the traffic-statistics-based optimization.

Chapter 14 goes over the application of repeaters and tower mounted amplifiers (TMA) in WCDMA systems. It looks at the engineering considerations regarding the usage of repeaters and tower mounted amplifiers from a practical point of view, and it also provides some guidelines for repeater deployment. It also examines the potential problems that may occur in the field and offers some practical solutions for them.

Chapter 15 explains intersystem interference, including interference between WCDMA and CDMA2000, between WCDMA and GSM, and between WCDMA and PHS systems. It also gives a brief description of adjacent channel performance.

Chapter 16 presents both a comparison of the similarities and a contrast of the differences of the two systems: WCDMA and CDMA2000.

References

[1] 3GPP TS25.101, v3.16.0, "User Equipment (UE) radio transmission and reception (FDD)," (Release 99).

Chapter 2

UMTS Fundamentals

The important fundamentals of UMTS include network topology, protocol stacks, channel mapping, frame timing, UE call states, and UE and subscriber identifiers. As such, the objectives of this chapter are to identify the topology of a UMTS network, explore the UMTS protocol stack, examine the functions performed by the access stratum, characterize the channels defined in a WCDMA system, understand the frame timing, depict the UE call states, and describe the user equipment and subscriber identities used in UMTS systems. Detailed information on the topics discussed in this chapter is given in various 3GPP technical specifications [1]–[7].

2.1 UMTS Network Topology

As shown in Figure 2.1, a UMTS system consists of three major parts: core network, access network, and user equipment. The core network contains all of the switching and routing elements that are capable of connecting the circuit-switched calls to the public switched telephone network (PSTN) and the packet-switched calls to a packet data supporting network (PDSN). The core network also performs the functions of mobility management, subscriber location management, and authentications. Basically, the functionalities of the core network in a WCDMA system are similar to that of a GSM/GPRS system. The difference is that a new interface to UTRAN is introduced. The access network and user equipment are totally new.

The access network contains all of the radio equipment that is necessary for accessing the core network. For WCDMA, the access network is UTRAN and for GSM/GPRS it is GSM/EDGE radio access network (GERAN). The user equipment could be a mobile, a stationary

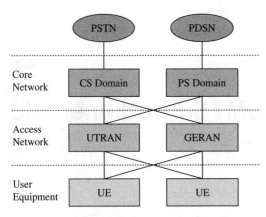

Figure 2.1 UMTS network topology.

terminal, or a laptop. It also includes a UMTS SIM card that contains the subscription information of a user.

2.2 UMTS Signaling Protocol Stack

As shown in Figure 2.2, there are two strata in the UMTS signaling protocol stack: the access stratum (AS) and the non-access stratum (NAS). The UMTS non-access stratum evolved from the GSM upper layers and is essentially the same as that of GSM. It consists of connection management (CM), session management (SM), mobility management (MM), and GPRS mobility management (GMM).

The connection management takes care of circuit-switched calls. It is responsible for call control, such as call establishment and call release, short message services, and supplementary services such as call forwarding and three-way calling. The session management deals

	Circuit-switched	Packet-switched
Non-Access Stratum	Connection Management	Session Management
	Mobility Management	GPRS Mobility Management
Access Stratum	L3, Radio Resource Control	
	L2, Radio Link Control	
	L2, Medium Access Control	
	L1, Physical Layer	

Figure 2.2 UMTS signaling protocol stack.

with packet-switched calls such as call establishment and call release. The mobility management governs circuit-switched domain mobility functions such as location area update. The GPRS mobility management manages packet-switched domain mobility functions such as routing area update.

The access stratum consists of three layers: Layer 1, Layer 2, and Layer 3. Layer 1 is the physical layer (PHY). Layer 2 consists of radio link control (RLC) and medium access control (MAC). Layer 3 is radio resource control (RRC).

The protocol stack can be further divided into four basic protocol stacks: circuit-switched control plane protocol stack, circuit-switched user plane protocol stack, packet-switched control plane protocol stack, and packet-switched user plane protocol stack. The control plane protocol stack deals with signaling protocols, whereas the user plane protocol stack deals with user protocols. To understand the functions of protocol stacks, we will examine each of these four protocol stacks individually.

2.2.1 Circuit-Switched Control Plane Protocol Stack

Figure 2.3 shows a circuit-switched control plane protocol stack between UE and MSC. In the non-access stratum, connection management protocols are defined between UE and MSC to take care of call establishment, call release, and other services. Similarly, mobility management protocols are defined between UE and MSC to deal with UE mobility. Since this is a circuit-switched control plane, neither session management nor GPRS mobility management protocol is required. In the access stratum, RRC, RLC, and MAC protocols are defined between UE and RNC. RRC controls establishment, release and configuration of radio resources. RLC performs segmentation, concatenation, reassembly, and other typical Layer 2 functions. MAC carries out transport format selection, multiplexing of control plane and user plane data, and so on.

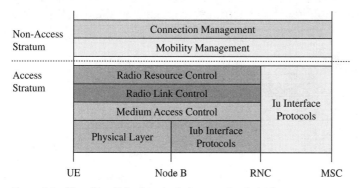

Figure 2.3 Circuit-switched control plane protocol stack.

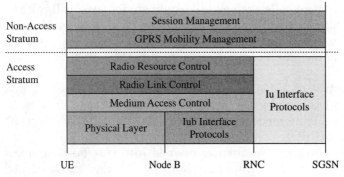

Figure 2.4 Packet-switched control plane protocol stack.

The physical layer protocol is defined between UE and Node B for transferring information over the air.

2.2.2 Packet-Switched Control Plane Protocol Stack

The packet-switched control plane protocol stack (Figure 2.4) is similar to the circuit-switched control plane protocol stack, except that it does not have CM or MM in the NAS. Instead, it uses SM to deal with the packet session establishment and release, and uses GMM to handle UE mobility. Additionally, the MSC is replaced with SGSN for packet-switched data. The access stratum of the packet-switched control plane protocol stack is identical to that of the circuit-switched control plane protocol stack.

2.2.3 Circuit-Switched User Plane Protocol Stack

Figure 2.5 illustrates the circuit-switched user plane protocol stack between UE and MSC. In the non-access stratum, the application layer could consist of many sub-layers. For instance, for a voice call, the uppermost sublayer is the actual voice heard by the users, while a

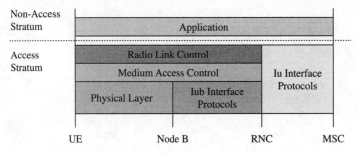

Figure 2.5 Circuit-switched user plane protocol stack.

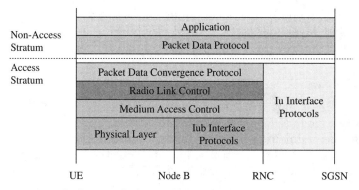

Figure 2.6 Packet-switched user plane protocol stack.

lower sublayer carries the digital bits of the voice. In the access stratum, the RLC, MAC and physical layer protocols for the user plane are the same as that for the control plane; but, there is no RRC in the user plane protocol stack. RRC just sets up the radio bearers (RB) and the channels and does not handle the user data.

2.2.4 Packet-Switched User Plane Protocol Stack

Figure 2.6 shows the packet-switched user plane protocol stack between UE and SGSN. In the non-access stratum, the application layer could be e-mail, database access, or data downloading. In the access stratum, the RLC, MAC, and physical layer protocols for the user plane are the same as that for the control plane. Again, RRC is not included in the user plane protocol stack. However, there is a packet data convergence protocol (PDCP) defined between UE and RNC. The PDCP provides protocol transparency for NAS protocols such as PPP and IP.

2.3 Access Stratum Data Flow

In access stratum, in addition to the protocol layers, there are both radio bearers and channels, which carry information between the layers. Radio bearers carry signaling data between RRC and RLC layers. They also carry user data between the application layer and Layer 2. Logical channels carry information between RLC and MAC layers. Transport channels carry information between MAC and physical layers. Physical channels transfer information over the air.

2.3.1 RRC Functions

Major functions of RRC include access stratum control, broadcast of system information, RRC connection management, radio bearer management, RRC mobility functions, paging and notification functions, routing of higher layer messages, control of ciphering and integrity protection, power control, and measurement control and reporting.

2.3.2 RLC Functions

RLC supports three data transfer modes: transparent mode (TM), unacknowledged mode (UM), and acknowledged mode (AM). Its major functions include segmentation, reassembly, concatenation, padding, retransmission control, flow control, duplicate detection, in-sequence delivery, error correction, and ciphering for UM and AM logical channels.

2.3.3 MAC Functions

Principal functions of MAC include mapping and multiplexing logical channels to transport channels, priority handling of data flow, UE identification on common channels, traffic volume measurements, random access channel procedures, and ciphering for TM logical channels.

2.3.4 Physical Layer Functions

The physical layer performs the following functions:

- cyclic redundancy check (CRC) attachment
- concatenation of transport blocks
- coding and decoding
- error detection
- interleaving and de-interleaving
- multiplexing and de-multiplexing
- discontinuous transmission (DTX) insertion
- rate matching
- power weighting and combining of physical channels
- modulation and demodulation
- spreading and scrambling
- measurements
- closed loop power control
- soft handover
- RF processing
- frequency and time synchronization
- physical layer procedures (e.g. initial system acquisition)

2.4 UMTS Channels

UMTS channels can be defined in terms of data flow direction, sharing, and functions. Downlink channels are transmitted by UTRAN and received by UE, while uplink channels are transmitted by UE and received by UTRAN. Common channels carry information to and from multiple UEs, while dedicated channels carry information to and from a single UE. Logical channels are defined by type of information that is transferred (either signaling or user data). Transport channels are defined by the way the data is transferred. Physical channels are defined by physical mappings, channel frame structure, and physical layer procedure used to transfer information over the radio link.

UMTS Release 99 defines the following channels:

Logical channels:

- BCCH—Broadcast control channel
- PCCH—Paging control channel
- CCCH—Common control channel
- DCCH—Dedicated control channel
- DTCH—Dedicated traffic channel

Transport channels:

- BCH—Broadcast channel
- PCH—Paging channel
- FACH—Forward access channel
- RACH—Random access channel
- DCH—Dedicated channel

Physical channels:

- PCCPCH—Primary common control physical channel
- SCCPCH—Secondary common control physical channel
- PICH—Paging indicator channel
- PRACH—Physical random access channel
- AICH—Acquisition indicator channel
- CPICH—Common pilot channel
- DPDCH—Dedicated physical data channel
- DPCCH—Dedicated physical control channel
- SCH—Synchronization channel

In addition to Release 99 channels, there exist Release 4 channels and Release 5 channels. Release 4 channels include common packet channel (CPCH), physical common packet channel (PCPCH), and downlink share channel (DSCH). CPCH and PCPCH are uplink transport and uplink physical channels, respectively, while DSCH is a downlink transport channel.

Release 5 supports high-speed downlink shared channels. It contains a few new channels: high-speed downlink shared channel (HS-DSCH), high-speed physical downlink shared channel (HS-PDSCH), downlink high-speed shared control channel (HS-SCCH), and uplink high-speed dedicated physical control channel (HS-DPCCH). HS-DSCH is a downlink transport channel. HS-PDSCH and HS-SCCH are downlink physical channels, while HS-DPCCH is an uplink physical channel.

2.5 Channel Mappings

The overall possible channel mappings of Release 99 channels are shown in Figure 2.7.

As shown in Figure 2.7, logical channels are mapped to transport channels, which in turn are mapped to physical channels. There could be one-to-one mappings or one-to-multiple mappings, depending on the type of channels involved. It should be noted that not all mappings occur at the same time. But, some mappings could occur simultaneously. For instance, three DTCHs carrying a voice call may map to three DCHs.

Depending on the logical channels, RLC may be configured in TM, UM, or AM mode. The physical channels CPICH, SCH, DPCCH, AICH, and PICH only carry the contents defined at the physical layer. They do not carry higher layer signaling or user data.

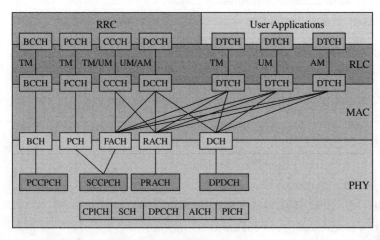

Figure 2.7 Release 99 channel mappings.

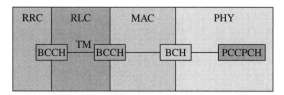

Figure 2.8 Channel mapping of BCCH.

To provide a detailed picture of channel mappings, we will discuss the mappings of individual channels in the following subsections.

2.5.1 Channel Mapping of BCCH

BCCH is a downlink common logical channel. It carries the system information required for the UE to access the system. Radio bearers carrying signaling from RRC are transferred to logical channels using one of the RLC data transfer modes. For BCCH, RLC is always configured in transparent mode. BCCH is broadcast continuously so that system information messages are repeatedly transmitted every 20 ms. The UE monitors this channel after initial system acquisition or when it camps on a new cell. It then monitors this channel periodically thereafter to get the most current system information.

Figure 2.8 shows the channel mapping of BCCH. At the UTRAN side, the logical channel BCCH maps to the transport channel BCH, which in turn maps to the physical channel PCCPCH. It is a one-to-one mapping. At the UE side, the mapping is in the reverse direction. That is, PCCPCH maps to BCH, which in turn maps to BCCH.

2.5.2 Channel Mapping of PCCH

PCCH is also a downlink common logical channel. It carries paging messages to page the UE and is associated with the physical paging indicator channel (PICH). For PCCH, RLC is also always configured in transparent mode.

Figure 2.9 shows the channel mapping of PCCH. At the UTRAN side, PCCH maps to PCH, which in turn maps to SCCPCH. It is also a one-to-one mapping.

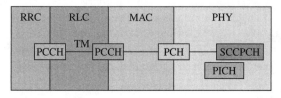

Figure 2.9 Channel mapping of PCCH.

PCH is broadcast continuously, but the UE only monitors the PICH during the paging occasions when it is in idle mode, Cell_PCH state, or URA_PCH state. The definition of idle mode, Cell_PCH state, and URA_ PCH state will be explained later in this chapter. The PICH carries indicator bits indicating whether a paging message for the UEs is sent on the associated SCCPCH. If an indicator indicates that there is a paging message, all UEs assigned to read that indicator at the assigned paging occasion must decode the associated SCCPCH to determine which UE should read the entire paging message.

2.5.3 Channel Mapping of CCCH to RACH/ FACH Common Channels in Idle Mode

In idle mode, there is no dedicated channel allocated to the UE. Channel mapping of CCCH to RACH/FACH is shown in Figure 2.10. At the UTRAN side and in the downlink direction, CCCH maps to FACH, which, in turn, maps to SCCPCH. In the uplink direction PRACH maps to RACH, which then maps to CCCH. At the UE side, SCCPCH maps to FACH, which in turn maps to CCCH in the downlink direction. In the uplink direction, CCCH maps to RACH, which then maps to PRACH.

For CCCH, RLC is configured in transparent mode on the uplink and unacknowledged mode on the downlink. Associated with RACH is an access process in which the UE sends PRACH preambles to UTRAN, according to the physical random access procedure we will discuss in Chapter 7. UTRAN responds on FACH when it receives a preamble from the UE. The FACH maps to an SCCPCH. The response is then sent in terms of a positive acquisition indicator or a negative acquisition indicator on AICH. Upon receiving a positive acquisition indicator from UTRAN, the UE may start to send the PRACH message. As mentioned earlier, the AICH does not carry the upper layer signaling or user data.

2.5.4 Channel Mapping of DCCH and DTCH to RACH/FACH Common Channels in Connected Mode

In connected mode, there is no CCCH; DCCH is used for signaling instead. DCCH and DTCH may map to RACH and FACH channels and

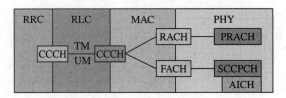

Figure 2.10 Channel mapping of CCCH/RACH/FACH.

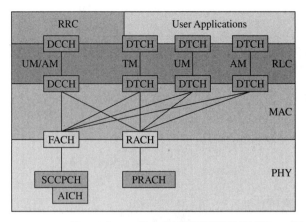

Figure 2.11 Channel mapping of DCCH/DTCH, FACH, and RACH.

their associated physical channels. The channel mapping of DCCH and DTCH to RACH and FACH is shown in Figure 2.11. At the UTRAN side, DCCH maps to FACH and then to SCCPCH in the downlink direction. In addition, DTCH may also map to FACH and then to SCCPCH for low-rate burst data transmission when no dedicated transport and physical channels are allocated. In the uplink direction, PRACH maps to RACH and then to DCCH for signaling and DTCH for data transmission.

At the UE side, DCCH and DTCH map to RACH and then to PRACH in the uplink direction. In the downlink direction, SCCPCH maps to FACH and then to DCCH for signaling and DTCH for user data.

For DCCH mapping to RACH and FACH, RLC may be configured in unacknowledged mode or acknowledged mode. For DTCH mapping to RACH and FACH, RLC may be configured in transparent mode, unacknowledged mode, or acknowledged mode. The RACH and FACH protocols in connected mode are the same as that in idle mode. The same PRACH random access procedure is followed if the UE wants to send signaling messages or data to UTRAN.

2.5.5 Channel Mapping of Dedicated Channels

The dedicated channels DCCH, DTCH, DCH, and DPDCH carry signaling information and user data between UTRAN and an individual UE. These dedicated channels are allocated when a voice call or a data session is active.

The channel mappings of these channels are shown in Figure 2.12. DCCHs and DTCHs may map to a single DCH, or may map to individual DCHs. In general, all DCHs map to a single DPDCH. However, in the case of the so-called multi-code, DCHs may map to multiple DPDCHs.

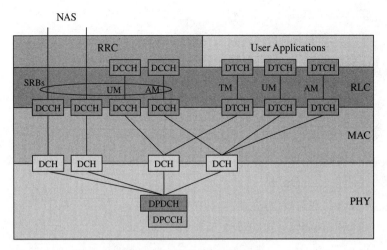

Figure 2.12 Channel mappings of dedicated channels.

Associated with DPDCHs is a single DPCCH which carries control information created at the physical layer, including pilot, power control, transport format combination index and feedback indicator bits.

In general, two DCCHs are allocated to carry RRC signaling with one configured for RLC acknowledged mode and the other configured for RLC unacknowledged mode. Another one or two DCCHs are allocated to carry NAS signaling. If two DCCHs are allocated, one carries high priority messages while the other carries low priority messages. These three or four DCCHs are also called signaling radio bearers (SRBs).

The number of DTCHs allocated for carrying user data depends on the type of applications. For example, three DTCHs are normally allocated to carry the three different classes of voice bits generated by the adaptive multi-rate (AMR) vocoder.

2.5.6 Channel Mappings of CPCH, DSCH, and HS-DSCH

Figure 2.13 shows the channel mappings of CPCH and DSCH. The CPCH is an uplink transport channel. The CPCH maps to the DCCH and DTCH logical channels above it, and the PCPCH physical channel below it. The PCPCH is an uplink physical channel carrying CPCH. The PCPCH access transmission consists of one or several access preambles, one collision detection preamble, a DPCCH power control preamble, and a message of variable length in multiples of 10 ms.

The downlink shared channel (DSCH) is a downlink transport channel shared by a number of UEs. It is associated with one or several downlink DCHs and is transmitted over the whole cell or over only a portion of the cell using intelligent antennas. It maps to DCCH and

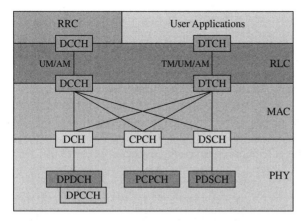

Figure 2.13 Channel mappings of CPCH and DSCH.

DTCH logical channels above it and physical downlink shared channel (PDSCH) below it. The PDSCH is used to carry the DSCH and is allocated to one and only one UE for each radio frame. Within one radio frame, UTRAN may allocate different PDSCHs under the same PDSCH root OVSF code to different UEs.

The HS-DSCH is a downlink transport channel shared by a number of UEs. It is a new channel in Release 5 and provides HSDPA services. It is associated with one downlink DPCH, and one or more high speed shared control channels (HS-SCCH). It is transmitted over the whole cell or over only a portion of the cell using smart antennas. Figure 2.14 shows the channel mapping of HS-DSCH. HS-DSCH maps to DCCH and DTCH logical channels above it and to an HS-PDSCH physical channel below it. The HS-PDSCH is used to carry HS-DSCH. The HS-SCCH is a downlink physical channel carrying downlink signaling for HS-DSCH transmission.

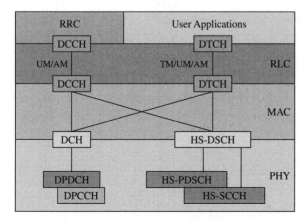

Figure 2.14 Channel mapping of HS-DSCH.

2.6 Protocol States

Figure 2.15 illustrates the idle mode, UTRAN connected mode, GSM connected mode, and GPRS packet transfer mode. The figure also illustrates the transitions between two modes. The transitions occur between idle mode and UTRAN connected mode. They may also occur between UTRAN RRC connected mode and GSM connected mode for CS domain services, and between UTRAN RRC connected mode and GSM/GPRS packet modes for PS domain services. In the following subsections, we will discuss in detail the idle mode and each protocol state within the UTRAN connected mode.

2.6.1 Idle Mode

In idle mode, the UE does not have any active CS or PS calls. It may be registered for services in the circuit-switched network and/or the packet-switched network. The UE sleeps most of the time in idle mode and wakes up during the paging occasions to monitor the PICH and the associated SCCPCH. If the network wants to deliver a call to the UE, UTRAN must first page the UE during the paging occasion assigned to the UE. Responding to the page, the UE requests UTRAN to establish an RRC connection. The UE may also request UTRAN to establish an RRC connection when it wants to set up a call. All UE requests for establishing an RRC connection are sent on RACH.

As shown in Figure 2.16, the UE may enter the idle mode when it is powered on and camped on a UTRAN cell, or when an RRC connection is released from a Cell_FACH state or a Cell_DCH state in the connected mode. In idle mode, the UE needs to perform a location or routing area

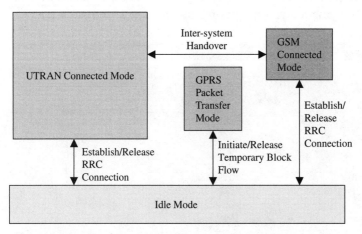

Figure 2.15 Modes and mode transitions (courtesy of ETSI).

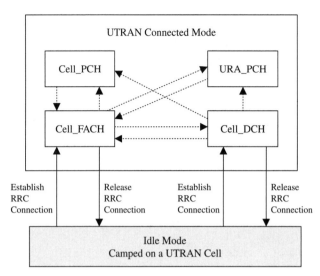

Figure 2.16 Idle mode.

update procedure whenever it moves to a new location or routing area. A location or routing area may contain many cells. Therefore, UTRAN must page the UE over all cells in the location or routing area where the UE most recently performed a location or routing area update.

In idle mode, UTRAN addresses the UE by using international mobile subscriber identity (IMSI), temporary mobile subscriber identity (TMSI), or packet-switched temporary mobile subscriber identity (P-TMSI).

2.6.2 Connected Mode

In connected mode, the UE has established an RRC connection with UTRAN for exchange of signaling messages. There are four states in connected mode specified in the 3GPP standard; namely, Cell_DCH state, Cell_FACH state, URA_PCH state, and Cell_PCH state. However, some network equipment vendors may not implement Cell_PCH state and regard it as a special case of URA_PCH state.

2.6.2.1 Cell_DCH State In the Cell_DCH state, a dedicated physical channel is allocated to the UE. As indicated in Figure 2.17, the Cell_DCH state may be entered from the idle mode when an RRC connection is established, or from the Cell_FACH state when a dedicated physical channel is allocated. Based on the active set information, UTRAN knows which cells are providing service to the UE. Since the UE and UTRAN are communicating on a dedicated physical channel when the UE is in Cell_DCH state, there is no need for addressing. If the UE moves from

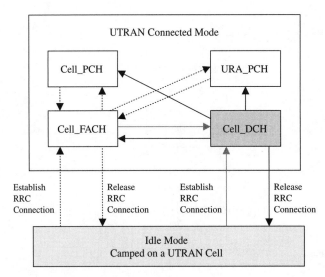

Figure 2.17 Cell_DCH state.

an old cell coverage area to a new cell coverage area, dedicated physical channels in the new cell are established while the dedicated physical channels in the old cell are released. The UE transitions to idle mode when the RRC connection is released. It may transition to other connected-mode states via explicit signaling.

A circuit-switched call needs a dedicated physical channel to deliver a guaranteed throughput. Therefore, it always operates in Cell_DCH state. A high data rate packet-switched call also operates in Cell_DCH state, especially if it is required to ensure a high throughput. However, low data rate packet-switched calls typically operate in Cell_FACH state to save the resource.

2.6.2.2 Cell_FACH State In the Cell_FACH state, no dedicated physical channel is allocated to the UE. As illustrated in Figure. 2.18, the Cell_FACH state may be entered from the idle mode or any of the other three states in the connected mode. When an RRC connection is established but no dedicated physical channel is setup, the UE moves from idle mode to the Cell_FACH state. The UE may also enter the Cell_FACH state from the Cell_DCH state when the dedicated physical channel is released due to low activity. In this case, dedicated logical channels are still allocated, but are mapped to common transport and physical channels. In addition, the UE may also move to the Cell_FACH state from the Cell_PCH and URA_PCH states when the UE needs to transmit signaling or user data from these two states.

In the Cell_FACH state, the UE continuously monitors the FACH. This is because UTRAN may send data or signaling to the UE at any time.

UMTS Fundamentals 27

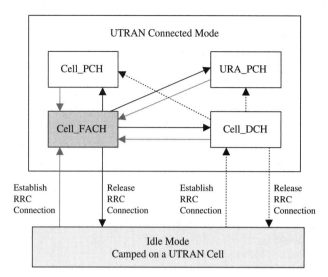

Figure 2.18 Cell_FACH state.

(For example, low rate packet-switched data.) There is no sleeping time for the UE in a Cell_FACH state. UTRAN knows the UE position at cell level based on the cell where the UE last performed a cell update. In the Cell_FACH state, UTRAN uses either a UTRAN radio network temporary identifier (U-RNTI) or a cell radio network temporary identifier (C-RNTI) to address the UE. A low rate packet-switched call generally operates in Cell_FACH state, because it does not require the high throughput provided by dedicated channels.

2.6.2.3 URA_PCH State In a URA_PCH state, no dedicated physical channel is allocated to the UE and no uplink activity can be carried out. As shown in Figure 2.19, the UE may enter a URA_PCH state from the Cell_FACH or Cell_DCH state. These transitions occur when there is a lack of activity during a packet-switched call. The UE sleeps most of the time in a URA_PCH state and only wakes up to monitor the PICH and the associated PCH during the paging occasions. If UTRAN wants to send data or signaling to the UE, UTRAN must first page the UE during the assigned paging occasion. Then, the UE moves to the Cell_FACH state and responds through the RACH. The UE may also move to the Cell_FACH state if it wants to send data or signaling on RACH.

In the URA_PCH state, UTRAN knows the UE location at the URA (UTRAN registration area) level because the UE needs to perform a URA update procedure when it moves to a new URA. In general, a URA may include a number of cells. UTRAN must page the UE in all cells of the URA. Similar to the Cell_FACH state, UTRAN addresses the UE using the UTRAN radio network temporary identifier (U-RNTI).

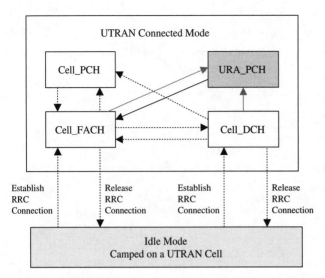

Figure 2.19 URA_PCH state.

2.6.2.4 Cell_PCH State The Cell_PCH state is similar to the URA_PCH state, except now the encompassed area is limited to just one cell. As shown in Figure 2.20, it may be entered from the Cell_FACH or Cell_DCH state. These transitions also happen when there is a lack of activity during a packet-switched call. The UE also sleeps most of the time in the Cell_PCH state and wakes up only to monitor the PICH and the associated PCH during the paging occasions. If UTRAN wants to send data or signaling to the UE, UTRAN must first page the UE

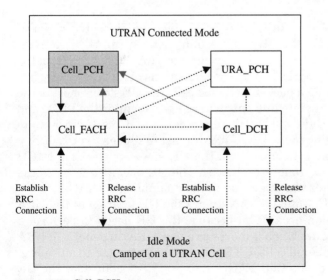

Figure 2.20 Cell_PCH state.

during the assigned paging occasion. UTRAN only needs to page the UE in one cell. The UE may also move from the Cell_PCH state to a Cell_FACH state if it wants to send data or signaling on RACH.

In Cell_PCH state, UTRAN knows the UE location at cell level according to the cell where the UE last performed a cell update in Cell_FACH state. UTRAN also use U-RNTI for addressing the UE in Cell_PCH state.

2.7 UE and Subscriber Identifiers

Identifiers are used for signalling messages exchanged between UTRAN and the UE on common channels. In other words, signalling messages on common channels must include an identifier. The identifier could be permanent or temporary and the type of identifier used depends on the message type. Some identifiers are related to the user, while others are related to the mobile. Identifiers related to the user are assigned by the core network and are stored in the USIM. These identifiers may be used with any mobile. Identifiers related to the mobile are stored in the memory inside the mobile. In the following subsections, we will discuss individual UE and subscriber identities, including IMSI, TMSI, radio network temporary identity (RNTI), and international mobile station equipment identity (IMEI).

2.7.1 International Mobile Subscriber Identity

IMSI is a permanent subscriber identifier stored in the USIM and the HLR. The UE uses IMSI to identify itself when it first registers with a network. Once the UE registers with the network, the network assigns a temporary identifier for the user to use from that point on. An IMSI is not sent over the air under normal conditions in order to prevent fraud and to protect a user's confidentiality.

As shown in Figure 2.21, an IMSI is not more than 15 digits and consists of three entities: mobile country code (MCC), mobile network

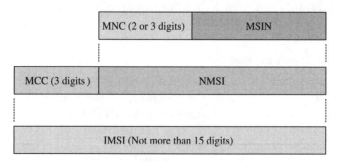

Figure 2.21 IMSI structure.

code (MNC), and mobile subscriber identification number (MSIN). The three-digit MCC identifies the country where the user subscribes to the networks. The MNC, which has two or three digits, identifies the home public land mobile network (PLMN) of the subscriber. The MSIN identifies a subscriber within a PLMN. A combination of MNC and MSIN is called a national mobile subscriber identity (NMSI).

The International Telecommunications Union (ITU), formerly known as the Consultative Committee on International Telegraphy and Telephony (CCITT), allocates the MCCs, while each individual country allocates the NMSI. If a country has many PLMNs, a unique MNC should be assigned to each PLMN.

2.7.2 Temporary Mobile Subscriber Identity

A TMSI or P-TMSI is assigned to a subscriber when the subscriber registers to the network for services. As mentioned earlier, using TMSI or P-TMSI instead of IMSI can prevent the IMSI from fraud and protect the user's identity. A VLR assigns the TMSI for circuit-switched services, while the SGSN assigns the P-TMSI for packet-switched services. TMSI is used by the UE and the VLR in the area where the VLR has sovereignty, while P-TMSI is used by the UE and the SGSN in the area controlled by SGSN. This is why the standards do not specify the identifier's internal structure and coding.

The length of a TMSI or a P-TMSI is 32 bits. These 32 bits cannot consist of only numeral 1s. The two most significant bits tell whether it is a circuit-switched TMSI or a packet-switched P-TMSI. For a circuit-switched TMSI, the two most significant bits could be 00, 01, or 10, while for a packet-switched P-TMSI, the two most significant bits are 11. All other bits are arbitrary and assigned by the VLR or SGSN.

If the UE uses a TMSI or a P-TMSI during its first contact with UTRAN, it must also provide UTRAN with a location area identity (LAI) or a routing area identity (RAI) in order to uniquely identify itself. In addition, the VLR or SGSN may also reassign a TMSI or a P-TMSI to the UE at any time. The reassignment may be carried out as part of the attach procedure, location area update procedure, or routing area update procedure.

2.7.3 Radio Network Temporary Identity

UTRAN always assigns the U-RNTI to the UE when it enters a Cell_FACH or Cell_DCH state. On top of the U-RNTI, UTRAN may also assign the C-RNTI to the UE at the same time. U-RNTI has 32 bits, among which 12 bits are for serving RNC, and 20 bits for serving RNC's RNTI. The C-RNTI has a length of 16 bits.

When the UE is in connected mode, the RRC uses the U-RNTI for all signaling messages exchanged between UTRAN and the UE on common logical channels. Typical examples include signaling messages in the cell update procedure, paging procedure, and all radio bearer assignment, release, and reconfiguration procedures. The MAC layer uses the C-RNTI when mapping DCCH and DTCH to FACH or RACH. However, the C-RNTI is used in the Cell_FACH state only.

2.7.4 International Mobile Station Equipment Identity

IMEI is a permanent identifier assigned to a mobile at the factory. It has a total of 15 hexadecimal digits. The first 6 digits constitute the type approval code (TAC). The TAC identifies the country where the mobile obtains the approval and the approval number. Type approval means that the mobile passes the tests defined in the mobile station conformance specification (MSCS). The next two digits, called the final assembly code (FAC), identify the place of manufacture or final assembly. The six digits that follow represent the serial number (SNR), which is assigned by the manufacturer for unique identification of the mobile. The last digit is a spare digit.

Similar to IMSI, IMEI is normally not sent over the radio link. But there are two exceptions. The first is that the network can request the UE to send the IMEI for the purpose of checking its validity. The second is that the IMEI may be used in the RRC Connection Request message for establishing an emergency call if the UE does not have a valid IMSI, TMSI, or P-TMSI.

2.8 System Frame Timing

Two frame timing numbers that are very important for UMTS system operations are the system frame number (SFN) and the connection frame number (CFN).

2.8.1 System Frame Number

The fundamental concept of system time within a cell is SFN, which increases one count every 10 milliseconds and repeats every 40.96 seconds. Therefore, its range is from 0 to 4095. UTRAN sends the SFN every 20 milliseconds on the BCH transport channel, which maps to PCCPCH physical channel.

UMTS is normally deployed as an asynchronous system, which means that frame timing among the Node Bs is not synchronized. The frame boundaries among Node Bs may be offset in time. They may also drift independently in time, and the SFN value for each Node B may be different.

SFN plays an important role in many procedures. For example, in the paging procedure, the paging indicator that the UE has to monitor depends on the UE's IMSI and the SFN of the serving Node B. Another example is handover procedure. The UE must report the measured time difference between the SFN of its serving cell and the SFN of the new cell before entering soft handover. As such, UTRAN can configure the dedicated channel for the new cell in such a way that the frame timing of the new cell is relatively close to that of the serving cell. This enables the UE to combine the symbols when the UE is in soft handover with multiple Node Bs.

2.8.2 Connection Frame Number

CFN specifies the frame timing of a dedicated physical channel (DPCH). Based on the SFN of the serving cell, the CFN is initialized at the moment when the DPCH is established. It then increases one count every 10 milliseconds and repeats every 2.56 seconds. Therefore, its range is from 0 to 255. CFN is never transmitted over the radio link and is maintained at the UE and UTRAN levels. CFN also plays an important role in many procedures, such as handover and ciphering.

2.9 Summary

The key elements that are indispensable for understanding a UMTS system include network topology, protocol stack, UMTS channels and channel mapping, frame timing, UE call states, and UE and subscriber identifiers. From the network topology viewpoint, a UMTS system consists of three major components: core network, access network and user equipment. The UMTS core network is similar to that of GSM/GPRS. The access network for UMTS is UTRAN, which employs WCDMA technology. The user equipment, which could be a mobile or a laptop, contains a USIM card for storing the user information.

The UMTS protocol stack is divided into two strata: the access stratum and the non-access stratum. The UMTS non-access stratum is basically the same as that of GSM, whereas the access stratum consists of physical, RLC, MAC, and RRC layers. The protocol stack is further divided into four basic protocol stacks: circuit-switched control plane protocol stack, circuit-switched user plan protocol stack, packet-switched control plane protocol stack, and packet-switched user plane protocol stack.

In access stratum, radio bearers and channels are used to carry signaling or data between the layers. Radio bearers carry signaling data between RRC and RLC layers, and user data between application layers and Layer 2. Logical channels carry information between RLC and MAC layers. Transport channels transfer information between MAC and physical layers. Physical channels transport information over the air.

UMTS channels are defined in terms of data flow direction, sharing, and functions. Downlink channels transport information from UTRAN to the UE, while uplink channels transport information from the UE to UTRAN. Common channels are shared by multiple UEs, while dedicated channels serve a single UE. Logical channels are defined by the type of information carried, either signaling or user data. Transport channels are defined by the method used for transferring the data. Physical channels are defined by physical mappings, channel frame structure, and physical layer procedure used to transport information over the air.

Channel mappings map logical channels to transport channels and transport channels to physical channels. Based on the types of channels, the mappings could be one-to-one mappings or one-to-multiple mappings. Some, but not all, mappings may occur at the same time.

A UE could be in idle mode or in connected mode. In idle mode, the UE is not active for any CS or PS call. In connected mode, the UE has an RRC connection with UTRAN and it could be in Cell_DCH, Cell_FACH, URA_PCH, or Cell_PCH state. It may enter the Cell_DCH state from idle mode when an RRC connection is established, or from the Cell_FACH state when a dedicated physical channel is established. It may enter the Cell_FACH state from idle mode or from any of the other three states in the connected mode. It enters the Cell_FACH state from idle mode when an RRC is established but no dedicated physical channel is setup. It enters the Cell_FACH state from the Cell_DCH state when the dedicated physical channel is released. Furthermore, when it is in the Cell_PCH or URA_PCH state and wants to send signaling or user data, it needs to transition to the Cell_FACH state. In addition, the UE may transition from the Cell_FACH or Cell_DCH state to the URA_PCH or Cell_PCH state when there is a lack of activity during a packet-switched call.

An identifier is required when UTRAN and UE exchange signalling messages on common channels. Depending on the message type, the identifier could be permanent or temporary, and it could be either user-related or mobile-related. A user-related identifier is assigned by the core network, stored in the USIM, and may be used with any mobile. A mobile-related identifier is stored in the memory within the mobile. The identifiers include IMSI, TMSI, RNTI, and IMEI.

SFN and CFN are very important for UMTS system operations. SFN is essential for system time within a cell, and it increases one count every 10 milliseconds and repeats every 40.96 seconds. Every 20 milliseconds, UTRAN broadcasts the SFN on the BCH transport channel. SFN is used in many procedures, such as paging and handover procedures.

CFN specifies the frame timing of a DPCH. When the DPCH is established, the CFN is initialized according to the SFN of the serving cell.

CFN increases one count every 10 milliseconds and repeats every 2.56 seconds. It is also used in many procedures, including handover and ciphering.

References

[1] 3GPP, TS21.905, v5.8.0, "Vocabulary for 3GPP Specifications" (Release 5).

[2] 3GPP, TS23.003, v5.8.0, "Numbering, Addressing and Identification" (Release 5).

[3] 3GPP, TS24.007, v5.1.0, "Mobile radio interface signaling layer 3; General Aspects" (Release 5).

[4] 3GPP, TS24.008, v5.12.0, "Mobile radio interface Layer 3 specification; Core network protocols; Stage 3" (Release 5).

[5] 3GPP, TS25.211, v5.3.0, "Physical channels and mapping of transport channels onto physical channels (FDD)" (Release 5).

[6] 3GPP, TS25.301, v5.2.0, "Radio Interface Protocol Architecture" (Release 5).

[7] 3GPP, TS25.331, v5.7.1, "Radio Resource Control (RRC) protocol specification" (Release 5).

Chapter

3

Overview of 3GPP and UMTS Standards

The 3rd Generation Partnership Project (3GPP) was established in December 1998. It is not a legal entity but it brings a number of telecommunications standards bodies together. These standards bodies are known as 3GPP organizational partners. The current organizational partners include the Association of Radio Industries and Business (ARIB) of Japan, the Alliance for Telecommunications Industry Solutions of North America (ATIS), the China Communications Standards Association (CCSA), the European Telecommunications Standards Institute (ETSI) of Europe, the Telecommunications Technology Association (TTA) of Korea, and the Telecommunications Technology Committee (TTC) of Japan.

In addition to these organizational partners, another category of partnership was also created within the project called market representation partners (MRP), in order to obtain a consolidated view of market requirements. Current MRPs include TD-SCDMA Forum of China, the Global Mobile Suppliers Association (GSA) of the UK, the GSM Association of Ireland, IPv6 Forum of the UK, UMTS Forum of the UK, 3G Americas of the US, and TD-SCDMA Industry Alliance of China. Furthermore, "observer" status may also granted to telecommunication standards bodies that have the potential to become organization partners but have not yet done so. Current observers include Telecommunications Industries Association (TIA) of the US, ICT Standards Advisory Council of Canada (ISACC), and the Australian Communications Industry Forum (ACIF).

One major responsibility of 3GPP is to create globally applicable technical specifications and technical reports for a 3rd generation mobile system based on GSM-evolved core network and the radio access technologies. The other responsibility of 3GPP is to maintain and develop GSM communication technical specifications and technical reports, including GPRS and EDGE. For a detailed description of the 3GPP, readers may refer to [1].

3.1 Technical Specification Groups

Within 3GPP, the organizational partners determine the general policy and strategy of 3GPP, while the Project Coordination Group (PCG) runs the projects. Under the PCG there are four Technical Specification Groups (TSGs): Services and System Aspects (SA), Core Network and Terminal (CT), Radio Access Networks (RAN), and GSM/EDGE Radio Access Network (GERAN). Each of these TSGs has a number of working groups made of individuals from companies that are members of one or more of the organizational partners, market representation partners, or observers.

3.1.1 Services and System Aspects Group

The responsibility of the Services and System Aspects group is in the areas of overall architecture and service capabilities of UMTS systems, definition and maintenance of the overall system architecture, definition of required bearers and services, development of service capabilities and a service architecture, charging, security, and network management.

3.1.2 Core Network and Terminals Group

The Core Network and Terminals group is responsible for specifying terminal interfaces, terminal capabilities, and the core network of UMTS systems. These include Layer 3 radio protocols between UE and CN, signalling between the core network nodes, interconnection with external networks, operation and maintenance requirements, generic user profile, descriptions of IP multimedia subsystem, and so forth.

3.1.3 GSM / EDGE Radio Access Network Group

The GSM/EDGE Radio Access Network group is responsible for the specification of the radio access part of GSM/EDGE. It includes the RF layer, Layers 1, 2, and 3, internal and external interfaces, conformance test specifications for GERAN base stations and terminals, and GERAN-specific operation and maintenance specifications for the GERAN nodes.

3.1.4 Radio Access Network Group

The Radio Access Network (TSG RAN) group is responsible for the definition of the functions, requirements, and interfaces of the UTRA network in its two modes, FDD & TDD. These include radio performance, physical layer, Layer 2 and Layer 3 radio resource specifications in UTRAN, specification of the access network interfaces (Iu, Iub, and Iur), definition of the O&M requirements in UTRAN, and conformance testing for base stations.

3.2 3GPP Specification Releases

The International Telecommunications Union (ITU) started to define the standard for third generation systems well before the 3GPP was established in 1998. At the same time, the European Telecommunications Standards Institute (ETSI) was responsible for the UMTS standards. In 1998, 3GPP took over the technical specification work.

3.3 3GPP Specification Numbering Scheme

There is a 3GPP specification number for all 3G and GSM specifications. The number consists of four or five digits (e.g. 04.03 or 25.101). The first two digits specify the specification series such as those listed in Table 3.1. For 01 to 13 series, two more digits follow the first two digits. For 21 to 55 series, three more digits follow the first two. The terms "3G" and "GSM" are used to mean a 3GPP system using a UTRAN radio access network and a 3GPP system using a GERAN radio access network, respectively. The term "GSM" also includes GPRS and EDGE features.

The full title, specification number and latest version number for every specification can be found at *ftp://ftp.3gpp.org/specs*.

For 21 to 35 series, if the third digit is a "0", then the specification applies to both 3G and GSM. If the third digit is not a "0", then the specification applies to 3G only. For example, while the Specification 29.001 applies to both 3G and GSM systems, Specifications 25.101 and 25.331 apply to 3G only.

3.4 3GPP Specification Series

The 3GPP specification series are summarized in Table 3.1. From an RF engineering standpoint, 3GPP radio aspects specification series are the most important and relevant specifications that need to be closely followed.

TABLE 3.1 3GPP Specification Series (Courtesy of ETSI)

Specification Series	3G/GSM R99 and Later	GSM Only (Release 4 and Later)	GSM Only (Before Release 4)
General information			00 series
Requirements	21 series	41 series	01 series
Service aspects ("stage 1")	22 series	42 series	02 series
Technical realization ("stage 2")	23 series	43 series	03 series
Signalling protocols ("stage 3")—user equipment to network	24 series	44 series	04 series
Radio aspects	25 series	45 series	05 series
CODECs	26 series	46 series	06 series
Data	27 series		07 series
Signalling protocols ("stage 3") —(RSS-CN)	28 series	48 series	08 series
Signalling protocols ("stage 3")—intra-fixed-network	29 series	49 series	09 series
Programme management	30 series	50 series	10 series
User Identity Module (SIM / USIM)	31 series	51 series	11 series
O&M	32 series	52 series	12 series
Access requirements and test specifications		13 series (1)	13 series (1)
Security aspects	33 series	(2)	(2)
SIM and test specifications	34 series	(2)	11 series
Security algorithms (3)	35 series	55 series	(4)

Note (1): The 13 series GSM specifications relate to European-Union-specific regulatory standards. On the closure of ETSI TC SMG, responsibility for these specifications was transferred to ETSI TC MSG, (Mobile Specification Group).
Note (2): The specifications of these aspects are spread throughout several series.
Note (3): Algorithms may be subject to export licensing conditions.
Note (4): The original GSM algorithms are not published and are controlled by the GSM Association.

3.5 Summary

3GPP develops and maintains the UMTS standards. It creates globally applicable technical specifications and technical reports for a 3rd generation mobile system based on GSM-evolved core network and the radio access technologies. It also maintains and develops GSM communication technical specifications and technical reports, including GPRS and EDGE.

A large number of technical specifications and reports have been published by 3GPP. From an RF engineering standpoint, 3GPP radio aspects specification series (25 Series) are the most important and relevant specifications that need to be closely followed by RF engineers.

Reference

[1] http://WWW.3gpp.org.

Chapter 4

Radio Resource Control

Radio Resource Control (RRC) is the captain of the access stratum. It performs overall control of the access stratum. As shown in Figure 4.1, RRC can configure the layers below it in the access stratum, through the control interface between the RRC layer and each of the other layers. Furthermore, RRC provides a control and signaling interface to the non-access stratum layer. More specifically, the RRC functions include access stratum control, system information processing, paging and notification, RRC connection management, ciphering and integrity protection control, radio bearer control, RRC management of UE mobility, power control, measurement control and reporting, and NAS layer message routing.

4.1 RRC Message Specifications

An RRC message is a piece of information exchanged between UTRAN and UE, when RRC is performing its functions. Typical examples include RRC Connection Request message, RRC Connection Setup message, RRC Connection Setup Complete message, RRC Connection Release message, Security Mode Command message, Radio Bearer Setup message, Radio Bearer Reconfiguration message, Radio Bearer Release message, and so forth.

RRC messages are defined and encoded by using Abstract Syntax Notation One (ASN.1) format. With the exception of the system information message, a message contains a message type field that allows the message to be correctly decoded. For system information messages, the message type is embedded in each segment of the message because system information messages are segmented across multiple frames. The contents of a message are a list of Information Elements (IEs). These information elements are marked as either mandatory present (MP), mandatory with default value (MD), conditional on value (CV),

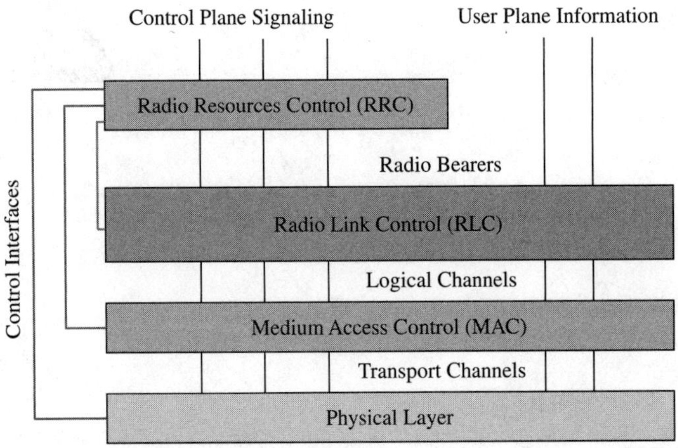

Figure 4.1 RRC control interfaces.

or conditional on history (CH). The meanings of each of these terms are described in Table 4.1, which is extracted from [1] with permission.

4.1.1 Protocol Extensions

As stated in [1], RRC messages may be extended in future versions of the protocol by adding values for choices, enumerated types, size constrained types, and information elements. Two types of extension are defined in [1]: *non-critical* and *critical extensions*.

For non-critical extensions, new values for choices, enumerated types, and size-constrained types may be added where spare values have been allocated. It is possible to indicate in the current protocol version how many non-critical spare values need to be reserved for future extensions. Additional information elements may be added to RRC messages by appending them at the end of the messages. If a receiver does not comprehend the non-critical extensions, the receiver shall accept the extensions and process the entire message as if the extensions were not included.

For critical extensions, the general approach is to define a new version of the message and then indicate it at the beginning of the message. No spare values are reserved since backward compatibility is not required. These messages may be modified completely. For example, information elements may be inserted, removed, or redefined. If a receiver does not comprehend the critical extensions, the receiver rejects the whole message. There is no partial rejection.

4.2 System Information

The UTRAN RRC broadcasts system information messages to all UEs. The messages contain information elements carrying information about access stratum and non-access stratum. The system information elements

TABLE 4.1 Meaning of Abbreviations Used in RRC Messages and Information Elements (Courtesy of ETSI)

Abbreviation	Meaning
Mandatory present (MP)	A value for this information is always needed, and no information is provided about a particular default value. If the transfer syntax allows absence (for example, due to extension), then absence leads to an error diagnosis.
Mandatory with default value (MD)	A value for this information is always needed, and a particular default value is mentioned (in the Semantical information column). This opens the possibility for the transfer syntax to use absence or a special pattern to encode the default value.
Conditional on value (CV)	The need for a value for that information depends on the value of some other IE or IEs, and/or on the message flow (for example, channel or SAP). The need is specified by means of a condition, the result of which may be that the information is mandatory present, mandatory with default value, not needed, or optional.
	If one of the results of the condition is that the information is mandatory present, the transfer syntax must allow for the presence of the information. If in this case the information is absent, an error is diagnosed.
	If one of the results of the condition is that the information is mandatory with default value, and a particular default value is mentioned (in the Semantical information column), the transfer syntax may use absence or a special pattern to encode the default value.
	If one of the results of the condition is that the information is not needed, the transfer syntax must allow encoding the absence. If in this case the information is present, it will be ignored. In specific cases however, an error may be diagnosed instead.
	If one of the results of the condition is that the information is optional, the transfer syntax must allow for the presence of the information. In this case, neither absence nor presence of the information leads to an error diagnosis.
Conditional on history (CH)	The need for a value for that information depends on information obtained in the past (for example, from messages received in the past from the peer). The need is specified by means of a condition, the result of which may be that the information is mandatory present, mandatory with default value, not needed, or optional.
	The handling of the conditions is the same as described for CV.
Optional (OP)	The presence or absence is significant and modifies the behavior of the receiver. However, whether the information is present or not does not lead to an error diagnosis.

are broadcast in system information blocks. A system information block (SIB) contains one or more system information elements. There are many types of system information blocks. The characteristics may be different for different system information blocks. For example, the repetition rate of two different system information blocks may be different.

System information is organized as a tree. The master information block (MIB) provides references and scheduling information to a number of system information blocks in a cell. The MIB may also contain reference and scheduling information for one or two scheduling blocks (SB). Scheduling blocks provide references and scheduling information for additional SIBs. Scheduling information for an SIB may only be contained in either the MIB or one of the scheduling blocks.

The MIB, SBs, and SIBs are broadcast on the BCH in a group of system information messages. When a UE first camps on a cell, it gets all system information messages for that cell from the BCH. The UE may store the system information messages for a given cell. So, if the UE moves out of the cell and then moves back in later on, it can use the stored system information messages instead of getting them from the BCH again.

4.2.1 System Information Blocks

There are many types of SIBs ranging from Type 1 to Type 18. With the exceptions of SIB Types 15.2, 15.3, and 16, the content is the same for each occurrence for all SIBs that use value tags. SIB Types 15.2, 15.3, and 16, may occur more than once with different contents. In this case, each occurrence of the system information block is provided with scheduling information. The content for each occurrence may be different for all SIBs that do not use value tags.

All SIBs and their characteristics are listed in Table 4.2, which is extracted from [1] with permission.

The Area Scope column in Table 4.2 defines the area where the value tag for a system information block is valid. If the area scope is defined as Cell, the system information block is valid only in the cell in which it is read. If system information blocks have been stored in the UE for a cell, the UE checks whether the value tag for the system information block of this cell is the same as the stored value tag. If so, the UE does not need to re-read the system information block. Otherwise, it re-reads the system information block.

If the area scope is PLMN or Equivalent PLMN, the system information block is valid only for the PLMN or equivalent PLMN. The UE checks the value tag for the system information block when it selects a new cell. If the value tag for the system information block in the new cell is different from that stored in the UE, the UE re-reads the system information block.

There may be multiple occurrences of system information block types 15.2, 15.3, and 16. Each occurrence has its own value tag. The UE re-reads the system information block of a particular occurrence if the value tag for this occurrence is different from that stored in the UE.

The UE Mode/State When Block is Valid column in Table 4.2 identifies the UE state or UE mode in which the IEs in a system information block are valid. In other words, the received system information block becomes invalid when UE transitions to a mode/state that is not included in this column. However, system information block type 16 remains valid when UE transitions to or from GSM/GPRS.

TABLE 4.2 System Information Blocks and Their Characteristics (Courtesy of ETSI)

System Information Block	Area Scope	UE Mode/State When Block is Valid	UE Mode/State When Block is Read	Scheduling Information	Modification of System Information
Master information block	Cell	Idle mode, CELL_FACH, CELL_PCH, URA_PCH	Idle mode, CELL_FACH, CELL_PCH, URA_PCH	SIB_POS = 0 SIB_REP = 8 (FDD) SIB_REP = 8, 16, 32 (TDD) SIB_OFF=2	Value tag
Scheduling block 1	Cell	Idle mode, CELL_FACH, CELL_PCH, URA_PCH	Idle mode, CELL_FACH, CELL_PCH, URA_PCH	Specified by the Scheduling information IE in MIB	Value tag
Scheduling block 2	Cell	Idle mode, CELL_FACH, CELL_PCH, URA_PCH	Idle mode, CELL_FACH, CELL_PCH, URA_PCH	Specified by the Scheduling information IE in MIB	Value tag
System information block type 1	PLMN	Idle mode CELL_FACH, CELL_PCH, URA_PCH, CELL_DCH	Idle, CELL_FACH, CELL_PCH, URA_PCH	Specified by the Scheduling information IE	Value tag
System information block type 2	Cell	URA_PCH	URA_PCH	Specified by the Scheduling information IE	Value tag
System information block type 3	Cell	Idle mode, CELL_FACH, CELL_PCH, URA_PCH	Idle mode, CELL_FACH, CELL_PCH, URA_PCH	Specified by the Scheduling information IE	Value tag
System information block type 4	Cell	CELL_FACH, CELL_PCH, URA_PCH	CELL_FACH, CELL_PCH, URA_PCH	Specified by the Scheduling information IE	Value tag
System information block type 5	Cell	Idle mode, CELL_FACH, CELL_PCH, URA_PCH, CELL_DCH (TDD only)	Idle mode, CELL_FACH, CELL_PCH, URA_PCH, CELL_DCH (TDD only)	Specified by the Scheduling information IE	Value tag
System information block type 6	Cell	CELL_FACH, CELL_PCH, URA_PCH, CELL_DCH (TDD only)	CELL_FACH, CELL_PCH, URA_PCH, CELL_DCH (TDD only)	Specified by the Scheduling information IE	Value tag
System information block type 7	Cell	Idle mode, CELL_FACH, CELL_PCH, URA_PCH, CELL_DCH (TDD only)	Idle mode, CELL_FACH, CELL_PCH, URA_PCH, CELL_DCH (TDD only)	Specified by the Scheduling information IE	Expiration timer = MAX(32, SIB_REP × Expiration TimeFactor)
System information block type 8	Cell	CELL_FACH, CELL_PCH, URA_PCH	CELL_FACH, CELL_PCH, URA_PCH	Specified by the Scheduling information IE	Value tag

(Cont.)

TABLE 4.2 System Information Blocks and Their Characteristics (Courtesy of ETSI)

System Information Block	Area Scope	UE Mode/State When Block is Valid	UE Mode/State When Block is Read	Scheduling Information	Modification of System Information
System information block type 9	Cell	CELL_FACH, CELL_PCH, URA_PCH	CELL_FACH, CELL_PCH, URA_PCH	Specified by the Scheduling information IE	Expiration timer = SIB_REP
System information block type 10	Cell	CELL_DCH	CELL_DCH	Specified by the Scheduling information IE	Expiration timer = SIB_REP
System information block type 11	Cell	Idle mode CELL_FACH, CELL_PCH, URA_PCH, CELL_DCH	Idle mode CELL_FACH, CELL_PCH, URA_PCH	Specified by the Scheduling information IE	Value tag
System information block type 12	Cell	CELL_FACH, CELL_PCH, URA_PCH, CELL_DCH	CELL_FACH, CELL_PCH, URA_PCH	Specified by the Scheduling information IE	Value tag
System information block type 13	Cell	Idle Mode, CELL_FACH, CELL_PCH, URA_PCH	Idle Mode, CELL_FACH, CELL_PCH, URA_PCH	Specified by the Scheduling information IE	Value tag
System information block type 13.1	Cell	Idle Mode, CELL_FACH, CELL_PCH, URA_PCH	Idle Mode, CELL_FACH, CELL_PCH, URA_PCH	Specified by the Scheduling information IE	Value tag
System information block type 13.2	Cell	Idle Mode, CELL_FACH, CELL_PCH, URA_PCH	Idle Mode, CELL_FACH, CELL_PCH, URA_PCH	Specified by the Scheduling information IE	Value tag
System information block type 13.3	Cell	Idle Mode, CELL_FACH, CELL_PCH, URA_PCH	Idle Mode, CELL_FACH, CELL_PCH, URA_PCH	Specified by the Scheduling information IE	Value tag
System information block type 13.4	Cell	Idle Mode, CELL_FACH, CELL_PCH, URA_PCH	Idle Mode, CELL_FACH, CELL_PCH, URA_PCH	Specified by the Scheduling information IE	Value tag
System information block type 14	Cell	Idle Mode, CELL_FACH, CELL_PCH, URA_PCH, CELL_DCH	Idle Mode, CELL_FACH, CELL_PCH, URA_PCH, CELL_DCH	Specified by the Scheduling information IE	Expiration timer = MAX(32, SIB_REP × Expiration TimeFactor)
System information block type 15	Cell	Idle Mode, CELL_FACH, CELL_PCH, URA_PCH	Idle Mode, CELL_FACH, CELL_PCH, URA_PCH	Specified by the Scheduling information IE	Value tag
System information block type 15.1	Cell	Idle Mode, CELL_FACH, CELL_PCH, URA_PCH	Idle Mode, CELL_FACH, CELL_PCH, URA_PCH	Specified by the Scheduling information IE	Value tag

(*Cont.*)

TABLE 4.2 System Information Blocks and Their Characteristics (Courtesy of ETSI)

System Information Block	Area Scope	UE Mode/State When Block is Valid	UE Mode/State When Block is Read	Scheduling Information	Modification of System Information
System information block type 15.2	Cell	Idle Mode, CELL_FACH, CELL_PCH, URA_PCH	Idle Mode, CELL_FACH, CELL_PCH, URA_PCH	Specified by the Scheduling information IE	Value tag
System information block type 15.3	PLMN	Idle Mode, CELL_FACH, CELL_PCH, URA_PCH	Idle Mode, CELL_FACH, CELL_PCH, URA_PCH	Specified by the Scheduling information IE	Value tag
System information block type 15.4	Cell	Idle Mode, CELL_FACH, CELL_PCH, URA_PCH	Idle Mode, CELL_FACH, CELL_PCH, URA_PCH	Specified by the Scheduling information IE	Value tag
System information block type 15.5	Cell	Idle Mode, CELL_FACH, CELL_PCH, URA_PCH	Idle Mode, CELL_FACH, CELL_PCH, URA_PCH	Specified by the Scheduling information IE	Value tag
System information block type 16	Equivalent PLMN	Idle Mode, CELL_FACH, CELL_PCH, URA_PCH, CELL_DCH	Idle Mode, CELL_FACH, CELL_PCH, URA_PCH	Specified by the Scheduling information IE	Value tag
System information block type 17	Cell	CELL_FACH, CELL_PCH, URA_PCH, CELL_DCH	CELL_FACH, CELL_PCH, URA_PCH, CELL_DCH	Specified by the Scheduling information IE	Expiration timer = SIB_REP
System information block type 18	Cell	Idle mode, CELL_FACH, CELL_PCH, URA_PCH, CELL_DCH	Idle mode, CELL_FACH, CELL_PCH, URA_PCH	Specified by the Scheduling information IE	Value tag

The UE Mode/State When Block is Read column in Table 4.2 identifies the UE state or UE mode in which the IEs in a system information block may be read by the UE. The UE must have the necessary information before it executes any procedure. This requires that information be obtained from the appropriate system information block. The time that the UE must read the system information may be derived from procedure specifications that set the required IEs for different UE modes/states and from performance requirements.

4.2.2 System Information Block Segmentation and Concatenation

As mentioned earlier, the system information blocks are broadcast on BCH in a set of system information messages. The size of the system information message has to fit the size of a BCH transport block. UTRAN RRC segments or concatenates the system information blocks to fit the size of a BCH transport block. If the encoded system information block is larger than the system information message, it is segmented and transmitted in several messages. If the encoded system information block is smaller than the system information message, UTRAN may concatenate several system information blocks or their segments into the same message. There are four different types of segments defined for a system information block: first segment, subsequent segment, last segment, and complete. The first segment, subsequent segment, and last segment are used to transfer segments of an MIB, an SB, or an SIB. The complete is used to transfer a complete MIB, a complete SB, or a complete SIB. Each segment contains a header and a data field. The data field carries the encoded system information elements. The header contains SIB type and segment type specific parameters. The SIB type identifies whether the segment belongs to an MIB, an SB, or an SIB. Segment type specific parameters are SEG_COUNT and segment index. The header contains SEG_COUNT (number of segments in the system information block) only if the segment type is first segment. Similarly, it contains segment index only if the segment type is subsequent segment or last segment.

UTRAN broadcasts a system information message on the BCH every 20 ms. The message contains the system frame number (SFN), which is always even because the message is transmitted on 20 ms boundaries, while SFN increases one count every 10 ms. UTRAN may combine one or several segments of different lengths in the same system information message. Each message contains one of the following 11 types of combinations [1]:

1. No segment
2. First segment
3. Subsequent segment
4. Last segment
5. Last segment + first segment
6. Last segment + one or several complete
7. Last segment + one or several complete + first segment
8. One or several complete
9. One or several complete + first segment
10. One complete of size 215 to 226
11. Last segment of size 215 to 222

If there is no MIB, SB, or SIB scheduled for a specific BCH transport block, the No segment combination is used.

4.2.3 Example of a System Information Message

Among the 11 types of combinations for system information messages, the Type 7 combination contains more variety than any other type. Therefore, we use the Type 7 combination as an example to illustrate the structure of a system information message. Figure 4.2 shows the structure of the Type 7 combination (Last segment + one or several complete + first segment).

As can be seen from Figure 4.2, the first segment consists of a header and a data field. The header contains an SIB type and a segment count, while the data field carries the SIB bits. The last segment also consists of a header and a data field. The only difference is that the header now carries a segment index instead of a segment count. Up to a total of 16 segments are allowed in this system information combination, and each segment can have a length of up to 214 bits. Based on segment count and segment index, the UE can reassemble the complete SIB and detect the missing segments. The complete SIB portion of the message may contain up to 16 complete SIBs, each of which may have a length of up to 214 bits.

4.2.4 Contents of System Information Blocks

System information blocks were summarized in Table 4.2 in Section 4.2.1. Each system information block has its unique function and structure. The system information blocks are described individually in the following subsections.

4.2.4.1 Master Information Block

UTRAN sends the MIB every 80 milliseconds. The MIB (Figure 4.3) consists of the following IEs: MIB value tag, supported PLMN types, references to SIBs and SBs, and PLMN identity. The MIB value tag is an integer ranging from 1 to 8. It increases by one number whenever the contents of the MIB change. The supported-PLMN-types IE identifies the types of supported PLMNs. One possible example could be GSM-MAP PLMN. The PLMN identity contains the

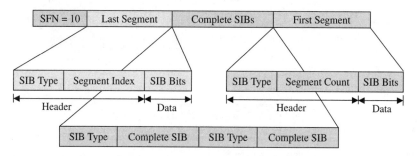

Figure 4.2 System information message–Combination Type 7.

48 Chapter Four

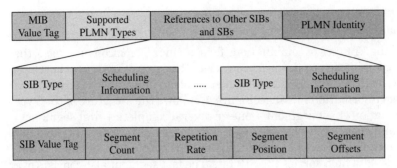

Figure 4.3 Master information block.

mobile country code (MCC) and mobile network code (MNC) of the GSM-MAP PLMN. The references-to-other-SIBs-and-SBs IE contains the SIB type and scheduling information for each SIB and SB. Scheduling information contains the SIB value tag, segment count, repetition rate at which the SIB is transmitted, the position of the first segment within one cycle of the cell SFN, and the offsets of subsequent segments.

4.2.4.2 System Information Block Type 1 SIB Type 1 (Figure 4.4) contains the common NAS information, core network domain information, and the UE constants and timers. The core network domain information includes domain identification, domain specific NAS information, and the domain specific DRX cycle coefficient. The common NAS information and domain-specific NAS information are defined in the CS and PS core network specifications. They include location area code (LAC), routing area code (RAC), and timers for location update. The UE uses the domain specific DRX cycle coefficient to determine the time period in which the UE monitors the paging channel when it operates in idle mode, URA_PCH state, or Cell_PCH state. The UE timers and constants are used

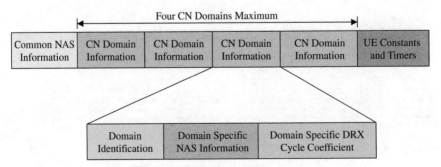

Figure 4.4 System information block type 1.

for various purposes. For example, T302 determines the time that the UE attempts to update cell or URA information, and N302 specifies the maximum number of retransmissions of the CELL UPDATE or URA UPDATE message.

4.2.4.3 System Information Block Type 2 SIB Type 2 consists of up to eight URA identities that are valid in the current cell. If there are multiple valid URA identities for a cell, UTRAN instructs the UE which URA identity to use when the UE enters the URA_PCH state.

4.2.4.4 System Information Block Type 3 SIB Type 3 contains information pertaining to cell reselection. It is used when the UE operates in idle mode. Figure 4.5 illustrates the structure of SIB Type 3. The IEs in SIB Type 3 include SIB Type 4 indicator, cell reselection parameters, cell access restriction parameters, and cell identity. The SIB Type 4 indicator tells whether SIB Type 4 is transmitted in the current cell. If SIB Type 4 is transmitted, the UE uses SIB Type 4 when it operates in connected mode. Otherwise, the UE still uses SIB Type 3 when it operates in connected mode.

The cell reselection parameters IEs provide the following information: quality measure, reselection thresholds, reselection timer, and maximum uplink transmit power. The quality measure is used to determine whether a cell is suitable. The reselection timer and reselection thresholds, including hysteresis, are used for cell reselection computations. The maximum uplink transmit power determines the maximum power the UE is allowed to transmit in the current cell.

The cell access restriction IEs consist of cell restricted indicators, cell-barred indicators, cell-barred timers, and access class barred indicators. These parameters decide whether a cell is barred or restricted and the duration for which the cell is barred. They also decide whether a cell is prohibited from being accessed by UEs of a particular access class.

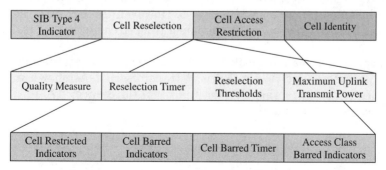

Figure 4.5 System information block type 3.

50 Chapter Four

4.2.4.5 System Information Block Type 4 SIB Type 4 is identical to SIB Type 3 except that it does not have the SIB Type 4 indicator IE. SIB Type 4 serves the same purpose as SIB Type 3. However, the UE uses the parameters from SIB Type 4 only when operating in connected mode.

4.2.4.6 System Information Block Type 5 SIB Type 5 contains information pertaining to the common physical channels PCCPCH, SCCPCH and PRACH. It is used when the UE operates in idle mode. Figure 4.6 shows the contents of SIB Type 5. The IEs in SIB Type 5 include SIB Type 6 indicator, power offsets, PCCPCH transmit diversity, PRACH information, SCCPCH information, and cell broadcast service discrete receiving (CBS DRX) information.

The SIB Type 6 indicator tells whether SIB Type 6 is transmitted in the current cell. If SIB Type 6 is transmitted, the UE uses SIB Type 6 when it operates in connected mode. Otherwise, the UE still uses SIB Type 5 when it operates in connected mode.

The power-offsets IE defines the power offsets for PICH and AICH. The PCCPCH transmit diversity IE uses a Boolean to indicate whether transmit diversity is used on the PCCPCH. The PRACH information includes many physical and transport channel parameters, such as available signatures, spreading factors, preamble scrambling code number, power offsets, puncturing limit, transport format set, transport format combination set, number of preamble cycles, and random back-off parameters.

The SCCPCH information also includes many physical and transport channel parameters, such as spreading factor, code number, timing offset relative to PCCPCH, PICH channelization code, number of paging indicators per frame, and transport format set and transport channel combination set for PCH and FACH. The last IE, CBS DRX information, contains the parameters for computing the DRX cycle of the cell broadcast service.

4.2.4.7 System Information Block Type 6 SIB Type 6 is identical to SIB Type 5 except that it does not have SIB Type 6 indicator IE. SIB Type 6 serves the same purpose as SIB Type 5. However, the UE uses the parameters from SIB Type 6 only when operating in connected mode.

4.2.4.8 System Information Block Type 7 SIB Type 7 contains fast changing parameters related to PRACH transmissions. As shown in Figure 4.7, it contains the IEs for uplink interference, dynamic persistence, and expiration time factor.

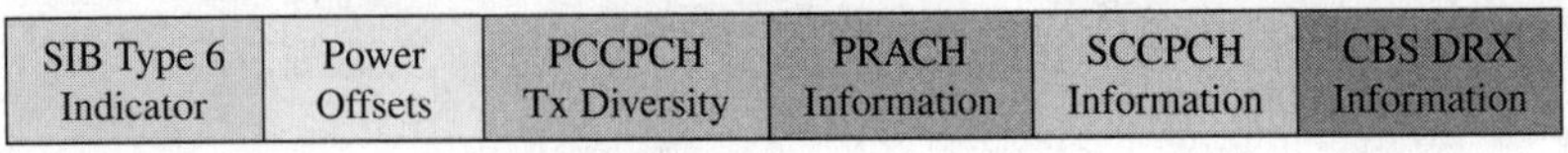

SIB Type 6 Indicator	Power Offsets	PCCPCH Tx Diversity	PRACH Information	SCCPCH Information	CBS DRX Information

Figure 4.6 System information block type 5.

| Uplink Interference | Dynamic Persistence | ..... | Dynamic Persistence | Expiration Time Factor |

Figure 4.7 System information block type 7.

The uplink interference IE is used for calculating the initial preamble power in PRACH transmissions. The dynamic persistence IE contains an assigned value N = 1, 2, 3, ..., 8; one dynamic persistence IE for one PRACH listed in SIB Type 5 and SIB Type 6. The assigned value maps to a persistence value given as

$$P(N) = 1/2^{(N-1)}. \qquad (4\text{-}1)$$

The MAC layer determines the priority of the attempts that try to access the PRACH using this persistence value.

The expiration time factor is used as a change control mechanism for ensuring that the UE always has the correct values for the parameters in SIB Type 7. The expiration timer is calculated as

Expiration timer = Max (32, SIB_REP × ExpirationTimeFactor) ms.

$$(4\text{-}2)$$

4.2.4.9 System Information Block Type 8 SIB Type 8 contains static CPCH information to be used in the cell, such as CPCH parameters. It is used only for FDD. The UE stores all relevant IEs included in this system information block when it is in connected mode. But, the UE does not use the values of the IEs in this system information block when it is in idle mode.

4.2.4.10 System Information Block Type 9 SIB Type 9 also contains CPCH information to be used in the cell. The information includes a CPCH set persistence levels list and CPCH set persistence levels. SIB Type 9 is also used only for FDD. When the UE is in connected mode, it stores all relevant IEs contained in this system information block. The UE starts a timer set to the value given by the repetition period (SIB_REP) for that system information block. When the UE is in idle mode, it does not use the values of the IEs in this system information block.

4.2.4.11 System Information Block Type 10 SIB Type 10 contains information to be used by the UEs with their DCH being controlled by a dynamic resource allocation control (DRAC) procedure. It is used only for FDD. When the UE is in Cell_DCH state, it stores all relevant IEs included in this system information block. The UE starts a timer set to the value given by the repetition period (SIB_REP) for that system information block and performs the DRAC procedure [1]. When the UE is in idle mode, Cell_FACH, Cell_PCH or URA_PCH state, it does not use the values of the IEs in this system information block.

4.2.4.12 System Information Block Type 11 SIB Type 11 contains measurement control information to be used in a cell. As shown in Figure 4.8, it consists of a SIB Type 12 indicator, FACH measurement occasion information, and measurement control information. The SIB Type 12 indicator indicates whether SIB Type 12 is transmitted or not. The FACH measurement occasion information contains IEs for the cycle length coefficient, FDD indicator, inter-RAT indicator, and TDD indicator; while the measurement control information contains IEs for UE internal, traffic volume, intra-frequency, inter-frequency, and inter-RAT measurements.

The UE uses the information contained in the FACH measurement occasion IEs to perform cell reselection measurements when it is in Cell_FACH state. The FDD indicator, inter-RAT indicator, and TDD indicator specify whether the UE should perform FDD, inter-RAT, and TDD measurements, respectively. The UE also uses the cycle length coefficient to calculate a periodic interval during which these measurements may be performed.

The UE internal measurements IE specifies whether the UE should measure UE transmit power, UTRA carrier RSSI, and/or UE Rx-Tx time difference.

The UE uses the information contained in the traffic volume measurements IEs to perform traffic volume measurements when it is in connected mode. These measurements tell UTRAN the available uplink data during a packet data call. They also help UTRAN to decide the channel resources needed to transfer the uplink data.

The intra-frequency, inter-frequency, and inter-RAT measurement IEs contain similar information. Each IE carries a list of cells to be added to the measurement procedure and a list of cells to be removed. If a cell is to be added, the IE will also carry the cell selection and reselection

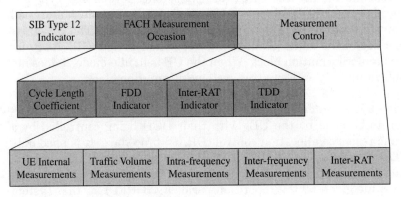

Figure 4.8 System information block type 11.

parameters of that cell, such as offset, minimum quality, and maximum allowed transmit power. These parameters will assist the UE to determine whether a neighbor cell is suitable for reselection based on the measurements taken prior to reselection of that cell.

4.2.4.13 System Information Block Type 12 SIB Type 12 also contains measurement control information. It is used when the UE is in connected mode. It is similar to SIB Type 11, except that there is no SIB Type 12 indicator. All other IEs contain the same information as those of SIB Type 11.

4.2.4.14 System Information Block Type 13 SIB Type 13 contains ANSI-41 system information. According to [1], no matter whether the UE is in idle mode or in connected mode, it has to store all relevant IEs included in SIB Type 13 except for a few IEs: CN domain specific DRX cycle length coefficient, UE timers and constants in idle mode, and capability update requirement. These IEs have to be stored only when the UE is in idle mode. Furthermore, if the PLMN type in the variable SELECTED_PLMN is set to ANSI-41 and the PLMN type IE in the MIB is set to ANSI-41 or GSM-MAP and ANSI-41, the UE must read SIB Type 13 and the associated SIB Types 13.1, 13.2, 13.3, and 13.4. In addition, the UE must also forward the content of CN domain specific NAS system information to the non-access stratum entity indicated by the CN domain identity, and use the CN domain specific DRX cycle length coefficient to calculate the frame number for the paging occasions and page indicator as specified in [2].

The SIB Types 13.1, 13.2, 13.3, and 13.4 contain the ANSI-41 RAND information, the ANSI-41 user zone identification information, the ANSI-41 private neighbor list information, and the ANSI-41 global service redirection information respectively.

4.2.4.15 System Information Block Type 14 SIB Type 14 is used only in 3.84 Mcps TDD. It contains parameters for common and dedicated physical channel uplink outer loop power control information for both idle and connected modes. The UE stores all relevant IEs contained in this system information block. The UE must use UL Timeslot Interference to compute the PRACH and DPCH transmit powers for the TDD uplink open loop power control.

4.2.4.16 System Information Block Type 15 SIB Type 15 contains information useful for UE-based or UE-assisted positioning methods. There are also SIB Types 15.1, 15.2, 15.3, 15.4, and 15.5 specified in the 3GPP specification [1]. We will not discuss each of these SIBs in detail, but just give a brief introduction for each of them instead.

If a UE supports GPS location services, it should store all relevant IEs included in SIB Type 15. When the GPS data ciphering information IE is included, the UE must act on the received reference position IE and GPS reference time IE. When the satellite information IE is included, the UE has to act on the list of bad satellites. Details on the action of UE can be found in [1].

SIB Type 15.1 contains information useful for UE positioning differential GPS (DGPS) corrections. The DGPS corrections message contents are based on a Type-1 message of DGPS specified in [3]. The UE should store all the relevant IEs included in this SIB in the variable UE_POSITIONING_GPS_DATA. The UE must act on DGPS information in the DGPS corrections IE and the received UE positioning GPS DGPS corrections IE.

SIB Type 15.2 contains information useful for the GPS navigation model [4]. It may have multiple occurrences, one for each satellite. Its scheduling information includes the SIB-occurrence-identity-and-value-tag IE to identify the different occurrences. The UE stores all the relevant IEs included in this SIB in UE_POSITIONING_GPS_DATA. For each occurrence, the UE must compare the value tag of the stored occurrence with the value tag included in the SIB-occurrence-identity-and-value-tag IE for the occurrence of the SIB with an identical occurrence identity. Details on the UE action based on SIB Type 15.2 can also be found in [1].

Similar to SIB Type 15.2, SIB Type 15.3 contains information useful for ionosphere delay, coordinated universal time (UTC) offset, and almanac. It may also have multiple occurrences, one for each set of satellite data. The UE actions based on SIB Type 15.3 are identical to that based on SIB Type 15.2. Interested readers may refer to [1] for more information.

SIB Type 15.4 contains ciphering information for SIB Type 15.5 and information useful for the observed-time-difference-of-arrival (OTDOA) UE-assisted positioning method. SIB Type 15.5 contains information useful for the OTDOA UE-based positioning method. Details on SIB Types 15.4 and 15.5 are described in [1].

4.2.4.17 System Information Block Type 16 SIB Type 16 contains radio bearer, transport channel, and physical channel parameters to be stored by UE in idle and connected modes. It may have multiple occurrences, one for each predefined configuration. Its scheduling information includes the predefined-configuration-identity-and-value-tag IE to identify the different predefined configurations. The UE stores all relevant IEs included in this system information block. For each predefined configuration, the UE must compare the value tag of the stored predefined configuration with the preconfiguration value tag in the predefined-configuration-identity-and-value-tag IE for the occurrence of the SIB with identical predefined configuration identity.

4.2.4.18 System Information Block Type 17
SIB Type 17 contains fast-changing parameters for the configuration of the shared physical channels to be used in connected mode. It is used only for TDD. Interested readers may refer to [1] for more information.

4.2.4.19 System Information Block Type 18
SIB Type 18 contains IEs that list the PLMN identities of neighboring cells. There are three lists, one for intra-frequency neighbors, another for inter-frequency neighbors, and still another for inter-RAT neighbors (GSM). Different sets of lists are used for idle mode and connected mode. These lists provide the information for the UE to perform cell reselection to equivalent PLMNs. The PLMN ID of a neighbor cell is compared with that of the listed equivalent PLMNs when the cell is considered for cell reselection. If the cell belongs to an equivalent PLMN, reselection of the cell will be allowed. Use of equivalent PLMNs enables a service provider to offer international seamless coverage. To achieve this, a service provider is allocated a number of PLMN IDs. A PLMN ID is made of a mobile country code (MCC) and a mobile network code (MNC).

4.3 Paging and Notification

In a mobile-terminated voice call setup there are several steps involved, including paging and notification, RRC connection setup, routing of NAS layer messages for mobility management, ciphering and integrity protection, routing of NAS layer messages for call control, and radio bearer management. Paging and notification is the first step for mobile-terminated voice call setup. There are two types of paging: Paging Type 1 and Paging Type 2.

4.3.1 Paging Type 1

Paging Type 1 is used when UE is in idle mode, Cell_PCH state, or URA_PCH state. When the UE is in idle mode, Paging Type 1 is used to establish an RRC connection to deliver a call. When UE is in Cell_PCH or URA_PCH state, it is used to resume user data transmission during a packet data session. It should be noted that when UE is in the Cell_PCH or URA_PCH state, the radio bearers are released.

Figure 4.9 shows the contents of a Paging Type 1 message. The message consists of a message type, an optional paging record list, a number of paging records, and an optional BCCH modification IE. The paging record list gives the number of paging records. A paging record could be for idle mode paging or connected mode paging. The idle mode paging record contains paging cause, CN domain identity, and UE identity. The connected mode paging record contains U-RNTI, paging cause, CN

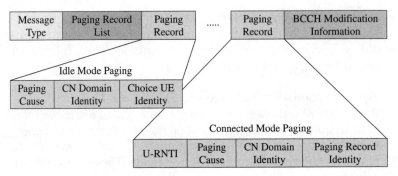

Figure 4.9 Paging Type 1 message.

domain identity, and paging record identity. The BCCH modification IE may contain an MIB value tag or modification time.

Upon request of upper layers, the UTRAN RRC starts the paging procedure by sending a Paging Type 1 message on the PCH mapped to SCCPCH at an appropriate paging occasion. This message may contain a number of paging records, each intended for an individual UE. The associated paging indicator is also set on the PICH. The UE monitors the PICH at its assigned paging occasion. If the paging indicator on the PICH indicates that the UE should read the associated SCCPCH, the UE will do so and check the Paging Type 1 message it receives. The UE performs address matching to determine whether the message contains a record that matches one of its addresses. IMSI, TMSI, or P-TMSI is used for addressing when the UE is in idle mode, while U-RNTI is used when the UE is in connected mode. If the Paging Type 1 message contains a paging record addressed to it, the UE RRC will inform the UE NAS of the CN domain identity (CS or PS) and paging cause when the UE is in idle mode. The UE NAS will start an RRC Connection Request. When the UE is in connected mode, the UE RRC will forward the paging cause, CN domain identity, and UE identity type to the upper layers if the optional IE, CN-originated page to connected mode UE, is included in the paging record. Otherwise, the UE RRC will perform a cell update procedure with the cause set to paging response.

4.3.1.1 Paging Discontinuous Reception (DRX) Cycle
The UE may use discontinuous reception (DRX) to prolong battery life when it is in idle mode, Cell_PCH state, or URA_PCH state. In idle mode, a DRX cycle is defined for each core network. In connected mode, a DRX cycle is defined for UTRAN. A paging occasion and a paging indicator are defined for each UE to determine for which occasion the UE must monitor a paging indicator within a DRX cycle. A DRX cycle is the time interval between paging occasions of a UE. Its duration is 2^k frames, where k is the DRX

cycle length coefficient. For the CN domain-specific DRX cycle length coefficient, the value of k is an integer ranging from 6 to 9, corresponding to a DRX cycle length of 64 to 512 frames or 640 milliseconds to 5.12 seconds. For the UTRAN DRX cycle length coefficient, the value of k is an integer ranging from 3 to 9, corresponding to a DRX cycle length of 8 to 512 frames or 80 milliseconds to 5.12 seconds.

As mentioned earlier in Section 4.2.4.2, the CN domain-specific DRX cycle coefficient for a circuit-switched CN is sent in SIB Type 1. For a packet-switched CN, the coefficient is also sent in SIB Type 1, but a NAS procedure may also set a new coefficient. In idle mode, if a UE attaches to both circuit-switched and packet-switched CN domains, it uses the shortest CN-domain specific DRX cycle length. When a UE enters connected mode, UTRAN sends a UTRAN DRX coefficient to the UE in the RRC Connection Setup message. The UE will use the shortest among the UTRAN DRX cycle length and the CN domain-specific DRX cycle lengths when it is in the Cell_PCH or URA_PCH state. In addition to using the RRC Connection Setup message, UTRAN may also use other messages such as Cell Update Confirm message, URA Update Confirm message, Radio Bearer Setup message, Radio Bearer Reconfiguration message, Radio Bearer Release message, Physical Channel Reconfiguration message, and Transport Channel Reconfiguration message to send a new UTRAN DRX coefficient to the UE.

4.3.1.2 Paging Occasion By definition, paging occasion is the occasion in which the UE monitors its paging indicator on the PICH. It is expressed in terms of system frame number (SFN). The paging indicator tells the UE whether it needs to read the associated SCCPCH to check for an incoming Paging Type 1 message.

Paging occasion is calculated using the UE's IMSI, DRX cycle length in frames, and the number of SCCPCHs that carry PCH. Specifically, it is given by

$$\text{Paging Occasion} = \{(IMSI \text{ div } K) \text{ mod DRX Cycle Length}\} + (n \times \text{DRX Cycle Length}) \quad (4\text{-}3)$$

where K = the number of available SCCPCHs that carry a PCH, n = 0, 1, 2, ... until the paging occasion exceeds 4095.

For example, assume K = 1, IMSI = 088613311065066, and DRX cycle length = 64 frames, then, paging occasion = 42, 106, ..., 4074.

4.3.1.3 Paging Indicator The paging indicator, carried by PICH, tells the UE to read the paging message on the appropriate frames of the SCCPCH. What RRC does for the paging indicator is to calculate a paging indicator value (PI value). This PI value, in conjunction with the paging occasion and the number of paging indicators per frame, is used by the physical layer to calculate the location of a set of paging indicator bits on the PICH. The PI value is calculated using the following formula:

$$PI = (IMSI \text{ div } 8192) \text{ mod } Np, \qquad (4\text{-}4)$$

where Np = number of paging indicators per frame. For FDD mode, Np could be 18, 36, 72, or 144 as given by SIB Type 5. If the paging indicator bits on the PICH frame (corresponding to the calculated PI value) are set to -1, the UE should read the associated SCCPCH to check for an incoming Paging Type 1 message. If more than one PCH and its associated PICH are defined in SIB Type 5, the UE has to select a SCCPCH from those listed in SIB Type 5 as

$$\text{Index of selected SCCPCH} = IMSI \text{ mod } K. \qquad (4\text{-}5)$$

4.3.2 Paging Type 2

Paging Type 2 is used when the UE is in the Cell_DCH or Cell_FACH state. The UTRAN RRC initiates the paging procedure by sending a Paging Type 2 message on the DCH or FACH channel. As shown in Figure 4.10, Paging Type 2 contains message type, RRC transaction identifier, paging cause, CN domain identity, and paging record type identifier. This message is sent on DCCH mapped to DCH or FACH.

UTRAN sends a Paging Type 2 signal to a UE for delivering a second call while the UE is already active in another call. Upon receiving the message, the UE RRC informs the UE NAS layer of the CN domain identity and paging cause. The UE NAS layer then takes appropriate actions.

4.4 RRC Connection Management

One of the RRC functions is to establish, maintain, and release an RRC connection between the UE and UTRAN. The RRC connection is always requested by the UE and set up and released by UTRAN. A UE may request an RRC connection for different reasons, such as response to a page, origination of a call, location area update, and routing area update. When an RRC connection is established, the UE goes through transitions from idle mode to either the Cell_DCH or Cell_FACH state. In an RRC connection establishment, normally four SRBs are set up between UTRAN and the UE for signaling purpose. Two SRBs are used to carry

| Message Type | RRC Transaction Identifier | Paging Cause | CN Domain Identity | Paging Record Type Identifier |

Figure 4.10 Paging Type 2 message.

RRC messages and the other two are used to carry upper layer (NAS) messages. User data RABs may or may not be needed depending on the reason for the RRC connection request. If the connection is requested for a voice or data call, RABs are required. If the connection is requested for location area or routing area update, no RAB is required.

There are three steps in the RRC connection process: First, the UE sends an RRC Connection Request message to UTRAN. Second, UTRAN responds with an RRC Connection Setup message. Third, the UE sends an RRC Connection Setup Complete message to UTRAN, completing the entire process.

4.4.1 RRC Connection Request

When a UE wants to request establishment of an RRC connection, it sends an RRC Connection Request message to UTRAN. The message is sent on logical channel CCCH mapped to transport channel RACH. As shown in Figure 4.11, the information elements in the RRC Connection Request message include RRC connection establishment cause, initial UE identity, protocol error indicator, and measured results on RACH.

The RRC connection establishment cause the IE to state the reason for requesting the RRC connection, such as responding to a page or originating a call.

In the initial UE identity IE, the UE may use, in decreasing order, TMSI, P-TMSI, IMSI, or IMEI to identify itself. If the UE does not have a valid TMSI, then a valid P-TMSI is used, and so forth.

The protocol error indicator IE could be set to TRUE or FALSE. When a UE wants to have an RRC connection established, it attempts to repeatedly transmit the RRC Connection Request message until it receives the RRC Connection Setup message with a valid configuration or until the number of retransmissions reaches the maximum number given by the counter N300, whichever comes first. If the maximum number of retransmissions is reached and the UE still cannot get an RRC connection established, the UE will stay in idle mode. For each transmission, the UE will wait for a response (for a duration given by the timer T300) before it attempts the next transmission. If the UE receives an invalid configuration in the RRC Connection Setup message, it will retransmit the RRC Connection Request message with the protocol error indicator IE set to TRUE.

RRC Connection Establishment Cause	Initial UE Identity	Protocol Error Indicator	Measured Results on RACH

Figure 4.11 RRC Connection Request message.

The measured-results-on-RACH IE contains the measurement results of the current cell and the monitored cells. UTRAN may use these measurement results in its calculations for handover decisions.

4.4.2 RRC Connection Setup

When UTRAN receives an RRC Connection Request message from the UE, it either accepts the request and responds with an RRC Connection Setup message, or rejects the request and responds with an RRC Connection Reject message.

The RRC Connection Setup message provides the UE with all necessary information to enter either the Cell_DCH or Cell_FACH state in the connected mode. UTRAN sends the RRC Connection Setup message on logical channel CCCH mapped to transport channel FACH. As shown in Figure 4.12, the Connection Setup message contains many IEs, including initial UE identity, specification mode, preconfiguration mode, predefined configuration identity, UTRAN DRX cycle length coefficient, default configuration mode, default configuration identity, reestablishment timer, UE capability requested, RBs to set up, UL transport channel information, DL transport channel information, frequency information, UL radio resources, and DL radio resources.

The initial UE identity IE provides the UE with the information for the UE to determine whether to read or ignore the rest of the RRC Connection Setup message. The UE compares the value of the initial UE identity IE with its own identity (that is, the value of the variable INITIAL_UE_IDENTITY). If the values are different, the UE ignores the rest of the message, otherwise it reads the rest of the message and acts accordingly.

The specification mode and pre-configuration mode IEs tell the UE to initiate the radio bearer and transport channel configuration in accordance with the predefined parameters or in accordance with the received information. Three different scenarios may occur:

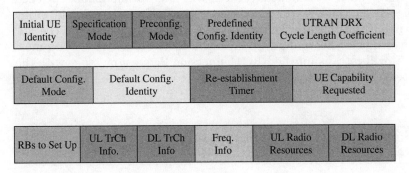

Figure 4.12 RRC Connection Setup message.

- **Scenario 1:** If the specification mode IE is set to pre-configuration and the pre-configuration mode IE is set to predefined configuration, the UE must initiate the radio bearer and transport channel configuration according to the predefined parameters identified by the predefined configuration identity IE and initiate the physical channels according to the received information in the UL radio resources and DL radio resources IEs.

- **Scenario 2:** If the specification mode IE is set to pre-configuration and the pre-configuration mode IE is set to default configuration, the UE must initiate the radio bearer and transport channel configuration according to the predefined parameters identified by the default configuration mode and default configuration identity IEs and initiate the physical channels according to the received information in the UL radio resources and DL radio resources IEs.

- **Scenario 3:** If the specification mode IE is set to complete specification, the UE must initiate the radio bearer, transport channel, and physical channel configuration according to the received information in the RBs-to-set-up UL transport channel information, DL transport channel information, UL radio resources, and DL radio resources IEs.

Normally, four signaling radio bearers (SRBs) may be configured; with two allocated to RRC and two allocated to NAS. Of the two allocated to RRC, one is configured as unacknowledged mode (UM) and the other is configured as acknowledged mode (AM). Of the two allocated to NAS, one is for high priority signaling and the other is for low priority signaling. The RB mapping information IE, which is embedded in the RBs-to-set-up IE, provides information on how to map the logical channels onto transport channels.

The UL transport channel information and DL transport channel information IEs contain information on permitted transport format combinations and other information specifically related to transport channels.

The UL radio resources and DL radio resources IEs provide physical dedicated channel information if the initial state for the UE to enter in the connected mode is Cell_DCH. These IEs also contain information on spreading factor, scrambling code number, and other physical channel parameters.

4.4.3 RRC Connection Setup Complete

As mentioned earlier, when the UE receives an RRC Connection Setup message, it will compare the value of the initial UE identity IE in the received RRC Connection Setup message with its own UE identity. If the values are different, the UE will ignore the rest of the RRC Connection

Setup message. If the values are identical, the UE will follow the following steps:

1. Take action, such as initiating the radio bearer, transport channel, and physical channel configurations, based on the information provided by the RRC Connection Setup message.
2. Stop timer T300.
3. Send an RRC Connection Setup Complete message to UTRAN after successfully entering the connected mode.

The UE sends the RRC Connection Setup Complete message to UTRAN on the uplink DCCH mapped to DCH or RACH. As shown in Figure 4.13, the RRC Connection Setup Complete message contains information elements including message type, RRC transaction identifier, start list, CN domain identity, UE radio access capability, UE radio access capability extension, UE system specific capability, and inter-RAT radio access capability.

For each CN domain, the start-list IE lists the START parameter, which is used in ciphering and integrity protection security procedures. If USIM or SIM is present, the UE must set the START for each CN domain in the start-list IE to the corresponding START value that is stored in the USIM or UE, and then set the START value stored in the USIM or UE for any CN domain to the value THRESHOLD of the variable START_THRESHOLD. If neither USIM nor SIM is present, the UE sets the START for each CN domain in the start-list IE to zero and sets the value of THRESHOLD in the variable START_THRESHOLD to the default value [1].

The UE radio access capability and UE radio access capability extension IEs contain information about the UE capabilities, including supported transport and physical channels, whether ciphering is supported, frequency bands, separation between transmit and receive frequencies, power class, positioning methods, and whether compressed mode is required for inter-frequency and inter-RAT measurements. The UE retrieves its UTRA UE radio access capability information element from the variable UE_CAPABILITY_REQUEST in the RRC Connection

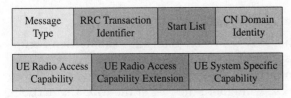

Figure 4.13 RRC Connection Setup Complete message.

Setup message and includes it in the UE radio access capability and UE radio access capability extension IEs.

The UE system specific capability IE contains information about inter-RAT-specific UE radio access capability. The UE retrieves its inter-RAT-specific UE radio access capability information element from the variable UE_CAPABILITY_REQUESTED in the RRC Connection Setup message and includes it in the UE system specific capability IE.

4.4.4 RRC Connection Release

RRC Connection Release message indicates the RRC connection is to be disconnected including all radio access bearers and all signaling radio bearers between the UE and UTRAN. By doing so, all established signaling connections are released. In RRC connection release, the NAS layers first carry out the release procedures and release the radio bearers when a call is terminated. UTRAN then releases the RRC connection when the last signaling radio bearer is released. In addition, UTRAN may at anytime initiate an RRC connection release by sending an RRC Connection Release message using UM RLC on DCCH when the UE is in the Cell_DCH or Cell_FACH state. UTRAN may also send an RRC Connection Release message on CCCH if the downlink DCCH is not available in UTRAN and the UE is in the Cell_FACH state.

The RRC Connection Release message contains a release cause IE that spells out the cause for the release. The causes for RRC connection release include normal event, unspecified pre-emptive release, congestion, re-establishment reject, user inactivity, and directed signaling connection re-establishment.

UTRAN may transmit several RRC Connection release messages to the UE to increase the probability of proper reception. In such a case, the RRC message sequence number (RRC SN) for these repeated messages should be the same. However, the number of repeated messages and the interval between the messages are optional.

Upon reception of the first RRC Connection Release message, the UE RRC passes the release cause to the NAS layer, which interprets the cause and then takes proper action. Normally, the UE will perform the RRC connection release procedure and send an RRC Connection Release Completion message to the lower layers for transmission on DCCH to UTRAN using AM RLC. For a detailed description of RRC connection release, the readers may refer to [1].

4.5 Ciphering and Integrity Protection Control

Ciphering is a security measure used to protect all user data and signaling from being overheard by unauthorized entities, while integrity protection is a message authentication method used to prevent a signaling message from being intercepted and altered by unauthorized devices. The UTRAN RRC may enable ciphering and/or integrity protection during call setup.

For ciphering, UTRAN sends a Security Mode Command message on the downlink DCCH in AM RLC using the most recent ciphering configuration to start or restart ciphering. The RRC configures the RLC and MAC layers to perform ciphering at the specified activation time.

For integrity protection, UTRAN also sends a Security Mode Command message on the downlink DCCH in AM RLC using the new integrity protection configuration to start or modify integrity protection. RRC computes a message authentication code (MAC-I) for each transmitted message by using a previously negotiated integrity key and inorder to validate the MAC-I that is received with each message. Thus, integrity protection provides a certain level of authentication on each message exchanged between UTRAN and the UE for preventing fraud and illegal interception of user data.

For both ciphering and integrity protection, when a UE receives a Security Mode Command from UTRAN, it normally responds with a Security Mode Complete message.

4.5.1 Security Mode Command Message

To start, stop, or modify ciphering and integrity protection, UTRAN needs to send a Security Mode Command message to the UE. The Security Mode Command message contains IEs, including security capability, ciphering mode information, integrity protection mode information, and CN domain identity. The security capability IE contains the algorithms supported. Currently, the algorithms defined for UMTS are UEA1 for ciphering and UIA1 for integrity protection, based on the Kasumi algorithm [1].

The ciphering-mode information IE provides information on start/stop of ciphering, ciphering algorithm, DCH TM activation time, and UM/AM activation time. An activation time is defined for each radio bearer. If transparent mode (TM) radio bearers exist for the CN domain identified by the CN domain identity IE, the activation time is a connection frame number (CFN) that is a multiple of 8 frames. For unacknowledged mode and acknowledged mode radio bearers, the activation time is an RLC sequence number.

The integrity-protection-mode information IE provides information on start and modification of integrity protection. For start of integrity protection, an integrity initialization number called FRESH [5] is provided. For modification of integrity protection, each signaling radio bearer is provided with an activation time. The activation time is an RRC sequence number (RRC SN). For a signaling radio bearer that has no pending integrity protection activation time, a new RRC sequence number is provided. For signaling a radio bearer that has a pending integrity protection activation time, the old RRC sequence number that was used in the previous security mode control procedure is reused.

4.6 Radio Bearer Control

As already mentioned earlier, RRC controls the establishment, reconfiguration, and release of radio bearers between UTRAN and the UE. To initiate these operations, UTRAN must send to the UE a Radio Bearer Setup message for radio bearer establishment, a Radio Bearer Reconfiguration message for radio bearer reconfiguration, and a Radio Bearer Release message for radio bearer release. These messages are sent on the downlink DCCH using AM or UM RLC. Radio Bearer Setup and Radio Bearer Reconfiguration messages contain parameters for the RLC layer, MAC layer, transport channel, physical channel, and mapping of logical channels to transport channels.

When the radio bearer establishment, reconfiguration, or release procedure is successful, the UE sends a Radio Bearer Setup Complete, Radio Bearer Reconfiguration Complete, or Radio Bearer Release Complete message to UTRAN. These messages are sent on the uplink DCCH using AM RLC. If the radio bearer establishment, reconfiguration, or release procedure is not successful, the UE sends a Radio Bearer Setup Failure, Radio Bearer Reconfiguration Failure, or Radio Bearer Release Failure message to UTRAN.

Radio bearers are used only in connected mode. The UE call state determines what radio bearers to establish. When the UE is in Cell_PCH or URA_PCH state, there are no SRBs or RABs. If the UE wants to transmit signaling information or originate a call, it must move to the Cell_FACH state and send a signaling message to UTRAN on the uplink CCCH, which maps to the RACH. UTRAN then initiates the radio bearer establishment procedure, which establishes SRBs but may or may not establish RABs depending on need. When the UE is in the Cell_DCH or Cell_FACH state, SRBs are established but RABs may or may not be established. Normally, as part of the RRC connection setup procedure, the SRBs are established when the UE first enters the Cell_DCH or Cell_FACH state, while the RABs may be established later. In radio bearer release, UTRAN may release RABs only while keeping SRBs. It may also release SRBs and RABs at the same time.

4.6.1 Radio Bearer Establishment

During a radio bearer establishment, UTRAN sends the Radio Bearer Setup message to the UE on logical channel DCCH mapped to transport channel FACH or DCH to configure the SRBs and RABs. The Radio

Bearer Setup message is very similar to the RRC Connection Setup message and contains many IEs that are the same as those in the RRC Connection Setup message. The differences exist in the UTRAN mobility information, RB information, and transport channel information fields. The Radio Bearer Setup message contains URA identity, RAB information for setup, RBs information to be affected, Deleted UL TrCH information, and Deleted DL TrCH information IEs, while the RRC Connection Setup message does not. The URA identity IE is optional and used to tell the UE to use a particular URA identity when there are multiple valid URAs listed in SIB Type 2. The RAB information for the setup IE is used for setting up an RAB, identifying the RAB, and specifying the core network domain to which the RAB belongs. The RBs information to be affected IE provides new RB mapping information for those previously established radio bearers that are affected when adding new SRBs and RABs. The Deleted UL TrCH information and Deleted DL TrCH information IEs are used for deleting previously established transport channels.

4.6.2 Radio Bearer Reconfiguration

For radio bearer reconfiguration, UTRAN sends the Radio Bearer Reconfiguration message to the UE on logical channel DCCH mapped to transport channel FACH or DCH. The Radio Bearer Reconfiguration message is similar to Radio Bearer Setup message. It is customarily used to make the UE transfer between the Cell_DCH and Cell_FACH states for supporting packet data services, and mapping dedicated logical channels to dedicated physical channels when the UE is in the Cell_DCH state, and to common physical channels when the UE is in the Cell_FACH state. The Radio Bearer Reconfiguration message is also used for reconfiguring the data rate of a dedicated channel. In the radio bearer reconfiguration procedure, UTRAN may indicate that uplink transmissions should be stopped or continued on certain radio bearers. However, uplink transmission on an SRB used for RRC signaling should not be stopped.

In addition to the Radio Bearer Reconfiguration message, there exist Transport Channel Reconfiguration and Physical Channel Reconfiguration messages. The Transport Channel Reconfiguration message is used by UTRAN to configure the transport channel of a UE. It can also be used to assign a TFC subset and reconfigure physical channels. The Physical Channel Reconfiguration message is used by UTRAN to assign, replace, or release a set of physical channels used by a UE.

4.6.3 Radio Bearer Release

For radio bearer release, UTRAN sends the Radio Bearer Release message to the UE to release radio bearers. The message can also contain modifications to the configurations of transport channels and physical channels. When one or more radio bearers are released, the remaining radio bearers along with the transport and physical channels may be reconfigured. In the case of concurrent services, the message can be used to indicate release of a signaling connection when a UE is connected to more than one CN domain. For example, if a UE has a simultaneous voice call and a packet data call, the Radio Bearer Release message can be used to release the voice call RABs, reconfigure the packet call RABs, and cause the UE to move from the Cell_DCH state to Cell_FACH state.

4.7 RRC Management of UE Mobility

Management of UE mobility is also one of the RRC functions. Several procedures are used to deal with the UE's mobility. Depending on the UE call state, whether a UE is in a CS call or a PS call, and the radio access technology of the target cell, different procedures are used. The cell reselection procedure is used when the UE is in idle mode, Cell_FACH, Cell_PCH, or URA_PCH state. The cell update and URA update procedures are used when the UE is in the Cell_FACH, Cell_PCH, or URA_PCH state. Active set update and inter-RAT handover from UTRAN procedures are used when the UE is in a CS call. Inter-RAT cell change order from UTRAN procedure is used when the UE is in a PS call.

4.7.1 Cell Reselection

When the UE is in idle mode, Cell_PCH, URA_PCH, or Cell_FACH state, it continues to search for a better cell. Cell reselection is the procedure of determining whether there is a neighbor cell that is better than the current serving cell. The UE RRC layer is in charge of this cell reselection procedure. The UE makes cell change decisions using the parameters provided by system information messages. Depending on the UE call state and the SIBs transmitted by UTRAN, cell reselection parameters from different SIBs may be used.

If the UE is in idle mode, the cell reselection parameters in SIB Type 3 and SIB Type 11 are used to determine whether a neighbor cell is better than the current serving cell. In idle mode, the UE does not need to tell UTRAN that a cell reselection has occurred. If the UE is in the Cell_PCH, URA_PCH, or Cell_FACH state, cell reselection parameters in SIB Type 4 and SIB Type 12 are used for cell reselection. In case these SIBs are not available, the UE still uses the information in SIB Type 3 and SIB Type 11. Contrary to the case of idle mode, in these states the UE has to notify UTRAN that it has moved to a new cell using the cell update or URA update procedures. Details on cell reselection will be discussed in Chapter 8.

4.7.2 Cell Update and URA Update Procedures

Cell update and URA update procedures are used by the UE to inform UTRAN that it has camped on a new cell while it is in connected mode. If the UE is in the Cell_FACH or Cell_PCH state, it uses the Cell Update message to notify UTRAN when it moves to a new cell. Similarly, if the UE is in the URA_PCH state, it uses the URA Update message to inform UTRAN when it moves to a new URA.

4.7.2.1 Cell Update Procedure The UE uses the cell update procedure to keep UTRAN informed of the cell on which it camps when it is in a Cell_PCH or Cell_FACH state. If the UE is in the Cell_PCH state, UTRAN may page the UE over a single cell only. To ensure that the UE can be paged, the UE must inform UTRAN whenever it moves to a new cell by sending a Cell Update message to UTRAN. It should be noted that when sending a Cell Update message to UTRAN, the UE must first transitions to the Cell_FACH state. When the UE is in a Cell_FACH state, its dedicated logical channels are rerouted to the new cell it camps on. Upon being informed of the new cell to which the UE has moved, UTRAN normally responds with a Cell Update Confirm message. However, UTRAN may also respond with an RRC Connection Release message. The UE will then perform the RRC Connection Release procedure, release all dedicated resources, and return to idle mode.

In addition to camping on a new cell due to cell reselection, the UE also needs to perform the cell update procedure first if it wants to transmit data or is paged by the network. Furthermore, in order to report possible radio link failures, the cell update procedure can also be configured to occur periodically.

4.7.2.2 URA Update Procedure The URA update procedure is used to keep UTRAN informed of the URA in which the UE is located. When the UE is in the URA_PCH state, UTRAN needs to know the UE's URA. The UE may move to a new cell due to cell reselection. If the UE moves to a cell belonging to a different URA, the UE must perform the URA update procedure by sending a URA Update message to UTRAN. Again, the UE must first transition to the Cell_FACH state before sending out the URA Update message. Upon receiving the URA Update message, UTRAN normally responds with the URA Update Confirm message. UTRAN may also respond with an RRC Connection Release message, causing the UE to perform the RRC Connection Release procedure, release all dedicated resources, and return to idle mode.

4.7.3 Active Set Update Procedure

The Active Set Update Procedure is used to add or remove cells from the active set of the connection between the UE and UTRAN. When a UE is in the Cell_DCH state, it is often in soft handover condition. That is, the UE establishes radio links with multiple cells at the same time. The set of cells with which the UE has established radio links is called the active set. The active set update procedure is initiated when UTRAN orders the UE to add cells to or remove cells from the active set. This is done by sending an Active Set Update message to the UE on downlink DCCH using AM or UM RLC. The Active Set Update message contains Radio link addition information and Radio link removal information IEs. The Radio link addition information IE contains downlink DPCH information such as primary scrambling code, frame offset (time difference between the PCCPCH and the DPCH), spreading factor, OVSF code number, and transmit power control combination index. The Radio link removal information IE contains the primary scrambling code. For cells to be added to the active set, the Radio link addition information IE is required. For cells to be removed from the active set, only the primary scrambling code is required to identify the cells. Upon reception of an Active Set Update message, the UE adds the cells indicated in the Radio link addition information IE and removes the cells indicated in the Radio link removal information IE.

4.7.4 Inter-RAT Mobility

Inter-radio-access-technology (Inter-RAT) mobility involves transfer of a connection between the UE and UTRAN from one access technology to another access technology. A typical example is transfer of a connection from UMTS to GSM. Inter-RAT transfer may occur due to cell reselection or handover. In the case of inter-RAT transfer due to cell reselection, the UE attempts to establish a connection with the inter-RAT cell without being directed by UTRAN. If the UE successfully connects to an inter-RAT cell, the core network informs UTRAN so that UTRAN can release all resources dedicated to the UE. More details on inter-RAT cell reselection will be discussed in Chapter 8.

In the case of inter-RAT transfer due to handover, the transfer is initiated by UTRAN. The procedures and messages for inter-RAT transfer are different for circuit-switched (CS) and packet-switched (PS) domains. In a CS domain, UTRAN sends the Handover from UTRAN Command message to the UE to transfer a CS connection from UTRAN to GSM when the UE is in the Cell_DCH state. In the PS domain, UTRAN sends the Cell Change Order from UTRAN message to the UE to transfer a PS connection from UTRAN to GPRS when the UE is in the Cell_DCH or Cell_FACH state.

4.7.4.1 Handover from UTRAN Command Message The Handover from UTRAN Command message is used for inter-RAT CS connection transfer. UTRAN sends this message to the UE in the Cell_DCH state to make a handover to an inter-RAT cell. This message carries information about the RABs to be transferred, the target cell identifiers, and radio parameters relevant to the target RAT, such as frequency band. The UE must be able to receive this message and carry out the inter-RAT handover, even if the UE has not previously made any measurements on the target cell.

Upon reception of a Handover from UTRAN Command message, the UE should establish the connection to the target cell using the information in the message. The UE should also connect the upper layer entities corresponding to the CS domain RABs indicated in the message to the radio resources provided by the target RAT.

If the UE successfully completes the inter-RAT handover, the CS core network informs the UTRAN so that UTRAN can release the radio connection and remove all context information for the UE. If the UE fails to complete the requested inter-RAT handover, the UE keeps the connection with UTRAN and sends a Handover from UTRAN Failure message to UTRAN, ending the inter-RAT handover procedure.

4.7.4.2 Cell Change Order from UTRAN Message The Cell Change Order from UTRAN message is used for inter-RAT PS connection transfer. UTRAN sends this message to the UE in the Cell_DCH or Cell_FACH state to make a cell change to an RAT other than UTRAN. The message carries information about the target cell identity and radio parameters relevant to the target RAT. The cell change order procedure may be used when no RABs are established or when the RABs are established only in the PS domain. This procedure may not be used if there is no PS signaling connection. The UE must be able to receive this message and carry out a cell change, even if the UE has not previously made any measurements on the target cell.

Upon reception of a Cell Change Order from UTRAN message, the UE should start timer T309 and establish the connection to the target cell using the information in the message. The range for T309 is from 1 to 8 seconds with a default value of 5 seconds.

The cell change order procedure is considered completed when the UE receives a successful response from the target RAT. If the UE successfully completes the inter-RAT cell change, the UE stops timer T309 and clears or sets UE variables upon leaving the UTRAN RRC connected mode. UTRAN then releases the radio connection and removes all context information for the UE. It should be noted that the release of the UMTS radio resources is initiated from the target RAT, not from the PS core network.

If timer T309 expires before the successful establishment of a connection to the target RAT, or if the UE fails to establish the connection to the other RAT due to other reasons such as access failure or lack of resources, the UE keeps the connection with UTRAN and sends a Cell Change Order from UTRAN Failure message to UTRAN, ending the cell change order procedure.

4.8 Measurements and Reporting

UTRAN may control the measurements UE performs either by broadcast of system information (SIB Type 11 and SIB Type 12) or by sending a Measurement Control message to the UE. UTRAN may send a Measurement Control message to the UE when the UE is in the Cell_DCH or Cell_FACH state. The message contains information on the measurements the UE has to perform. There are many types of measurements that the UE performs. These include intra-frequency measurements, inter-frequency measurements, inter-RAT measurements, quality measurements such as transport channel block error rate, UE internal measurements such as UE transmit power, traffic volume measurements, and UE position measurements. The UE sends a Measurement Report message to the UTRAN on a periodical basis or when a requested measurement event is triggered.

4.8.1 Measurement Control Message

UTRAN sends the Measurement Control message to the UE on logical channel DCCH mapped to transport channel DCH or FACH. Through this message, UTRAN may request the UE to set up, modify or release a measurement in the UE. A Measurement Control message must contain measurement identity, measurement command, and measurement type. Depending on the measurement type, a Measurement Control message may also contain any of the following control information: measurement objects, measurement quantity, reporting quantities, measurement reporting criteria, measurement validity, measurement reporting mode, additional measurement identities, and compressed mode information.

Measurement identity is a reference number used by UTRAN when setting up, modifying, or releasing the measurement, and by the UE in the measurement report. Measurement command tells the UE to set up, modify, or release a measurement. Measurement type defines the measurement type and all of the related parameters associated with the measurement identity. Measurement objects list the objects on which the UE performs the measurements. Measurement quantity specifies the quantity the UE has to measure on the measurement object.

Reporting quantities indicate the quantities the UE must include in the report in addition to the mandatory ones. Measurement reporting criteria could be periodical or event-triggered reporting. Measurement validity defines the UE states in which the measurement is valid. Measurement reporting mode specifies whether the report should be sent in AM or UM mode. Additional measurement identities contain reference numbers for other measurements that should be reported whenever a measurement report for the current measurement is triggered. Compressed mode information carries information to activate or deactivate compressed mode patterns.

4.8.2 Quality Measurement

Quality measurement is one of the measurement types included in the Measurement Control message. It contains information about measurements of downlink quality parameters, such as transport channel block error rate. Quality measurement contains quality reporting quantity and may also contain quality reporting criteria, periodical reporting criteria, or no-reporting. Quality reporting quantity gives a list of transport channels that the UE needs to report block error rates. Quality reporting criteria contain the parameters that reporting event 5A uses to determine when a quality report should be sent. These parameters include total CRC, bad CRC, and pending-after-trigger. Periodical reporting criteria define the number of reporting instances and reporting intervals. No-reporting is used to define a measurement that will be referenced in the additional measurement identities of another measurement later on. For details on quality measurement, readers may refer to [1].

4.8.3 UE Internal Measurements

The UE internal measurements IE contains information about UE internal measurements. It includes measurement quantity and reporting quantity. It may also include UE internal reporting criteria, periodical reporting criteria, or no-reporting. Measurement quantity specifies which quantity should be measured, such as UE transmission power, UE received signal strength power (RSSI), or UE receive-transmit time difference. Reporting quantity defines reporting quantities on top of those quantities that are required for the event. UE internal reporting criteria define reporting events from 6A to 6G and the related thresholds and time-to-trigger. Definitions of reporting events 6A through 6G are as follows [1]:

- Reporting event 6A – The UE transmit power becomes larger than an absolute threshold.

- Reporting event 6B – The UE transmit power becomes less than an absolute threshold.

- Reporting event 6C – The UE transmit power reaches its minimum value.
- Reporting event 6D – The UE transmit power reaches its maximum value.
- Reporting event 6E – The UE RSSI reaches the UE's dynamic receiver range.
- Reporting event 6F – The UE Rx-Tx time difference for a radio link in the active set becomes greater than an absolute threshold.
- Reporting event 6G – The UE Rx-Tx time difference for a radio link in the active set becomes less than an absolute threshold.

4.9 NAS Message Routing

NAS message routing is also one of the RRC functions. The UE RRC and UTRAN RRC use the Initial Direct Transfer, Uplink Direct Transfer and Downlink Direct Transfer messages to route messages between the UE NAS and UTRAN NAS through the access stratum protocols.

Initial Direct Transfer is used on the uplink to establish a signaling connection and to carry an initial UE NAS message over the radio interface. When the UE NAS requests for the transfer of a NAS message, the UE RRC initiates the initial direct transfer procedure. The UE RRC sends the message to the UTRAN RRC without changing the contents of the message. The UTRAN RRC then routes the message to the CN. Figure 4.14 shows an example of initial direct transfer.

Downlink Direct Transfer carries NAS messages over the radio interface in the downlink direction. It is initiated in the UTRAN RRC. A typical example of downlink direct transfer is transferring an authentication or ciphering request from the CN to the UE.

Similar to Downlink Direct Transfer, Uplink Direct Transfer carries NAS messages over the radio interface in the uplink direction. It is initiated in the UE RRC when the upper layers request a transfer of a NAS message on an existing signaling connection. For example,

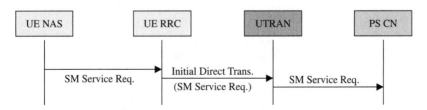

Figure 4.14 Illustration of initial direct transfer.

the uplink direct transfer can transfer an authentication or ciphering response from the UE to the CN.

4.10 Summary

The RRC controls and configures all other layers below it in the access stratum through the control interfaces. It also provides a control and signaling interface to the non-access stratum layer. Overall, the RRC functions include access stratum control, system information processing, paging and notification, RRC connection management, ciphering and integrity protection control, radio bearer control, RRC management of UE mobility, power control, measurement control and reporting, and NAS layer message routing.

The UTRAN RRC broadcasts system information messages to all UEs on the BCH. A UE gets all system information messages for a cell from the BCH when it first camps on that cell. The messages contain information elements carrying access stratum and non-access stratum information. The system information elements are broadcast in system information blocks (SIB). A system information block may contain one or more system information elements. The master information block (MIB) provides references and scheduling information to a number of SIBs in a cell, and may also contain reference and scheduling information for one or two scheduling blocks (SB). Scheduling blocks provide references and scheduling information to additional SIBs. Scheduling information for an SIB may only be contained in either the MIB or one of the SBs.

The SIBs range from Type 1 to Type 18. With the exceptions of SIB Types 15.2, 15.3, and 16, the content does not change for each occurrence for any SIBs that use value tags. SIB Types 15.2, 15.3, and 16, may occur more than once with different contents.

Each type of SIB has its own unique function and structure. SIB Type 1 contains the core network domain information and the UE timers and constants. SIB Type 2 contains a list of up to eight URA identities that are valid in the current cell. SIB Types 3 and 4 contain information pertaining to the cell reselection procedure, while SIB Types 5 and 6 contain information pertaining to common physical channels. SIB Type 7 carries fast changing parameters related to uplink PRACH transmissions. SIB Types 8, 9, and 10 are used only in FDD for a UE in connected mode. The UE should store all relevant information in these SIBs. SIB Types 11 and 12 carry measurement control information to be used in the cell. SIB Type 13 contains ANSI-41 system information. SIB Type 14 is used only in 3.84 Mcps TDD. SIB Type 15 contains information useful for UE-based or UE-assisted positioning methods. SIB Type 16 carries radio bearer, transport channel, and physical channel parameters to be stored by the UE in idle and connected modes. The SIB Type 17 contains fast changing parameters for configuration of the shared physical channels to be used in connected mode for TDD, while SIB Type 18 contains IEs listing PLMN identities of neighboring cells.

The RRC also performs paging and notification. There are two types of paging: Paging Type 1 and Paging Type 2. Paging Type 1 is used when UE is in idle mode, the Cell_PCH state, or URA_PCH state, while Paging Type 2 is used when the UE is in the Cell_DCH state or Cell_FACH state.

For RRC connection management, the RRC is responsible for establishing, maintaining, and releasing an RRC connection between the UE and UTRAN. The RRC connection is always requested by the UE, but it is set up and released by UTRAN.

The UTRAN RRC may enable ciphering and/or integrity protection during call setup. Ciphering is a security measure to protect user data and signaling from being overheard by unauthorized entities, while integrity protection is a message authentication method to prevent a signaling message from being intercepted and altered by unauthorized devices.

The RRC also controls the establishment, reconfiguration, and release of radio bearers between UTRAN and UE. To initiate these operations, UTRAN must send to the UE a Radio Bearer Setup message for radio bearer establishment, a Radio Bearer Reconfiguration message for radio bearer reconfiguration, and a Radio Bearer Release message for radio bearer release. When the radio bearer establishment, reconfiguration, or release procedure is successful, the UE sends a Radio Bearer Setup Complete, Radio Bearer Reconfiguration Complete, or Radio Bearer Release Complete message to UTRAN.

The RRC is also responsible for UE mobility procedures. Depending on the UE call state, whether UE is in a CS call or a PS call, and the radio access technology of the target cell, different procedures are used. The mobility procedures include cell reselection, cell update, and URA update, active set update, and inter-RAT mobility.

For measurement control, UTRAN controls the UE measurements either by broadcast of system information (SIB Type 11 and SIB Type 12) or by sending a Measurement Control message to the UE. UTRAN may send a Measurement Control message to the UE when the UE is in the Cell_DCH or Cell_FACH state. The message carries information on the measurements that the UE has to perform. The UE may perform many types of measurements, including intra-frequency measurements, inter-frequency measurements, inter-RAT measurements, quality measurements such as transport channel block error rate, traffic volume measurements, UE position measurements, and UE internal measurements such as UE transmit power. The UE sends a Measurement Report message to the UTRAN when a requested measurement event is triggered. It may also send the message to the UTRAN on a periodical basis.

For NAS message routing, the UE RRC and UTRAN RRC use the Initial Direct Transfer, Uplink Direct Transfer, and Downlink Direct Transfer messages to route messages between the UE NAS and UTRAN NAS through the access stratum protocols.

References

[1] 3GPP, TS25.331, v5.7.1, "Radio Resource Control (RRC) protocol specification," (Release 5).

[2] 3GPP TS 25.304, v5.3.0, "User Equipment (UE) procedures in idle mode and procedures for cell reselection in connected mode," (Release 5).

[3] RTCM-SC104: "RTCM Recommended Standards for Differential GNSS Service (v.2.2)".

[4] ICD-GPS-200: "Navstar GPS Space Segment/Navigation User Interface".

[5] 3GPP TS 33.102, v5.5.0, "3G Security; Security Architecture," (Release 5).

Chapter
5
Radio Link Control

Layer 2 in the access stratum protocol stack consists of four sublayers: Packet Data Convergence Protocol (PDCP), Broadcast/Multicast Control (BMC), Radio Link Control (RLC), and Medium Access Control (MAC).

Each of these sublayers has its unique functions. Packet data convergence protocol compresses and decompresses IP data stream headers such as TCP/IP and RTP/UDP/IP headers. Broadcast/multicast control provides cell broadcast functions. Radio link control supports a variety of functions such as segmentation, concatenation, reassembly, padding, transfer of user data, and so forth. Medium access control maps logical channels onto transport channels, takes care of traffic volume reporting, and selects transport format.

This chapter focuses on RLC, starting with a review of Layer 2 fundamentals and then addressing the RLC functions, architectures, and RLC data transfer modes. It will also cover RLC ciphering and RLC configurable parameters.

5.1 Layer 2 Fundamentals

To understand RLC, you need to be familiar with the key terms: SDU, PDU, TTI, and CCTrCh. These terms play important roles in both RLC and MAC layers.

The service data unit (SDU) is the basic data unit for layer-to-layer communication. It flows between layers and sublayers of the protocol stack. On the transmitting side, each layer receives SDUs from the layer just above it, while on the receiving side, each layer receives SDUs from the layer below.

A protocol data unit (PDU) is the basic data unit for peer-to-peer communication. It is exchanged between peer layers in the UE and UTRAN. A PDU is generated from an SDU. For example, on the transmitting end,

RLC transforms the SDU received from the RRC layer into one or multiple PDUs through segmentation, concatenation, and by adding a header. This PDU now becomes an SDU of the MAC layer. On the receiving end, RLC transforms the SDU received from the MAC layer into a PDU through reassembly and by removing the header. This PDU now becomes an SDU of the RRC layer.

Transmission time interval (TTI) is the time interval in which MAC produces a set of transport blocks for a transport channel. In other words, TTI is the time duration for transmitting a set of data on a transport channel. For Release 99, each transport channel is configured to support a TTI of 10, 20, 40, or 80 ms. The transport channel for HSDPA (Release 5) is configured to support a TTI of 2 ms.

Coded composite transport channel (CCTrCh) is a multiplexed channel that consists of the coded transport blocks from one or more transport channels. On the transmitting side, the MAC layer transfers a set of transport blocks in a TTI to the physical layer. The physical layer then adds a CRC to each transport block and sends the transport blocks through a series of processing steps such as transport block concatenation and code block segmentation, channel coding, rate matching, DTX insertion, interleaving, and radio frame segmentation. Radio frame segmentation breaks a TTI into 10-ms radio frames if it is larger than 10 ms. After radio frame segmentation, the transport channels are serial multiplexed using the 10-ms radio frames to form a CCTrCh.

In the protocol stack, on the transmitting side, each layer provides services to the layer above and uses the services provided by the layer below. On the receiving side, each layer provides services to the layer below and uses the services provided by the layer above. The peer layers in the UE and UTRAN communicate with each other using PDU as the basic data unit.

When the UTRAN RRC sends a signaling message to the UE RRC, the message must be sent down the UTRAN's protocol stack from the RRC layer to the physical layer where it is sent over the air to the UE. On the UE side, the message climbs up the UE's protocol stack from the physical layer to the RRC layer. Figure 5.1 illustrates an example of this peer-to-peer communication. On the UTRAN side, the signaling message is packaged into an RRC PDU and then sent to the RLC layer as an RLC SDU. The RLC layer transforms this RLC SDU into an RLC PDU by adding an RLC header and sends it to the MAC layer as an MAC SDU. The MAC layer transforms this MAC SDU into a MAC PDU by adding a MAC header and sends it to the physical layer.

On the UE side, the physical layer sends the data to the MAC layer, which removes the MAC header and sends the data to the RLC layer. The RLC layer then removes the RLC header and sends the data to the RRC layer. The data received by the UE RRC is exactly the same as that originally sent by the UTRAN RRC. It contains the original RRC signaling message sent by UTRAN.

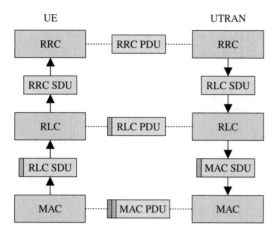

Figure 5.1 Communication between the UE and UTRAN peer layers.

5.2 RLC Functions

RLC functions include segmentation and reassembly, concatenation, padding, user data transfer, error correction, in-sequence delivery of upper layer PDUs, duplicate detection, flow control, sequence number check, protocol error detection and recovery, ciphering, and SDU discard. RLC provides services to RRC in the control plane and the application layer in the user plane.

RLC supports three data transfer modes: transparent mode (TM), unacknowledged mode (UM), and acknowledged mode (AM). Depending on the data transfer mode, some or all of the previously mentioned functions are needed to support a data transfer service. For TM data transfer service, the functions needed are segmentation and reassembly, transfer of user data, and SDU discard. Supporting UM data transfer service requires additional functions, which include concatenation, padding, ciphering, and sequence number check. For AM data transfer service, all the functions needed for UM data transfer service are still needed with the exception of sequence number check. On top of that, AM data transfer service also requires error correction, in-sequence delivery of upper layer PDUs, duplicate detection, flow control, and protocol error detection and recovery.

5.3 RLC Entities

As shown in Figure 5.2, there exist TM, UM, and AM RLC entities in the RLC layer for supporting transfers of signaling and user data. A TM and a UM RLC entity can be configured as a transmitting RLC entity or as a receiving RLC entity. An AM RLC entity consists of a transmitting

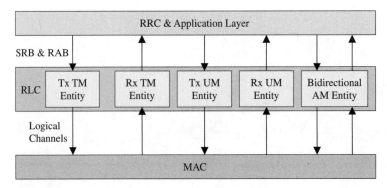

Figure 5.2 RLC entities.

side and a receiving side. Data flows between RLC and RRC are called signaling radio bearers (SRBs) while data flows between RLC and user application layers are called radio access bearers (RABs). SRBs carry signaling SDUs between the RLC and RRC layers, while RABs carry user data SDUs between the RLC and application layers. Data flows between RLC and MAC are called *logical channels*, which carry data and signaling SDUs between RLC and MAC. Each SRB and RAB is mapped to a logical channel. When the RRC maps an SRB or an RAB to a logical channel, it allocates an RLC entity and configures the data transfer mode as TM, UM, or AM. Each TM and UM RLC entity uses one logical channel to send or receive signaling and data SDUs. An AM RLC entity can be configured to use one or two logical channels to send or receive signaling and data SDUs. If two logical channels are configured, they must be of the same type, either DCCH or DTCH.

5.4 RLC Data Transfer Modes

As mentioned previously, the RLC entities may be configured as TM, UM, or AM. For TM and UM, the transmit entity and the receive entity are separately configured. For AM, a single entity is configured for both transmitting and receiving.

BCCH, PCCH, uplink CCCH, and circuit-switched DTCH use TM. Downlink CCCH uses UM. Of the two DCCHs allocated to RRC in connected mode, one uses UM and the other uses AM. DCCHs allocated to NAS in connected mode use AM. Packet-switched DTCH use both UM and AM.

5.4.1 RLC Transparent Mode

When operating in transparent mode, service is unreliable. Figure 5.3 shows the model of two transparent mode peer RLC entities. The transmitting

Radio Link Control 81

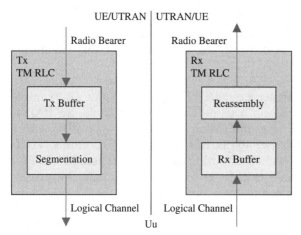

Figure 5.3 Transparent mode RLC entities.

TM entity receives RLC SDUs from upper layers. All received RLC SDUs must be of a length that is a multiple of one of the valid RLC PDU lengths. RLC PDUs are transferred to the logical channel without adding an RLC header. The TM RLC entity may only perform segmentation, reassembly, and SDU discard functions.

On the transmitting side (for example, UTRAN), if the RRC has configured segmentation and the RLC SDU size is larger than the RLC PDU size used by the MAC layer for that TTI, the transmitting TM RLC entity will segment the RLC SDUs received from the upper layers to fit the RLC PDU size. Note that an RLC PDU is a MAC SDU. All the RLC PDUs carrying one RLC SDU are sent in one TTI, and no segment from another RLC SDU can be sent in the same TTI. If the RRC has not configured segmentation, then no segmentation occurs. In this case, more than one RLC SDU can be sent in one TTI by placing one RLC SDU in one RLC PDU. All RLC PDUs in one TTI must be of equal length.

On the receiving side (for example, UE), the receiving TM RLC entity receives RLC SDUs from the MAC layer through logical channels. Note that these RLC SDUs received on the receiving side are those RLC PDUs transmitted on the transmitting side. If the RRC has configured segmentation, the receiving TM entity reassembles all received RLC SDUs to form SDUs of the upper layers and sends them to the upper layers. If segmentation is not configured, each received RLC SDU is treated as an upper layer SDU. These upper layer SDUs are exactly the same as those RLC SDUs received by the transmitting TM entity on the transmitting side.

SDU discard may be performed if the transmitting TM entity is configured to do so. Normally, if the transmitting TM entity is configured to perform timer-based SDU discard, an SDU that is not transmitted after a configurable period of time is discarded. However, if SDU discard is not configured for the transmitting TM entity and a new SDU arrives before the previous one is transmitted, then the new SDU will simply overwrite the old one.

5.4.1.1 RLC TM Data Transfer Example Figure 5.4 shows an example of RLC transparent mode data transfer. In this example, an RAB is configured for RLC TM. The application layer generates a PDU with a maximum size of 103 bits per 40 ms. RLC maps the PDU onto a DTCH logical channel without adding anything to the PDU. Then, MAC maps the DTCH logical channel to a DCH transport channel. There is no need to add a MAC header in this case because the logical-to-transport mapping is one-to-one. The transport block size is configured to be the same as that of the RLC PDU. The transmission time interval is configured to 40 ms. The physical layer encodes, interleaves, segments, and multiplexes the transport block onto a CCTrCh, which contains four 10-ms radio frames. One quarter of each transport block is sent in the first 10-ms radio frame, one quarter in the second 10-ms radio frame, another quarter in the third 10-ms radio frame, and still another quarter in the fourth 10-ms radio frame. It should be noted that after interleaving, the data bits in the transport block are redistributed. The first radio frame does not carry all the data bits in the first quarter of the transport block.

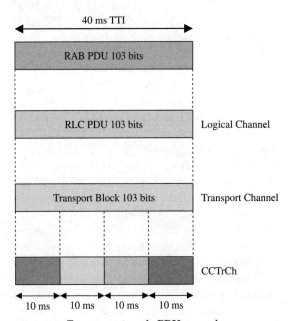

Figure 5.4 Transparent mode PDU example.

5.4.2 RLC Unacknowledged Mode

When operating in unacknowledged mode, the SDU can be arbitrary sizes. But, the service is not reliable. In UM operation, RLC adds a header, to each PDU which carries segmentation, concatenation, and the sequence number information. Figure 5.5 shows the unacknowledged mode operation. On the transmitting side, the UM RLC transmitting entity performs segmentation and concatenation, RLC header addition, ciphering, padding, and SDU discard. On the receiving side, the UM RLC entity performs deciphering, RLC header removal, sequence number check, and reassembly.

Similar to TM RLC on the transmitting side, if the RLC SDUs received from the upper layers are larger than the RLC PDU size, the transmitting UM RLC entity segments the RLC SDUs to fit the RLC PDU size. On the receiving side, all the RLC PDU carrying one RLC SDU are reassembled to form an SDU of the upper layers. On the contrary, if the RLC SDUs are very small, a number of RLC SDUs may be combined into a single RLC PDU. If the RLC PDU, still has a vacancy after multiple RLC SDUs are placed in it, padding bits are added to the RLC PDU.

The sequence number check is used by the receiving entity during reassembly. It can detect the corrupted SDUs.

In UM operation, RLC ciphers RLC PDU, except the header if ciphering is enabled. This is different from the TM operation where RLC does not perform ciphering. Ciphering is done in the transmitting entity while deciphering is done in the receiving entity.

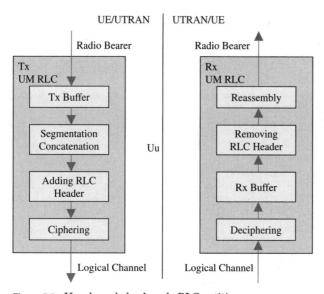

Figure 5.5 Unacknowledged mode RLC entities.

SDU discard is a function that may be configured in the transmitting entity using a timer. If timer-based SDU discard is configured, any RLC SDU that is not transmitted after the timer expires is discarded. On the other hand, if timer-based SDU discard is not configured in the transmitting entity, the new RLC SDUs will overwrite the old ones when the transmission buffer is full.

5.4.2.1 RLC UM PDU Format Having discussed the RLC unacknowledged mode, we will take a look at the RLC UM PDU format. As shown in Figure 5.6, an RLC UM PDU consists of a number of different fields including sequence number, extension bit, length indicator, data, and pad.

The sequence number, if any, occupies the first 7 bits of the first byte of the PDU. It is used to detect missing PDUs and as an input to the ciphering procedure. The sequence number is required if multiple PDUs are transmitted.

The extension bit is the last bit of the first byte and is used to indicate whether the next byte is the beginning of data or is a length indicator plus an extension bit. If the extension bit is 0, the next byte is data. If it is 1, the next byte is a length indicator plus an extension bit.

The length indicator occupies 7 or 15 bits. It is used to indicate the last byte of each RLC SDU that ends within the PDU. Whether 7 or 15 bits should be used is determined by the maximum size of a UM PDU. If the largest UM data PDU size is smaller or equal to 125 bytes, the 7-bit length indicator is used. Otherwise, the 15-bit length indicator is used.

The data is all or part of an RLC SDU, which has a length of multiple bytes. A number of RLC SDU segments may be concatenated within an RLC PDU. As such, there are boundaries between the RLC SDUs. It is the length indicator that tells the locations of these boundaries.

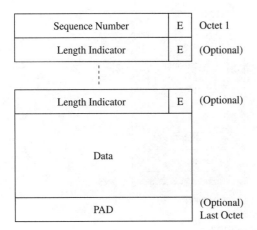

Figure 5.6 RLC UM data PDU format (courtesy of ETSI).

When the RCL PDU is smaller than the configured PDU size, padding bits are added to the end of the RLC PDU to make the PDU size equal to one of the PDU sizes that have been configured.

5.4.2.2 RLC UM Data Transfer Example Figure 5.7 shows an example of RLC UM data transfer. In this example, one SRB carrying RRC signaling messages is mapped to a logical channel in UM with a maximum RLC SDU size of 160 bits. RLC segments, concatenates, and pads the signaling messages to fit into this RLC SDU size. The RLC then adds an 8-bit RLC header to form a 168-bit RLC PDU. This 168-bit RLC PDU becomes a MAC SDU.

The logical channel is then mapped to a DCH transport channel. Normally, four SRB logical channels are mapped into a transport channel. Therefore, MAC adds a 4-bit MAC header to form a 172-bit transport channel. The MAC header contains a logical channel identifier. In this example, the transport block size is configured to be 172 bits and the TTI of the transport block is configured to be 20 ms.

The physical layer maps the 172-bit transport block onto a CCTrCh, which contains two 10-ms radio frames. As such, the entire transport block is transmitted in two 10-ms radio frames. Note that depending on the RRC SDU size, multiple transport blocks may be required for the entire RRC SDU.

5.4.3 RLC Acknowledged Mode

The RLC acknowledged mode operation is more complex than that of the unacknowledged mode. It supports both positive and negative acknowledgement (ACK and NACK), and hence provides reliable services as opposed to unreliable services.

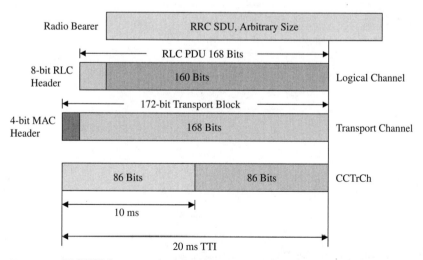

Figure 5.7 RLC UM data transfer example.

Acknowledged mode RLC performs more functions than unacknowledged mode RLC. The functions that AM RLC performs include segmentation and reassembly, concatenation and padding, error correction, in-sequence delivery, duplicate detection, flow control, ciphering, protocol error detection and recovery, and SDU discard. Many of these functions are the same as that for unacknowledged mode, such as segmentation and reassembly, concatenation and padding, ciphering, and SDU discard. (These functions are not discussed here again.) The other functions including error correction, in-sequence delivery, duplicate detection, flow control, and protocol error detection and recovery are AM-specific and are briefly discussed next.

Error correction requires that the PDUs received in error be retransmitted. The transmitter may poll the receiver to remind it to send an acknowledgement status report. In-sequence delivery guarantees that PDUs are delivered to the upper layers from the receiving side RLC in the same order as they were passed to the transmitting side RLC from the upper layers. Duplicate detection ensures that a PDU is delivered to the upper layers only once. Flow control function controls the data flow in dedicated channels by properly configuring the transmitting and the receiving window sizes. Protocol error detection and recovery means that AM RLC may be reset to recover from protocol errors.

5.4.3.1 RLC AM PDU Format Figure 5.8 shows the RLC data PDU format for acknowledged mode. Similar to RLC UM data PDU, an RLC AM data PDU contains many fields, including data/control (D/C), sequence number, polling bit (P), header extension (HE), extension bit (E), length indicator, data, pad, and piggybacked status PDU.

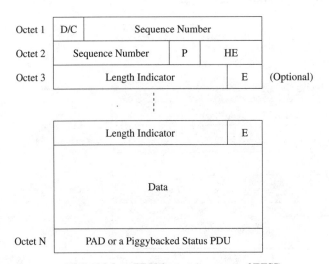

Figure 5.8 RLC AM data PDU format (courtesy of ETSI).

The data/control (D/C) field contains one bit to indicate whether the PDU is a data PDU or a control PDU. If the D/C bit is 0, the PDU is a control PDU. If the D/C bit is 1, then the PDU is a data PDU.

The sequence number field provides a 12-bit sequence number, which is used to detect missing PDUs, to identify a PDU in an ACK or NACK, and as an input to the deciphering procedure.

The polling bit is used to ask for a status report from the receiver. A polling bit of 0 indicates that no status report is requested, while a polling bit of 1 indicates that a status report is requested.

The header extension field (2 bits) and extension bit (1 bit) are used to indicate whether the next byte is the beginning of data or is a length indicator plus extension bit field. For the header extension field, the bits 00 indicate that the next byte contains data, and 01 indicate that the next byte contains a length indicator and an extension bit, while 10 and 11 are reserved. For the extension bit, 0 indicates that the next byte is data, piggybacked status PDU, or padding, and 1 indicates that the next byte is a length indicator plus an extension bit.

The length indicator, data, and pad fields are the same as that in the RLC UM data PDU format. However, for AM, the 7-bit length indicator is used if the AM PDU size is less than or equal to 126 bytes. Otherwise, the 15-bit length indicator is used. For AM, there is an additional field called piggybacked status PDU that is sent at the end of data PDU if there is space available.

An RLC AM control PDU may be a status PDU, a reset PDU, or a reset ACK PDU. Figure 5.9 shows a status PDU format. A status PDU can be sent from transmitter to receiver, or from receiver to transmitter. Figure 5.10 illustrates a reset PDU and a reset ACK PDU format. A reset PDU is only sent from transmitter to receiver, while a reset ACK PDU is only sent from receiver to transmitter. All of these three types of PDU have the common fields: Data/Control, PDU type, and PAD. For control PDU, the D/C bit is 0. The PDU type field contains three bits with 000 indicating that the PDU is a status PDU, 001 indicating that the PDU is a reset PDU, and 010 indicating that the PDU is a reset ACK PDU.

D/C	PDU Type	SUFI$_1$	Octet 1
SUFI$_1$			Octet 2
....			
SUFI$_k$			
PAD			Octet N

Figure 5.9 Status PDU format (courtesy of ETSI).

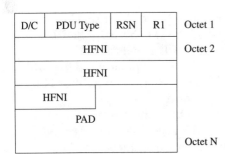

Figure 5.10 Reset PDU and reset ACK PDU format (courtesy of ETSI).

The status PDU contains a super field (SUFI) in addition to the three fields mentioned previously. The SUFI is a variable length field that reports receiver status back to the transmitter. It may include three subfields: type information, length information, and a value.

The size of the type subfield is non-zero, but the size of the other subfields may be zero. The type subfield specifies the status information including no more data (NO_MORE), window size (WINDOW), positive acknowledgments (ACK), a list of sequence numbers that were not correctly received (LIST), a bitmap of sequence numbers (BITMAP), and a request to move the reception window (MRW).

The size and presence of the subfield length and value depend on the superfield type and are specified for each super field separately.

In the reset PDU and reset ACK PDU, there are reset sequence number (RSN), reserved field (R1), and hyper frame number indicator (HFNI) fields.

The RSN is 1 bit used to indicate the sequence number of the transmitted reset PDU. If a reset PDU is a retransmission of the original reset PDU, it will have the same RSN value as the original reset PDU. Otherwise it will have the next RSN value. The initial value of this field is zero. The value of this field is reinitialized when the RLC is reestablished. It is not reinitialized when the RLC is reset.

The reserved field R1 is always coded to 000. HFNI occupies 20 bits. It is used to indicate the hyper-frame number (HFN) to the peer entity. With the help of HFNI, the HFN in UE and UTRAN can be synchronized.

5.4.3.2 RLC AM Data Transfer Example Figure 5.11 shows an RLC AM data transfer example. The radio bearer is a packet-switched RAB carrying user data. It is mapped to an AM logical channel with a maximum RLC SDU size of 400 bits. The RLC adds a 16-bit header to form a 416-bit PDU. RLC segments, concatenates, and pads the data as needed to fit into this PDU size. If there is a status PDU to send, it may be piggybacked at the end of the data, provided that there is room for it. It is obvious that to make room for piggybacked status PDU, the SDU passed to the RLC must be less than 400 bits in size in this example.

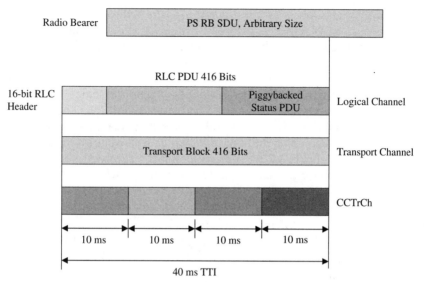

Figure 5.11 RLC AM data transfer example.

The logical channel, which is the 416-bit RLC PDU, is then mapped to a DCH transport channel. In this example, there is no MAC header added because the logical to transport channel mapping is one-to-one, and it is a dedicated transport channel. The transport block size is configured to be equal to the RLC PDU size, and the TTI of the transport block is configured to be 40 ms.

The physical layer maps the 416-bit transport block onto a CCTrCh, which contains four 10-ms radio frames. The entire transport block is transmitted in four 10-ms radio frames. Depending on the packet-switched radio bearer SDU size, multiple transport blocks may be required for the entire radio bearer.

5.5 RLC Ciphering

As mentioned in Chapter 4, ciphering is a security measure to protect all user data and signaling from being overheard by unauthorized entities. On the transmitting side, a ciphering method called ciphering algorithm f8 can be used to encrypt a plain text block by applying a key stream block using a bit per bit binary addition of the plain text and the key stream. This ciphering process generates a cipher text block. On the receiving side, the cipher text block is deciphered by applying the same key stream block using a bit per bit binary addition, recovering the original plain text.

The input parameters to this algorithm include the cipher key, a time dependent input COUNT-C, the bearer identity, the direction of transmission, and the length of the key stream required. For detailed descriptions of these parameters, interested readers may refer to [3].

On the transmitting side, RLC performs ciphering on UM and AM logical channels, while MAC performs ciphering on TM logical channels. All data PDUs are ciphered. But, the first byte of a UM PDU and the first two bytes of an AM PDU are not ciphered. This is because these bytes contain the RLC sequence number (SN), which is used as an input to the deciphering procedure. At the receiving end, the receiving entity must extract the sequence number from the received PDU to construct the COUNT-C, which is an input parameter to the deciphering function.

For each uplink and downlink UM and AM logical channel, a 32-bit COUNT-C is maintained between the UE and UTRAN. COUNT-C consists of two parts. For RLC UM, the first part of COUNT-C is the 25-bit RLC UM hyper-frame number (HFN) while the second part is the 7-bit RLC sequence number, which is a part of the RLC UM PDU header. For RLC AM, the first part of COUNT-C is the 20-bit RLC AM HFN, while the second part is the 12-bit RLC sequence number, which is a part of the RLC AM PDU header. The HFN is initialized using the security parameter START. The sequence number is initialized to zero and incremented when a new PDU is transmitted.

5.6 RLC Configurable Parameters

Many of the RLC parameters that control the functionalities of the RLC entity used by a setup or configured radio bearer, are configurable. Typical examples are the SDU discard parameter and many AM parameters. These configurable parameters are addressed in the following subsections.

5.6.1 SDU Discard

The SDU discard parameter determines how RLC deals with an SDU that cannot be successfully transmitted after a period of time or after a number of transmissions. The transmission failure may be due to poor radio link conditions. Also, data on a low-priority channel may not have a chance to be transmitted because a higher priority channel occupies all of the available bandwidth. SDU discard may be specified as Timer-based discard, Discard after MaxDat, No discard, and Discard not configured.

Timer-based discard is supported by all RLC modes. With timer-based discard, a timer is started when an SDU is received from higher layers and the SDU is discarded if the timer expires before it is successfully transmitted. For acknowledged mode, timer-based discard with explicit signaling is used for discarding SDUs and transferring the discard information between two peer entities. The transmitter sends a Move Receiving Window (MRW) in the type subfield of the SUFI in a status PDU to the receiver. Based on the MRW SUFI, the receiver discards the acknowledged mode data (AMD) PDUs carrying that SDU and updates the reception window. For transparent mode and unacknowledged mode, the transmitter discards the SDU without explicit signaling.

Discard after MaxDAT and No discard are only applicable for acknowledged mode RLC. With Discard after MaxDAT, each SDU's transmission attempts are counted. The SDU is discarded after it is retransmitted MaxDAT times. Just as with the timer-based discard, the transmitter uses explicit signaling to inform the receiver. With No discard, if the number of times an AMD PDU is scheduled for transmission reaches MaxDAT, the transmitter initiates the RLC Reset procedure.

Discard not configured is applicable to both TM RLC and UM RLC. For TM RLC, upon reception of new SDUs from the upper layer, the transmitter discards all SDUs (received from the upper layer in previous TTIs) that are not yet submitted to the lower layer and submits the new SDUs in the first possible TTI. For UM RLC, SDUs in the transmitter are not discarded unless the transmission buffer is full. When the transmission buffer in an UM RLC entity is full, the transmitter may discard the SDU without explicit signaling if segments of the SDU to be discarded have been submitted to a lower layer. If no segments of the SDU to be discarded have been submitted to a lower layer, the transmitter removes the SDU from the transmission buffer.

5.6.2 AM Configurable Parameters

There are many configurable parameters for RLC AM. Some of them are optional or are disabled by default. They are used for flow control, error recovery, and in-sequence delivery.

The parameters related to flow control are transmission window size and reception window size, which respectively specify the maximum number of transmitted RLC PDUs waiting to be acknowledged and the maximum number of RLC PDUs that can be received.

The parameters related to error recovery are maximum reset attempts and reset timer, which respectively specify the maximum number of transmissions of a reset PDU before an unrecoverable error is declared and the time interval between retransmissions of a reset PDU.

The in-sequence delivery parameter specifies whether data delivered to the upper layers on the receiving side is required to be in the same order as it was submitted to the RLC from the upper layers on the transmitting side.

On the receiving side, the receiver sends status reports to the transmitter to ACK or NACK. A status report may be sent either as a standalone status PDU or as a piggybacked status PDU along with a data PDU. Status reports may be sent periodically, whenever a missing PDU is detected, or when polled by the transmitter. The parameters associated with status reports are periodic status timer, status prohibit timer, missing PDU indicator, and estimated PDU counter (EPC) mechanism. The periodic status timer controls the period at which the receiver sends a status report. The status prohibit timer specifies the minimum time interval between transmissions of status reports. The missing PDU indicator determines whether RLC should report every time a PDU is missing. The estimated PDU counter mechanism is used to minimize the delay of the status report retransmission.

On the transmitting side, RLC AM may be configured to request the receiver to send a status PDU. In other words, the transmitter may poll for a status report from the receiver. The transmitter does the polling by setting the polling bit in the RLC header of a data PDU to 1. Polling may be triggered when a timer expires, when no further new PDUs are available to be sent, when no further PDUs are available to be retransmitted, when a PDU or SDU counter is reached, or when a certain percentage of the transmission window size is reached.

5.7 Summary

RLC is a sublayer of Layer 2. RLC provides services to RRC and application layers. It supports three data transfer modes: transparent mode, unacknowledged mode, and acknowledged mode. Its functions include transfer of user data and signaling, segmentation and reassembly, concatenation, padding, error correction, in-sequence delivery of upper layer PDUs, duplicate detection, flow control, sequence number check, protocol error detection and recovery, ciphering, and SDU discard.

An SDU is the basic data unit that flows between layers and sublayers of the protocol stack, while a PDU is the basic data unit exchanged between peer layers of UTRAN and UE.

Each radio bearer is mapped to a logical channel through an RLC entity, which may operate in TM, UM, or AM. The RRC allocates an RLC entity and configures the data transfer mode. In TM operation, PDUs are transferred with little intervention from the RLC. In UM operation, the RLC adds a header to each PDU, which carries segmentation, concatenation, and the sequence number information. In AM operation, RLC performs more functions than in UM operation.

The data PDU formats for TM, UM and AM are somewhat different. TM PDUs contains only data, while UM and AM PDUs contain headers and data. In AM operation, in addition to data PDUs, there are status

PDUs, reset PDUs, and reset ACK PDUs. A status PDU can be sent either from transmitter to receiver or from receiver to transmitter. A reset PDU is only sent from transmitter to receiver while a reset ACK PDU is only sent from receiver to transmitter.

References

[1] 3GPP TS 25.301, v5.5.0, "Radio interface protocol architecture," (Release 5).

[2] 3GPP TS 25.322, v5.7.0, "Radio link control protocol specification," (Release 5).

[3] 3GPP TS 33.102, v5.3.0, "3G security; security architecture," (Release 5).

Chapter

6

Medium Access Control

Medium Access Control (MAC) is a sublayer of Layer 2. It serves the RLC layer and maps logical channels to transport channels. The functions of MAC include prioritizing data flows from RLC, multiplexing and de-multiplexing RLC PDUs when multiple logical channels are mapped to a single transport channel, and selecting the transport format for transport channels. MAC performs UE identification when dedicated logical channels are mapped to common transport channels. It also performs other functions such as traffic volume measurements, directing the RACH procedures, and ciphering of RLC TM data PDUs.

This chapter will cover MAC functions and architecture, explain the contents of a MAC header, examine the headers that are required for common and dedicated channels, preview transport format combination (TFC) selection, investigate how MAC performs traffic volume measurements, explain how ciphering works in MAC and how MAC controls access to the RACH, and describe the MAC-configurable parameters.

6.1 MAC Architecture

Figure 6.1 describes the MAC architecture, which consists of four parts: a broadcast part (MAC-b), a dedicated part (MAC-d), a common part (MAC-c/sh) and a high-speed part (MAC-hs). MAC-b is the MAC entity that handles the broadcast channel (BCH). MAC-c/sh is the MAC entity that handles PCH, FACH, RACH, uplink CPCH, and DSCH. MAC-d is the MAC entity that handles DCH. MAC-hs is the MAC entity that handles the high-speed downlink shared channel (HS-DSCH) for high-speed downlink packet access (HSDPA), which will be discussed in Chapter 11.

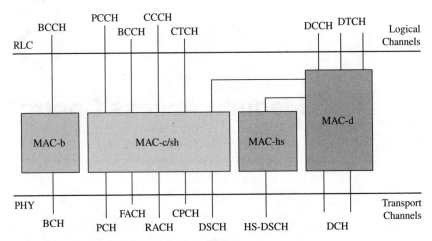

Figure 6.1 MAC architecture (courtesy of ETSI).

6.2 Logical to Transport Channel Mappings

Channel mappings have been discussed in detail in Chapter 2. For convenience, logical to transport channel mappings are shown again in Figure 6.2. Logical to transport channel mappings include downlink and uplink mappings. Downlink mappings include the following:

- BCCH mapped to BCH
- PCCH mapped to PCH
- CCCH mapped to FACH
- DCCH mapped to FACH
- DTCH mapped to FACH
- DCCH mapped to DCH
- DTCH mapped to DCH
- DCCH mapped to DSCH
- DTCH mapped to DSCH
- DCCH mapped to HS-DSCH
- DTCH mapped to HS-DSCH.

Uplink mappings include the following:

- CCCH mapped to RACH
- DCCH mapped to RACH

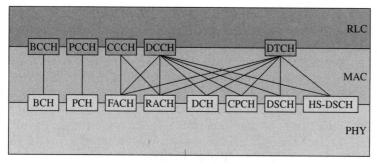

Figure 6.2 Logical to transport channel mappings.

- DTCH mapped to RACH
- DCCH mapped to DCH
- DTCH mapped to DCH
- DCCH mapped to CPCH
- DTCH mapped to CPCH.

6.3 MAC Header

On the transmitting side, MAC may add a header to the SDU coming from the RLC layer. The SDU becomes a MAC PDU after the MAC header is added to it. Depending on channel mapping, a MAC header may include one or more of the following fields: Target channel type field (TCTF), C/T field, UE-Id field, and UE-Id type field. Figure 6.3 shows a non-HS-DSCH MAC PDU containing all four fields of a MAC header and MAC SDU. For HS-DSCH, there are MAC-d PDUs and HS-DSCH MAC PDUs. The MAC-d PDU format is the same as the non-HS-DSCH MAC PDU format. An HS-DSCH MAC PDU consists of one MAC-hs header and one or more MAC-hs SDUs, with each MAC-hs SDU equaling one MAC-d PDU. For details on MAC-hs headers, interested readers may refer to [2].

The TCTF is used when multiple types of logical channels are mapped to a single transport channel. In other words, it is a flag that provides

Figure 6.3 MAC PDU (non-HS-DSCH) (courtesy of ETSI).

identification of the logical channel type on FACH and RACH transport channels. For example, CCCH, CTCH, DCCH, and DTCH logical channels may be mapped to FACH and RACH transport channels. In these mappings the TCTF field of the MAC header is used to indicate the type of logical channel from which the MAC got the SDU. This way, the MAC at the receiving end can route the SDU to the correct logical channel. The TCTF field size of FACH is either 2 bits or 8 bits depending on the value of the two most significant bits. If the two most significant bits are 01 or 10, then the TCTF field size is 8 bits. Otherwise, the TCTF field is 00 or 11. The TCTF field size of RACH is 2 bits.

The C/T field is used when multiple logical channels of the same type are mapped to a single transport channel. It provides identification of the logical channels that are carried on the same transport channel (other than HS-DSCH) or same MAC-d flow (HS-DSCH). A typical example is that four DCCHs are mapped to a single DCH. In this case, the C/T field in the header is used to indicate the channel number of the logical channel that sent this SDU. The size of the C/T field is fixed to 4 bits for both common and dedicated transport channels.

The UE-Id field is used to identify UEs on common transport channels. The UE-Id field contains an RNTI. As mentioned earlier in Chapter 2, the RNTI is assigned when the RRC connection is established. For the receiving UE, the RNTI is used to identify the SDUs on the FACH that are addressed to it. For the transmitting UE, the RNTI is used to identify its SDUs on the RACH to UTRAN. When DCCH is mapped to FACH, either U-RNTI or C-RNTI may be used in the MAC header of DCCH. U-RNTI is never used in the uplink direction. When DCCH and DTCH are mapped to RACH, C-RNTI is used in the MAC header. C-RNTI is also used on DTCH in downlink when mapped onto common transport channels, except when mapped onto the DSCH transport channel. DSCH-RNTI is used on DTCH and DCCH in downlink when mapped onto the DSCH transport channel.

The UE-Id type field is used to identify which type (C-RNTI, U-RNTI or DSCH-RNTI) of UE-Id is used. It is required to ensure correct decoding of the UE-Id field in MAC headers.

6.3.1 MAC Header for Dedicated Logical Channels

As illustrated in Figure 6.4, depending on mapping configuration, five types of MAC headers are used for dedicated logical channels.

The first configuration is a single DCCH or DTCH mapped to a single DCH. In this configuration, only a single UE is using the channel; no UE-Id field or UE-Id type field is required. Also, the logical channel is one-to-one mapped to the transport channel; neither a TCTF nor a C/T field is needed. Therefore, in this configuration, no header is required.

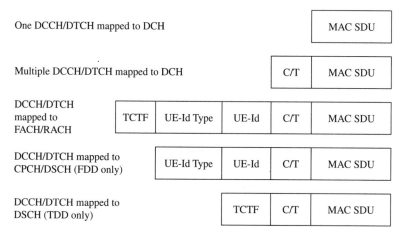

Figure 6.4 MAC PDU formats for DTCH and DCCH (courtesy of ETSI).

The second configuration is multiple DCCHs or multiple DTCHs mapped to a single DCH. In this case, a C/T field is needed to identify the DCCH or DTCH from which the SDU comes. Since there is only one UE using the channel and there is only one type of logical channel involved, no TCTF, UE-Id, or UE-Id type field is required. Therefore, the MAC header for this configuration contains only the C/T field.

The third configuration is multiple DCCHs and multiple DTCHs mapped to FACH or RACH. In this case, the UE-Id and UE-Id type fields are needed for the necessary addressing because multiple UEs are transmitting/receiving on common RACH/FACH channels. The TCTF field is also required to identify the logical channel type as either DCCH or DTCH. Therefore, for this configuration, all four fields of the MAC header are required.

The fourth configuration is DCCH or DTCH mapped to DSCH (FDD only) or CPCH. The UE-Id type field used is DSCH-RNTI for mapping to DSCH and C-RNTI for mapping to CPCH.

The fifth and last configuration is DCCH or DTCH mapped to DSCH (TDD only).

6.3.1.1 Dedicated Logical Channels Mapping Example
Figure 6.5 shows a typical example of mapping three signaling DCCHs onto a single DCH, using 20-ms TTI. The three SRBs are configured in the following ways: RRC UM signaling with priority 1 for SRB 1 (DCCH 1), RRC AM signaling with priority 2 for SRB 2 (DCCH 2), and NAS signaling with priority 3 for SRB 3 (DCCH 3).

The mapping process is briefly described here. First, RLC segments, concatenates, and pads the signaling SDUs as necessary to fit the RLC

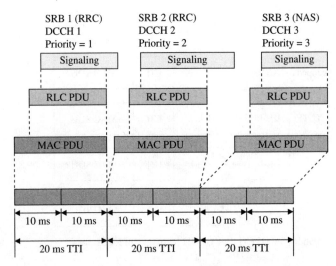

Figure 6.5 Example of dedicated channel mapping for standalone SRBs.

PDU size that is configured. Second, for each TTI (20 ms), MAC checks with RLC to see if there is data to send from any DCCH. If there are multiple DCCHs that have data to send, MAC decides the RLC PDU to send based on the configured logical channel priorities. Third, MAC adds a header containing the C/T field to identify the logical channel that sent the data. Finally, the physical layer maps the transport block onto a CCTrCh, which contains two 10-ms radio frames.

6.3.2 MAC Header for Common Transport Channels

MAC headers for common transport channels are relatively simple. Depending on mapping configuration, a MAC header may or may not be required. Figure 6.6 shows the two different scenarios.

No MAC header is required for the mappings from BCCH to BCH and from PCCH to PCH. This is because these are broadcast channels that carry signaling information for all UEs and the mappings are one to one.

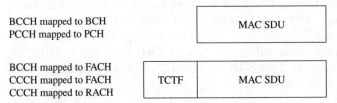

Figure 6.6 MAC PDU formats for common channels (courtesy of ETSI).

For mappings where BCCH is mapped to FACH and CCCH is mapped to FACH or RACH, a TCTF field is required to identify the type of logical channels on which the information was sent. There is no need to have C/T in the header because a single BCCH or CCCH is mapped to FACH for the downlink, and a single CCCH is mapped to RACH for the uplink. Furthermore, address fields are not needed for mapping of CCCH to FACH or RACH, though multiple UEs share the same transport channel. This is because a U-RNTI addressing field is defined in all RRC signaling messages that are transmitted on CCCH. The receiving MAC passes all CCCH SDUs to RLC, which in turn passes them to RRC. RRC then matches the address.

6.3.2.1 Common Channels Mapping Example Figure 6.7 illustrates an example of mapping common and dedicated logical channels to a common transport channel.

The mappings shown in this example include BCCH mapped to BCH, BCCH mapped to FACH, PCCH mapped to PCH, uplink CCCH mapped to RACH, downlink CCCH mapped to FACH, uplink DCCH mapped to RACH, downlink DCCH to FACH, uplink DTCH mapped to RACH, and downlink DTCH mapped to FACH.

Mapping of BCCH to BCH is used for UTRAN to broadcast the system information to UEs, while mapping of BCCH to FACH is used when UTRAN sends a message on BCCH for system information changes. PCCH is always mapped to PCH, representing the paging channel.

In the uplink direction, CCCH is mapped to RACH when a UE attempts to access the network. On the UE side, the signaling message on CCCH is mapped to RACH. On the UTRAN side, the message on RACH is mapped to CCCH since this channel is shared by many UEs. In the downlink direction, CCCH is mapped to FACH when UTRAN responds to a UE's access attempt.

When a UE is in the Cell_FACH state, it is allocated dedicated logical control channels (DCCH) for transferring a signaling message. In the

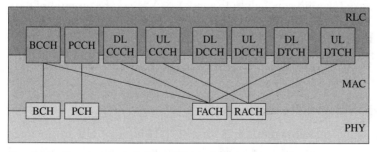

Figure 6.7 Common and dedicated logical channel mappings to common transport channels.

uplink direction, the DCCH is mapped to RACH, while in the downlink direction the DCCH is mapped to FACH.

Dedicated logical traffic channels (DTCH) may also be allocated to the UE for transferring user data when it is operating in the Cell_FACH state. For example, a PS call may operate in the Cell_FACH state, because the burst nature of a packet data service can tolerate the lower throughput of a common channel. Uplink DTCH is mapped to RACH, while downlink DTCH is mapped to FACH.

6.4 Transport Format Combination Selection

One of the important functions MAC performs is transport format combination (TFC) selection, which involves the selections of transport block sizes and a valid combination of transport block sizes. For every transport time interval, the physical layer asks for data from the MAC layer. The MAC layer checks with the RLC layer to see whether there is any data to send, and also determines the amount of data it can deliver to the physical layer. There may be more data available to send than the physical channel can carry. In addition, PDU sizes on logical channels may vary with time. Therefore, the MAC layer must select a valid TFC that maximizes utilization of physical channel resources.

Details on TFC will be discussed in Chapter 7. The following paragraph will briefly discuss how the MAC layer behaves in transferring data from the RLC layer to the physical layer.

Assume that the physical layer channel is configured to carry radio frames with different amounts of data up to a certain data rate. The MAC layer will select a TFC based on available data on logical channels. Once the MAC picks a TFC, that TFC will determine the data rate of the physical channel radio frame by radio frame. Since signaling data is bursting, there may be no data available to send on the logical channels at one moment and data available to send on multiple logical channels the next. If there are data to send on multiple logical channels, MAC will use logical channel priorities to determine which data to send. In addition, the amount of data available to send may vary from frame to frame. In this case, MAC may choose a higher rate when more data is available. If both signaling and user data are available, MAC has to choose a TFC that will maximize the amount of data sent on the highest priority channel.

6.5 Traffic Volume Measurement

Traffic volume measurement is a task UTRAN asks UE to perform during a PS call. UTRAN sends the traffic volume measurement information elements in SIB Type 11 and SIB Type 12 through BCCH mapped to BCH and the Measurement Control on DCCH mapped to FACH or DCH. Traffic volume is measured in the UE MAC layer and the results are reported to UTRAN. Basically, UTRAN may configure the UE to measure and report the amount of traffic available in the uplink. UTRAN will use the measurement reports to determine when to transfer the UE from one connected-mode state to another connected-mode state, for example, from the Cell_DCH state to Cell_FACH state. UTRAN may also use the measurement reports to determine the maximum DCH data rate and to arrange resource allocation.

The 3GPP specifications do not specify the UTRAN resource allocation and UE call state transition algorithms. It is up to network equipment vendors to develop their own algorithms. However, the performance of these algorithms is important to the overall performance of a packet data system. More specifically, user data throughput, overall system throughput, and system capacity for both voice and data will be affected.

6.6 MAC Ciphering

As mentioned in Chapter 5, RLC does not perform ciphering of TM dedicated logical channels—the MAC does. Ciphering is performed in the MAC-d entity. The MAC header is not ciphered. When a dedicated logical channel is mapped to a common transport channel, the MAC header contains a UE-Id address field. All UEs receiving the MAC PDU can read the UE-Id field, but only the UE that the PDU is addressed to can decipher the rest of the PDU.

The ciphering and deciphering process in the MAC layer is the same as that in the RLC layer, except that the 32-bit COUNT-C is different. For MAC ciphering of TM dedicated logical channels, a common COUNT-C applies to all radio bearers for both uplink and downlink directions and is maintained between UTRAN and UE. More specifically, the same COUNT-C is independently maintained in the mobile equipment MAC-d entity and the serving RNC MAC-d entity. COUNT-C consists of two parts: a long sequence number and a short sequence number. The long sequence number of COUNT-C is a 24-bit hyperframe number (HFN), and the short sequence number is an 8-bit connection frame number (CFN). HFN is initialized to the ciphering START value and incremented when CFN rolls over. The CFN is initialized to the CFN at which ciphering is activated and incremented every radio frame. For details on MAC ciphering, interested readers may refer to [4].

6.7 MAC RACH Functions

MAC RACH functions include determination of access service class (ASC) and control of RACH transmissions. MAC picks the access service class, computes the persistence delay, and then starts a RACH transmission by asking the physical layer to perform a preamble ramping cycle. During a preamble ramping cycle, if the UE receives an acknowledgment on AICH, the UE MAC asks the physical layer to transmit the RACH message. If the UE receives no acknowledgment or a negative acknowledgment, the UE MAC tries another preamble ramping cycle if the maximum preamble ramping cycle number configured in SIB Type 5 or SIB Type 6 has not been reached. In addition, in the case when a negative acknowledgment is received, the UE MAC calculates a randomized back off interval before trying another preamble ramping cycle. The detailed procedure of MAC RACH is described in the following subsection.

6.7.1 MAC RACH Procedure

A flow diagram of the MAC RACH procedure is illustrated in Figure 6.8. On the UE side, the MAC layer performs the following steps for controlling the transmission operations on the RACH.

- **Step 1: Persistency check**

    Before a UE attempts to transmit on RACH, the UE MAC carries out a persistency check. It generates a random variable between 0 and 1 and compares the random variable with the persistence value corresponding to the access service class. If the persistency check passes,

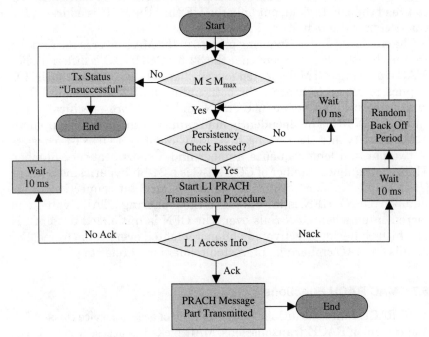

Figure 6.8 MAC RACH procedure flow chart (courtesy of ETSI).

the MAC starts a RACH transmission by requesting the physical layer to perform a PRACH transmission procedure (preamble ramping cycle).

- **Step 2: Physical layer PRACH transmission procedure**

  In this step, the UE transmits a preamble at increasing power levels until it receives either a positive or negative acquisition indicator on the AICH.

- **Step 3: Transmission of PRACH message or maximum preamble ramping cycle check**

  During the preamble ramping cycle, if the UE receives an acknowledgement on AICH, the UE MAC requests the physical layer to transmit the RACH message. If the UE receives no acknowledgment, the UE MAC will wait 10 ms before going to maximum preamble cycle check, where the UE MAC checks whether the maximum preamble cycle number, M_{max}, configured in SIB Type 5 or SIB Type 6 has been reached or not. If a negative acknowledgment is received on AICH, the UE MAC will wait 10 ms and then compute a randomized back off interval before going to maximum preamble cycle check. In both no acknowledgement and negative acknowledgement situations the UE MAC will determine that transmission is unsuccessful and will stop the access attempt if the maximum preamble ramping cycle number has been reached. Otherwise, it will go back to persistency check again and continue the MAC RACH procedure.

Reception of a negative acknowledgement may be due to various reasons, including poor RF condition, insufficient UE power, or an unavailability of resources at the UTRAN side. UTRAN may send a negative acknowledgement to the UE if its PRACH resources are currently not available. In addition, if UTRAN detects that many UEs are trying to access the system simultaneously, it may send a positive acknowledgement to one UE and negative acknowledgements to the rest of the UEs.

6.7.2 Access Class and Access Service Class

An access class (AC) ranging from 0 to 9 is assigned to every UE. The AC is stored in the USIM. A UE will use AC 0 if it has no USIM. Special ACs ranging from 11 to 15 may also be programmed into the USIM for high priority users, such as public utilities, security services, and emergency services (fire fighting, rescue missions, and so forth).

The access service class (ASC) must be determined before RACH transmits. The ASC ranging from 0 to 7 determines certain RACH procedure parameters, such as PRACH signature, AICH sub-channel, and persistence value.

When a UE sends an RRC Connection Request message to establish an RRC connection, the UE RRC determines the ASC to be used for RACH transmission by using the mapping between AC and ASC specified in SIB Type 5.

If an RRC connection has already been established, MAC uses the logical channel priorities to determine the ASC for all subsequent RACH transmissions while the UE is in the Cell_FACH state. The logical channel priorities are assigned when the radio bearers are established. The ASC for a message is determined by the priority of the logical channel from which the message is sent. A higher priority channel has a lower ASC (ASC 0 = highest priority, ASC 7 = lowest priority). ASC 0 is used in case of an emergency call or for other reasons with equivalent priority.

6.7.3 Persistency Check

Persistency check is the first step of the MAC RACH procedure. For each ASC, there is a corresponding persistence value. The persistence values P_i associated with each ASC can be derived from the following formula by using the dynamic persistence level N = 1, ... , 8, which is broadcast in SIB Type 7, and the persistence scaling factors, s_i, broadcast in SIB Type 5 or SIB Type 6.

$$P_i = s_i P(N) = 2^{-(N-1)} s_i \qquad (6\text{-}1)$$

where i is a dummy variable running from 2 to 7. For ASC 0, the associated persistence value is $P_0 = 1$. For ASC 1, the associated persistence value is $P_1 = P(N)$. For ASC 2 to ASC 7, the associated persistence value is given by (6-1) with $i = 2, ... , 7$. A lower ASC has a higher associated persistence value.

Figure 6.9 shows a flow chart of persistency checking. Before each physical layer PRACH preamble ramping cycle, MAC performs a persistency check. The persistency check involves picking a random number between 0 and 1 and then comparing the random number with the persistence value. If the random number is smaller than the persistence value, the physical layer starts a preamble ramping cycle. Otherwise, MAC waits 10 ms and carries out another persistency check. This process is repeated until the persistency check passes.

Before each persistency check, MAC must check to see whether the RACH control parameters that are sent in the system information messages have been changed. This is because the fast changing parameters for the PRACH such as dynamic persistence level may have changed between persistency checks. In addition, UTRAN may use the dynamic persistence level to regulate PRACH transmissions to avoid system overload.

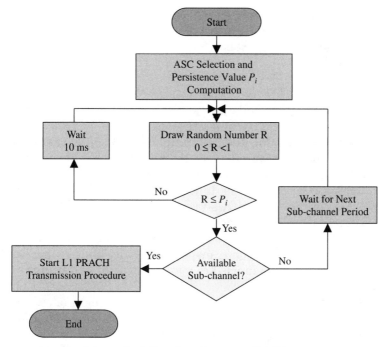

Figure 6.9 Persistency check flow chart (courtesy of ETSI).

6.8 MAC Configurable Parameters

MAC is configured by RRC based on system information, radio bearer configuration, and measurement control information. The key MAC configurable parameters include UE identity, radio bearer configuration, transport format combination set, traffic volume measurements, RACH transmission control, and ciphering.

For UE identity, MAC may use U-RNTI and C-RNTI in the MAC header when the UE is in connected mode. Radio bearer configuration allows MAC to map logical channels onto transport channels. Transport format set and transport format combination set are required for MAC to select respectively the transport format for every active transport channel and the transport format combination for multiple transport channels being active at the same time. Traffic volume measurements allow MAC to know the parameters configured for traffic volume measurements and reporting. RACH transmission control parameters such as ASC, persistence values, the number of preamble ramping cycles, and the maximum and minimum time between ramping cycles allow

MAC to perform the RACH procedure. Ciphering parameters such as ciphering key and the initial CFN are required for MAC to cipher the transparent mode logical channels.

6.9 Summary

Major functions of MAC include logical and transport channel mapping, identification of UEs on common transport channels, prioritizing logical channels, multiplexing and de-multiplexing of logical channels, transport format combination selection, traffic volume measurements, ciphering of TM logical channels, and RACH procedures.

When mapping logical channels to transport channels, MAC adds appropriate headers to MAC SDUs (such as RLC PDUs) to become MAC PDUs. A complete MAC header includes TCTF, UE-Id type, UE-Id, and C/T fields. A complete MAC header is used when dedicated logical channels are mapped to common transport channels, while no header is needed when dedicated channels are mapped to dedicated transport channels.

MAC performs transport format combination (TFC) selection, which involves the selections of transport block sizes and a valid combination of transport block sizes. MAC also determines how much data it can deliver to the physical layer. As such, it must select a valid TFC that maximizes utilization of physical channel resource.

Traffic volume measurement is performed by the UE during a PS call. UTRAN may configure the UE to measure and report the amount of traffic available in the uplink. UTRAN uses the measurement reports to determine when to transfer the UE from one connected-mode state to another connected-mode state. UTRAN may also use the measurement reports to determine the maximum DCH data rate and to arrange resource allocation.

MAC performs ciphering of TM dedicated logical channels. When the MAC performs ciphering, it does not cipher the MAC header because the header may contain a UE-Id address field. All UEs receiving the MAC PDU can read the UE-Id field, but only the UE that the PDU is addressed to can decipher the rest of the PDU.

MAC RACH procedures include determination of access service class (ASC) and control of RACH transmissions. MAC selects the access service class and computes the persistence delay and then starts a RACH transmission by requesting the physical layer to perform a preamble ramping cycle. During a preamble ramping cycle, if the UE receives an acknowledgment on AICH, the UE MAC will request the physical layer to transmit the RACH message. If the UE receives no acknowledgment or a negative acknowledgment, the UE MAC will try another preamble ramping cycle if the maximum preamble ramping cycle number has not been reached.

References

[1] 3GPP TS 25.301, v5.5.0, "Radio interface protocol architecture," (Release 5).

[2] 3GPP TS 25.321, v5.7.0, "Medium Access Control (MAC) protocol specification," (Release 5).

[3] 3GPP TS 25.331, v5.7.1, "Radio Resource Control (RRC) protocol specification," (Release 5).

[4] 3GPP TS 33.102, v5.7.1, "3G security; security architecture," (Release 5).

Chapter 7

Physical Layer

The physical layer is the bottom layer of the protocol stack that handles the air interface. It is also called Layer 1 of the protocol stack. It offers different transport channels to MAC and provides services to higher layers. A variety of physical channels are defined in the physical layer. There are two duplex modes that can be employed in the physical layer. They are the frequency division duplex (FDD) and time division duplex (TDD) modes. FDD mode uses code, frequency, and relative phase (in-phase or quadrature phase on the uplink) to characterize the physical channels. TDD mode uses time slots to characterize the physical channels.

The physical layer performs the following functions [1]:

- Error detection on transport channels and indication to higher layers
- FEC encoding/decoding of transport channels
- Multiplexing of transport channels and de-multiplexing of coded composite transport channels (CCTrCh)
- Rate matching of coded transport channels to physical channels
- Mapping of coded composite transport channels on physical channels
- Power weighting and combining of physical channels
- Modulation and spreading, as well as demodulation and de-spreading of physical channels
- Frequency and time (chip, bit, slot, frame) synchronization
- Radio characteristics measurements including FER, SIR, interference power, and so on, as well as indication to higher layers
- Macro-diversity distribution/combining and soft handover execution
- Inner-loop power control

- RF processing
- Synchronization shift control
- Beam-forming

Operating in either FDD or TDD mode, UMTS allows for efficient utilization of the available spectrum based on the frequency allocation in different nations. In the physical layer, the information spreads over either 5-MHz bandwidth (for FDD and 3.84 Mcps TDD) or 1.6 MHz (for 1.28 Mcps TDD). For FDD, uplink and downlink transmissions use two separate radio frequencies in different frequency bands. A pair of frequency bands is assigned for the system.

For TDD, uplink and downlink transmissions are carried over the same radio frequency by using synchronized time intervals. Time slots in a physical channel are divided into transmission and receiving parts. Uplink and downlink information are transmitted reciprocally.

The basic transmission unit in the physical layer is a radio frame, which is 10 ms in length (for non-HSDPA) and divided into 15 slots with 2560 chips per slot for FDD. The information rate of physical channels varies with symbol rate, which in turn depends on spreading factor. The spreading factor ranges from 4 to 256 for FDD uplink, from 4 to 512 for FDD downlink, and from 1 to 16 for 3.84 Mcps TDD uplink and downlink. As such, the corresponding symbol rate varies from 960 ksps to 15 ksps for FDD uplink, from 960 ksps to 7.5 ksps for FDD downlink, and from 3.84 Msps to 240 ksps for 3.84 Mcps TDD.

For the 1.28 Mcps TDD mode, a 10-ms radio frame is divided into two 5-ms subframes. There are 7 normal time slots and 3 special time slots in each subframe. With a spreading factor ranging from 16 to 1, the corresponding symbol rate varies from 80 ksps to 1.28 Msps.

This chapter will focus on the FDD mode. Major topics to be discussed in this chapter include orthogonal spreading codes, scrambling codes, synchronization codes, physical layer timing, downlink procedures, uplink procedures, physical channel structures, physical channel timing, and physical layer procedures.

7.1 Orthogonal Spreading Codes

Just like narrow band CDMA systems, WCDMA is a code division multiple access system. In CDMA, the data streams are spread using Walsh codes, while in WCDMA the data streams are spread using orthogonal variable spreading factor (OVSF) codes. OVSF codes are also known as channelization codes or Walsh codes. In its general form, an OVSF code can be written as $C_{ch,SF,k}$, where C_{ch} stands for channelization code, SF stands for spreading factor, and k stands for code number with $0 \leq k \leq SF-1$. OVSF codes are used to spread data symbols to

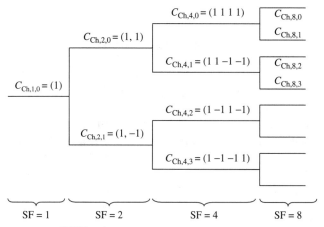

Figure 7.1 OVSF code tree.

chips on both the uplink and downlink. OVSF specifies the number of chips per symbol.

As shown in Figure 7.1, OVSF codes form a code tree. It begins with the first generation one-bit code $C_{ch,1,0} = (1)$, where the subscript ch stands for channel, the subscript 1 stands for spreading factor 1, and the subscript 0 stands for code number 0. The second generation consists of two codes: $C_{ch,2,0}$ and $C_{ch,2,1}$. They are two-bit codes with a spreading factor of 2. The third generation consists of four codes: $C_{ch,4,0}$, $C_{ch,4,1}$, $C_{ch,4,2}$, and $C_{ch,4,3}$. They are four-bit codes with a spreading factor of 4. The fourth generation has 8 eight-bit codes with a spreading factor of 8. The code tree can go up to 10 generations with the 10th generation having 512 codes. For a given code tree generation, the spreading factor is equal to the number of codes.

Figure 7.2 illustrates how the OVSF codes are generated. The code $C_{ch,2,0}$ is generated by using $C_{ch,1,0}$ for both its first and second bits, while the code $C_{ch,2,1}$ is generated by using $C_{ch,1,0}$ as its first bit and the negative

$$C_{ch,1,0} = 1$$

$$\begin{bmatrix} C_{ch,2,0} \\ C_{ch,2,1} \end{bmatrix} = \begin{bmatrix} C_{ch,1,0} & C_{ch,1,0} \\ C_{ch,1,0} & -C_{ch,1,0} \end{bmatrix} = \begin{bmatrix} 1 & 1 \\ 1 & -1 \end{bmatrix}$$

$$\begin{bmatrix} C_{ch,4,0} \\ C_{ch,4,1} \\ C_{ch,4,2} \\ C_{ch,4,3} \end{bmatrix} = \begin{bmatrix} C_{ch,2,0} & C_{ch,2,0} \\ C_{ch,2,0} & -C_{ch,2,0} \\ C_{ch,2,1} & C_{ch,2,1} \\ C_{ch,2,1} & -C_{ch,2,1} \end{bmatrix} = \begin{bmatrix} 1 & 1 & 1 & 1 \\ 1 & 1 & -1 & -1 \\ 1 & -1 & 1 & -1 \\ 1 & -1 & -1 & 1 \end{bmatrix}$$

Figure 7.2 Generation of OVSF codes.

of $C_{ch,1,0}$ as its second bit. The third-generation codes are generated from the second-generation codes by following the same process.

The OVSF code tree makes it possible to use multiple spreading factors. For a given spreading factor, all OVSF codes are orthogonal to one another. On the downlink, different OVSF codes separate UEs in a cell. On the uplink, different OVSF codes separate dedicated physical channels, such as DPCCH and DPDCH, in UE.

Spreading factors can range from 4 to 512 for the downlink and 4 to 256 for the uplink. Codes with different spreading factors have different lengths. For example, codes with a spreading factor of 128 have 128 chips, while codes with a spreading factor of 256 have 256 chips. During spreading, one data symbol is represented by one OVSF code. Therefore, high-rate calls use low spreading factors, while low rate calls use high spreading factors because the chip rate is kept constant at 3.84 Mcps.

7.1.1 Orthogonal Sequences

Two sequences are said to be orthogonal to each other if they have zero cross-correlation. Zero correlation means that the product of two sequences, summed over a period of time, is zero. The values of 1 and −1 are of opposite polarity. If the product of two binary sequences, made up of individual 1s and −1s, is a new sequence with equal numbers of 1s and −1s, the cross-correction of the two sequences is zero. As you have seen, OVSF codes are made up of individual 1s and −1s. Two OVSF codes are orthogonal to each other if their cross-correlation is zero. OVSF codes within the same generation of the code tree are orthogonal, and so are non-related codes from different generations. However, related-codes from different generations are not orthogonal. For example, $C_{ch,4,2}$ and $C_{ch,2,1}$ are not orthogonal.

7.1.2 Spreading and De-spreading

The fundamental concept of spreading and de-spreading is that when a symbol is multiplied with a given binary sequence, and the product is again multiplied with the same binary sequence, the original symbol is recovered. That is, the effect of multiplication is null when performed twice using the same code.

WCDMA does the orthogonal spreading by multiplying each encoded symbol with an OVSF code. For example, in Figure 7.3, an original symbol of value 1 is orthogonally spread with OVSF code $C_{ch,4,2}$, thus yielding a 4-chip representation of the symbol.

For voice with a symbol rate of 60 ksps, the 64-bit OVSF code used for spreading runs 64 times faster than the symbol, making the spread symbol run at 3.84 Mcps. For data with a symbol rate of 960 ksps, the

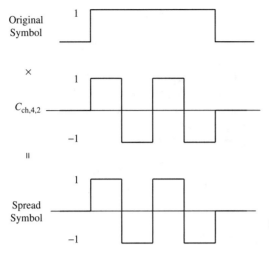

Figure 7.3 Orthogonal spreading.

4-bit OVSF code used for spreading runs four times faster than the symbol, resulting in again a spread symbol chip rate of 3.84 Mcps.

The spread symbols can be de-spread by using the same OVSF code used for spreading. As shown in Figure 7.4, the original symbol of value 1 can be recovered by multiplying the spread symbol with the same OVSF code $C_{ch,4,2}$. Note that under ideal (no noise) conditions, the symbols are completely recovered without errors. However, in practice, the WCDMA channels are normally not noise-free, some errors may occur during the spreading and de-spreading process. As such, WCDMA systems a employ forward error correction (FEC) technique to alleviate the effects of noise and improve the system performance.

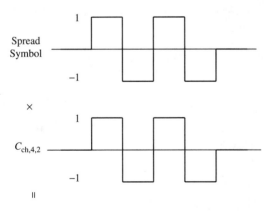

Figure 7.4 Recovery of the original symbol using the correct code.

The original symbol cannot be recovered if an incorrect orthogonal sequence is used for de-spreading. As shown in Figure 7.5, if the spread symbol is de-spread by using the wrong code, $C_{ch,4,1}$, the resulting symbol is a distorted symbol, not the original symbol of value 1. This demonstrates the advantage of the orthogonal property of OVSF codes. If someone attempts to decode the signal using a wrong code, the original data cannot be recovered, protecting the privacy of the original data.

In real systems, multiple users' spread symbols are combined and transmitted. At the receiving end, each individual's original symbols are recovered by using each individual's OVSF code. Figure 7.6 shows an example of spreading with three users designated as A, B, and C. In this example, the data symbols and OVSF codes of the users are given as follows:

User A: data symbols = (1 –1), OVSF Code = (1 –1 1 –1).

User B: data symbols = (1 1), OVSF Code = (1 1 –1 –1).

User C: data symbols = (–1 1), OVSF Code = (1 –1 –1 1).

The top three waveforms in the figure are the spread symbols of users A, B, and C. The bottom waveform is the composite signal of the three spread symbols.

At the receiving end, the data symbol of each user can be recovered by multiplying the composite signal with each user's OVSF code. For example, as shown in Figure 7.7, at User A's receiver, the composite

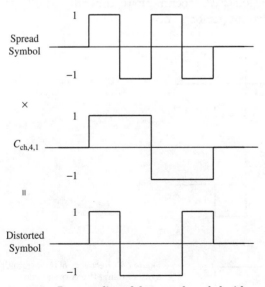

Figure 7.5 De-spreading of the spread symbol with a wrong code.

Physical Layer 117

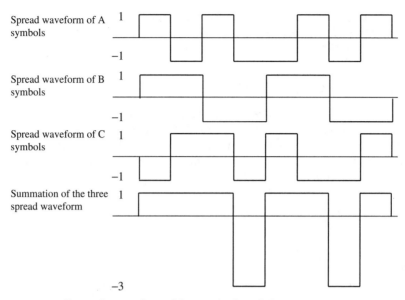

Figure 7.6 Composite waveform of three spread symbols.

signal is multiplied by User A's OVSF code and the result is then averaged over one symbol period. The average value over the first symbol period is 1 and the average value over the second symbol period is −1.

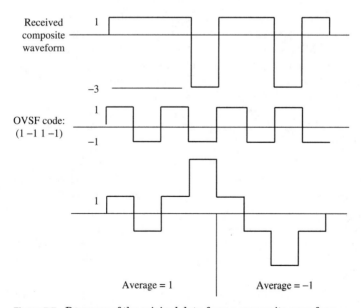

Figure 7.7 Recovery of the original data from a composite waveform.

Thus, the original data symbols (1 − 1) transmitted by User A are recovered. The original data symbols of User B and User C can also be recovered by using the corresponding OVSF codes.

7.2 Scrambling Codes

In WCDMA, Gold codes are used to scramble the data, separate the cells (sectors) on the downlink, and distinguish the users on the uplink. Gold codes are pseudorandom noise sequences that simulate a random noise process. They have good cross-correlation properties that are excellent for separating cells and UEs. Therefore, Gold codes with truncation are used as scrambling codes. In the following subsections, we will discuss how Gold codes, and thus scrambling codes, are generated.

7.2.1 Maximum Length Pseudorandom Binary Sequences

Maximum length pseudorandom binary sequences, also called M-sequences, are random binary number sequences generated by a deterministic process. The maximum achievable period of a generated sequence is given by 2^n-1, where n is the number of the shift registers. M-sequence is a periodical sequence and has 2^{n-1} ones and $2^{n-1}-1$ zeros. Half of the runs of consecutive zeros or ones are of length one, one quarter are of length two, one-eighth are of length three, and so forth. The chip by chip sum of an M-sequence C_m and any shift of itself C_{m+t} is a time-shifted version of the same sequence. The autocorrelation of an M-sequence is a periodic two-valued function equaling to 2^n-1 or −1. When two identical M-sequences are exactly aligned, the autocorrelation reaches the peak value of 2^n-1; with any other offset, the autocorrelation drops dramatically to −1. Therefore, when an M-sequence is correlated with itself over the entire 2^n-1 period, the autocorrelation value is 1.

7.2.2 Gold Codes

A Gold code is the product of a pair of preferred M-sequences. The generation of Gold codes can be carried out using two preferred M-sequence generators. If the two preferred M-sequence generators are of degree n, they can generate 2^n Gold codes by using a fixed non-zero seed in the first generator, and changing the seed of the second generator from 0 to 2^n-1. The cross-correlation of Gold codes is $R_{xy} = \{-t(n), -1, t(n)-2\}$, where $t(n)$ is equal to $2^{(n+1)/2}+1$ for n odd and $2^{(n+2)/2}+1$ for n even. The Gold code sequences are of length $2^{18}-1$ for downlink and $2^{25}-1$ for uplink.

7.2.3 Scrambling Code Generation

Gold codes have low cross-correlation, minimizing the possibility of mistaking one Gold code for another. As such, Gold codes are used for scrambling the channelized signals. In WCDMA, the Gold code sequence is truncated at 38,400 chips to match the 10-ms radio frame.

On the downlink, a total of $2^{18}-1 = 262,143$ scrambling codes, numbering 0, 1, 2, ..., 262,142, can be generated. The code length is $2^{18}-1$, but is truncated at 38,400 chips.

However, not all of them are used. In the normal mode, 8192 scrambling codes numbering 0, 1, 2, ..., 8191 are used. These 8192 codes are divided into 512 sets with each set containing a primary scrambling code (PSC) and 15 secondary scrambling codes (SSC). The PSCs are scrambling codes $16 \times m$, where $m = 0, 1, ..., 511$ (that is, 0, 16, 32, ..., 8176). The SSCs are scrambling codes $16 \times m + k$, where $k = 1, 2, ..., 15$ (for example, the SSCs associated with first PSC are codes 1, 2, ..., 15). The 512 PSCs are further divided into 64 scrambling code group, each with 8 PSCs.

Scrambling codes are used to identify the cells (sectors). Each cell is assigned a unique primary scrambling code (one of 512 codes). The cross-correlation between any two scrambling codes is very low regardless of the timing offset between them, allowing asynchronous deployment of the cells and the use of secondary scrambling codes. The PCCPCH, primary CPICH, PICH, AICH, and SCCPCH carrying PCH are always transmitted using PSC. The other downlink physical channels can be transmitted with either the PSC or a SSC from the set associated with the PSC of the cell. The mixture of PSC and SSC for one CCTrCh is allowable. Normally, secondary scrambling codes are used when there is a need to reuse the OVSF codes that are scrambled by the PSC.

On the uplink, there are 2^{24} scrambling codes. The code length is $2^{25}-1$ but is truncated at 38,400 chips. UTRAN informs the UE to use a particular uplink scrambling code when a dedicated physical channel is assigned. Each UE is assigned a unique scrambling code (one of 2^{24} scrambling codes), thus identifying the users.

7.3 Synchronization Codes

There is one primary synchronization code and 16 secondary synchronization codes. They are complex-valued sequences with a length of 256 chips. The primary synchronization code, C_p, is generated as follows:

Define $a = \{x_1, x_2, ..., x_{16}\} = \{1,1,1,1,1,1,-1,-1,1,-1,1,-1,1,-1,-1,1\}$, the primary synchronization code is given by

$$C_p = (1+j) \times \{a,a,a,-a,-a,a,-a,-a,a,a,a,-a,a,-a,a,a\}.$$

The secondary synchronization codes, $C_{s,k}$ ($k = 1,...,16$), are generated as follows:

Define z = {b,b,b,–b,b,b,–b,–b,b,–b,b,–b,–b,–b,–b,–b}

where $b = \{x_1, x_2, ..., x_8, -x_9, -x_{10}, ..., -x_{16}\} = \{1,1,1,1,1,1,-1,-1,-1,1,-1,1,-1,1,1,-1\}$.

Obtain Hadamard sequence from matrix H_8, which is a 256 × 256 matrix. The secondary synchronization codes are given by

$$C_{s,k} = (1+j) \times \{h_m(0) \times z(0), h_m(1) \times z(1), ..., h_m(255) \times z(255)\},$$

where $h_m(i)$ is the symbol at ith column and mth row of H_8, $z(i)$ is the ith symbol of z sequence and $m = 16(k-1)$ with $k = 1, 2,...,16$.

A Hadamard sequence is a row of a matrix H_8 constructed recursively by:

$$H_0 = 1 \qquad (7\text{-}1)$$

$$H_k = \begin{bmatrix} H_{k-1} & H_{k-1} \\ H_{k-1} & -H_{k-1} \end{bmatrix} \quad \text{for } k \geq 1. \qquad (7\text{-}2)$$

The primary synchronization code is used to identify the time slot boundaries, while the secondary synchronization codes are used to identify the frame boundaries. Details on the usage of PSC and SSC will be discussed in Subsection 7.8.1.

7.4 Physical Layer Timing

The physical layer timing is very straightforward. A physical channel radio frame is 10 ms in length. The physical channels are on radio frame boundaries. The transport channels are on 10 ms, 20 ms, 40 ms, and 80 ms boundaries. Therefore, the transport channels are broken into 10-ms radio frames to align with the physical channels. A 10-ms radio frame is divided into 15 time slots, each with a length of 2/3 ms. WCDMA operates at a chip rate of 3.84 Mcps, therefore there are 2560 chips per time slot.

Time slots are used not only for initial system acquisition, but also for physical layer control. As will be discussed in the physical channel structure in Subsections 7.5.10 and 7.7.8, power control bits are embedded in each time slot, indicating that power control in WCDMA performs at 1500 Hz.

7.5 Downlink Procedure

Generically, the downlink procedure starts from receiving the transport channel data from the MAC layer and ends with transmitting the physical channels into the air. As shown in Figure 7.8, the downlink procedure consists of many steps, including attachment of cyclic redundancy check (CRC), concatenation of transport block, segmentation of the data stream into code blocks, channel coding, rate matching, 1st discontinuous transmission (DTX), 1st interleaving, segmentation of transport channel into 10-ms radio frames, multiplexing of transport radio frames into a coded composite transport channel (CCTrCh), 2nd DTX insertion, 2nd interleaving, mapping to physical channels, spreading of physical channels using OVSF codes, scrambling of physical channels using downlink scrambling codes, and modulation. After the modulation step, the signal is transmitted into the air, completing the downlink physical procedure. In the following subsections, the steps of the downlink procedure are described individually.

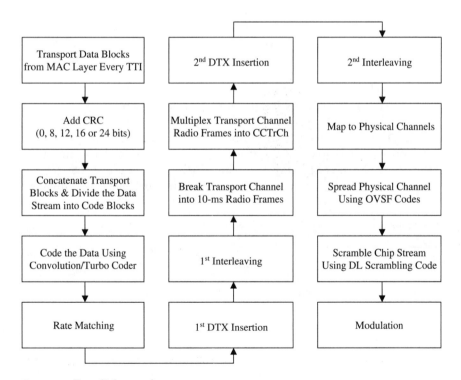

Figure 7.8 Downlink procedure.

7.5.1 Transport Channel Data Delivery to Physical Channels

The first step of downlink procedures is delivery of transport channel data to physical channels. Every transport time interval, the physical layer asks for data from the MAC layer. MAC in turn must query the RLC to determine how much data is available to send, and then determine how much data MAC can deliver to the physical layer.

7.5.1.1 Transport Block and Transport Block Set When transport channel data is delivered to the physical layer, it is delivered in transport blocks. A transport block is equivalent to a MAC PDU and is the basic data unit exchanged between MAC and the physical layer. The length of a transport block in time is called a transport (or transmission) time interval (TTI). A TTI could be 10, 20, 40, or 80 milliseconds. For a given transport channel, the physical layer asks for data from the MAC layer every TTI.

A transport block set consists of zero or multiple transport blocks that are exchanged between MAC and the physical layer at the same time using the same transport channel. Figure 7.9 illustrates these definitions.

The purpose of using transport blocks is to avoid the loss of large amounts of data. When a large data block is broken into a set of smaller blocks, each of the smaller blocks can have its own CRC. An error that occurs in a block results in the loss of that block only; other blocks are not affected. If a large data block is not broken into smaller ones and an error occurs in the block, all data in the block could be lost.

7.5.1.2 Transport Format and Transport Format Set A transport block set delivered to the physical layer must follow a transport format (TF) offered by the physical layer to the MAC layer. A transport format defines the format for MAC to deliver a transport block set during a

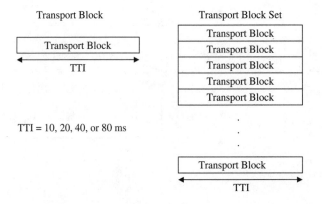

Figure 7.9 Transport Block and Transport Block Set.

TTI on a transport channel. It specifies the transport block size (in other words, bits per block), and the number of transport blocks in a transport block set. For example, assuming a transport block size of 640 bits, the transport format could be defined as TF$n = n \times 640$, where n, representing the number of transport blocks, could be 0 or any positive integer.

There may be many choices of transport formats that could be used for delivering transport blocks. A set of valid transport formats that can be used for a given transport channel is called a transport format set (TFS). In other words, the transport format set defines all of the valid transport formats for each transport channel. For example, a TFS could consist of TF1, TF2, TF4, and TF8. To support a 64 kbps circuit-switched radio access bearer (CS RAB) for streaming data with a transport block size of 640 bits, TF1 = 1×640 is chosen if TTI = 10 ms, TF2 = 2×640 is chosen if TTI = 20 ms, TF4 = 4×640 is chosen if TTI = 40 ms, and TF8 = 8×640 is chosen if TTI = 80 ms.

7.5.1.3 Transport Format Combination and Transport Format Combination Set As mentioned before, a number of transport channels may be multiplexed onto a CCTrCh. A TFS is defined for a transport channel and the TFS may be different for different transport channels. There are many possible valid combinations of transport formats (one for each transport channel) that could be chosen for mapping multiple transport channels onto a CCTrCh. Each valid combination is called a transport format combination (TFC).

A set of TFCs is called a transport format combination set (TFCS). A TFCS that contains all valid TFCs for a given CCTrCh is offered to the MAC layer, which then selects a TFC from the TFCS at every radio frame boundary. As mentioned in Chapter 6, the MAC layer must select a valid TFC that maximizes utilization of physical channel resources. To achieve this purpose, the MAC layer normally selects the TFC based on the following factors: the amount of data to be sent on each logical channel, the logical channel priorities, and the Quality of Service parameters for each logical channel. In addition, the MAC layer may treat those data that could not be sent in a particular TTI differently for different logical channels. For instance, time sensitive data may be discarded while non-time sensitive data may be saved for retransmission later on.

To represent a TFC, a transport format combination indicator (TFCI) is indexed into the TFCS. TFCI is transmitted on every radio frame on the dedicated physical control channel (DPCCH), allowing the receiver to identify the TFC that was used in each radio frame.

Figure 7.10 illustrates the relationships among TF, TFS, TFC, and TFCI. Each entry with color fill is a TF, each color column represents

TFCI \ TTI	TrCh 1 10 ms	TrCh 2 20 ms	TrCh 3 40 ms	TrCh 4 80 ms
1	1 × 100	2 × 50	2 × 100	4 × 100
2	1 × 200	4 × 50	4 × 100	4 × 100
3	1 × 200	2 × 50	2 × 100	4 × 100
4	1 × 100	1 × 100	2 × 100	4 × 100
5	2 × 100	2 × 50	4 × 100	4 × 100

Figure 7.10 Example of TTI, TF, TFS, TFC, TFCS, and TFCI.

the TFS for one transport channel, each color row represents a TFC, the first row represents TTI for each transport channel, and the numbers in the first column are TFCIs. The five color rows constitute a TFCS.

7.5.1.4 TFC Selection At every radio frame boundary (every minimum TTI = 10 ms), MAC selects a TFC from the TFCS for delivering transport blocks to the physical layer. Once a TFC is selected, the transport format for a given transport channel cannot change within the TTI of that channel.

To illustrate the TFC selection, the example in Figure 7.11 performs TFC selection over 90 ms using the TFCS in Figure 7.10.

At time 0, TFCI 1 is chosen. For the next 20 ms, the transport format for TrCh 2 cannot change. For the next 40 ms, transport format for TrCh 3 cannot change. For the next 80 ms, transport format for TrCh 4 cannot

TFCI	1	3	4	4	2	2	5	5	3
TrCh 1 TTI = 10 ms	1 × 100	1 × 200	1 × 100	1 × 100	1 × 200	1 × 200	2 × 100	2 × 100	1 × 200
TrCh 2 TTI = 20 ms	2 × 50	2 × 50	1 × 100	1 × 100	4 × 50	4 × 50	2 × 50	2 × 50	2 × 50
TrCh 3 TTI = 40 ms	2 × 100	2 × 100	2 × 100	2 × 100	4 × 100	4 × 100	4 × 100	4 × 100	2 × 100
TrCh 4 TTI = 80 ms	4 × 100	4 × 100	4 × 100	4 × 100	4 × 100	4 × 100	4 × 100	4 × 100	4 × 100

0   10   20   30   40   50   60   70   80   90
Time (ms)

Figure 7.11 TFC selection versus time.

change. Between 0 and 10 ms, data delivery from transport channels to the physical layer is as follows:

- Transport Channel 1: delivers 1 block of 100 bits to the physical layer
- Transport Channel 2: delivers 2 blocks of 50 bits each to the physical layer
- Transport Channel 3: delivers 2 blocks of 100 bits each to the physical layer
- Transport Channel 4: delivers 4 blocks of 100 bits each to the physical layer

At 10 ms, TFCI 3 is chosen. Between 10 ms and 20 ms, data delivery from transport channels to the physical layer is as follows:

- Transport Channel 1: delivers 1 block of 200 bits to the physical layer
- Transport Channels 2, 3, and 4 do not change from the previous 10 ms

At 20 ms, TFCI 4 is chosen. Between 20 ms and 30 ms, data delivery from transport channels to the physical layer is as follows:

- Transport Channel 1: delivers 1 block of 100 bits to the physical layer
- Transport Channel 2: delivers 1 block of 100 bits each to the physical layer
- Transport Channel 3 and 4 do not change from the previous 10 ms

At 30 ms, TFCI 4 is chosen again. Between 30 ms and 40 ms, there is no change from the previous 10 ms for all channels.

At 40 ms, TFC 2 is chosen. Between 40 ms and 50 ms, data delivery from transport channels to the physical layer is as follows:

- Transport Channel 1: delivers 1 block of 200 bits to the physical layer
- Transport Channel 2: delivers 4 blocks of 50 bits each to the physical layer
- Transport Channel 3: delivers 4 blocks of 100 bits each to the physical layer
- Transport Channel 4 does not change from the previous 10 ms

At 50 ms, TFC 2 is chosen again. Between 50 ms and 60 ms, there is no change from the previous 10 ms for all channels.

At 60 ms, TFC 5 is chosen. Between 60 ms and 70 ms, data delivery from transport channels to the physical layer is as follows:

- Transport Channel 1: delivers 2 blocks of 100 bits to the physical layer
- Transport Channel 2: delivers 2 blocks of 50 bits each to the physical layer
- Transport Channels 3 and 4 do not change from the previous 10 ms

At 70 ms, TFC 5 is chosen again. Between 70 ms and 80 ms, there is no change from the previous 10 ms for all channels.

At 80 ms, TFC 3 is chosen. Between 80 ms and 90 ms, data delivery is the same as that between 10 ms and 20 ms.

7.5.2 CRC Attachment

Upon reception of the transport blocks from the MAC layer in each TTI, the physical layer attaches a CRC to each of the transport blocks. Figure 7.12 shows an example illustrating CRC attachment. In this example, for TrCh 1, there is only one transport block in each TTI. For TrCh 2, there are two transport blocks in each of the first two TTIs and one transport block in the third TTI.

The size of the CRC could be 0, 8, 12, 16, or 24 bits. The CRC size that should be used for each transport channel is signaled from higher layers. A CRC size of zero bits means no CRC is attached. Through CRC, error detection can be provided for transport blocks.

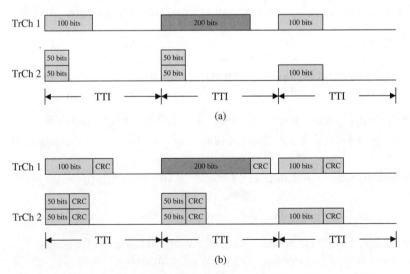

Figure 7.12 CRC attachment (a) MAC delivers transport block sets every TTI (b) Physical layer attaches CRC to each transport block.

7.5.3 Transport Block Concatenation and Code Block Segmentation

After CRC attachment, the physical layer performs the concatenation of all transport blocks in a TTI for each transport channel. Figure 7.13 illustrates transport block concatenation using the transport blocks shown in Figure 7.12. For TrCh 1, no concatenation is necessary because there is only one transport block in a TTI. For TrCh 2, in each of the first two TTIs, two transport blocks are concatenated to form a larger data block.

For a given transport channel, the number of bits including CRC in a TTI after concatenation is given by

$$X = M B, \tag{7-3}$$

where M is the number of transport blocks and B is the number of bits per block. Coding (convolutional or turbo) is performed on a code block of Y bits in size. The maximum code block size is 5114 bits for turbo coding and 504 bits for convolutional coding. The total number of bits in a TTI after concatenation, X, may be larger than a coding block size. If $X > Y$, it needs to segment the X bits into C coding blocks, where $C = X/Y$, rounding up to the nearest integer. Each coding block is then input to the encoder for channel coding. For turbo coding, if the last coding block does not have enough bits, filler bits are added to it.

7.5.4 Channel Coding

WCDMA employs convolutional coding and turbo coding. Convolutional coding has been widely used for decades. Turbo coding came out in the early nineties and is being used in many applications.

Convolutional coding is implemented on common and dedicated transport channels with data rates smaller than or equal to 32 kbps. For a convolutional coder having a constraint length of 9, eight tail bits with binary value 0 are added to the end of the code block before coding.

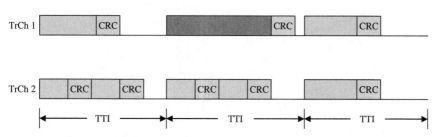

Figure 7.13 Transport block concatenation.

TABLE 7.1 Usage of Channel Coding Scheme and Coding Rate (Courtesy of ETSI)

Type of TrCh	Coding Scheme	Coding Rate
BCH	Convolutional coding	1/2
PCH		
RACH		
CPCH,DCH,DSCH,FACH		1/2, 1/3
	Turbo coding	1/3

The purpose of zero tailing of the data is to flush the coder. Turbo coding is used on dedicated transport channels with data rates equal to or greater than 64 kbps. Table 7.1 lists the channel coding scheme and coding rate for various transport channels.

7.5.4.1 Convolutional Coder For WCDMA, the convolution codes with constraint 9 and coding rates 1/2 and 1/3 are defined in 3GPP TS25.212 [2] with different generating functions. The configuration of the convolutional coder is presented in Figure 7.14. Output from the rate 1/2 convolutional coder is done in the order: Output A, Output B, Output A, ..., Output B. Output from the rate 1/3 convolutional coder is done in the order: Output A, Output B, Output C, Output A, Output B, Output C, Output A,...,Output C. Since the convolutional coder has eight shift registers, an input bit will influence many coded output bits, depending on the number of outputs and taps. The number of output bits influenced is equal to the number of outputs multiplied by the number of taps. For example, the numbers of output bits influenced are 20 bits and 45 bits for rate 1/2 and rate 1/3 convolutional coders, respectively.

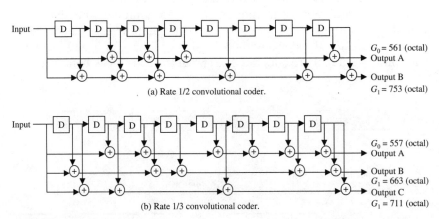

Figure 7.14 Rate 1/2 and rate 1/3 convolutional coders (courtesy of ETSI).

7.5.4.2 Turbo Coder The turbo coder uses a parallel-concatenated-convolution code (PCCC) with two 8-state constituent encoders and one turbo code internal interleaver. Figure 7.15 shows the structure of a turbo coder. The coding rate of a turbo coder is 1/3. Turbo coding is most efficient for blocks sizes greater than 1000 symbols. This corresponds to transport channel data rates of 64 kbps or higher.

7.5.5 Rate Matching

Rate matching matches the number of transport channel bits to the bits available in a physical channel radio frame by repeating or puncturing bits in the bit stream on a transport channel. Only some of the bits may be repeated or removed. During rate matching, the higher layers assign a semi-static rate-matching attribute for each transport channel. This attribute can only be changed through higher layer signaling and is used in the algorithm for calculating the number of bits to be repeated or punctured.

Another way of rate matching for the downlink is discontinuous transmission. The number of bits on a transport channel may not be the same for different transmission time intervals. If the number of bits is lower than the maximum, the transmission is interrupted. This is called discontinuous transmission (DTX). During DTX, DTX indication bits with no energy are added to the bit stream. These bits are interleaved but are not transmitted through the air. They do not contribute to transmitted signal energy. DTX is used only on the downlink.

There are two schemes of placing transport channels in a CCTrCh: fixed position transport channels and flexible position transport channels.

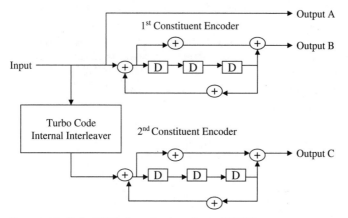

Figure 7.15 Rate 1/3 Turbo coder (courtesy of ETSI).

7.5.5.1 Fixed Position Transport Channels For fixed position transport channels, the location of a transport channel in a CCTrCh is fixed, and the starting location for each transport channel is based on the largest transport format. However, combination of the largest transport format for each transport channel could be an invalid TFC. Thus, fixed position channels may result in an increase of DTX. For example, as shown in Figure 7.16, the upper TFC, in which DTX is not needed in any of its transport channels, is not a valid TFC, while the lower TFC is a valid TFC, which requires DTX in transport channels 1 and 2.

7.5.5.2 Variable Position Transport Channels Variable position transport channels are used to minimize the number of bits sent. Not all transport channels send their largest TF at the same time, thus reducing the amount of DTX. The example shown in Figure 7.17 illustrates how transport channels are placed in a CCTrCh. In this example, there is no need to add any DTX bits in any of the transport channels.

7.5.6 First DTX Insertion

As discussed in rate matching, DTX indication bits are inserted to fill up the radio frames. The DTX insertion depends on whether a fixed position scheme or a flexible position scheme is used for placing a transport channel in a CCTrCh. UTRAN decides for each CCTrCh whether a fixed position or a flexible position scheme is used during the connection. The DTX indication bits are used to signify when the transmission should be turned off; they are not really transmitted.

Insertion of DTX indication bits is used only if the positions of the transport channels in the radio frame are fixed. With a fixed position transport channel scheme, a fixed number of bits equal to the number of bits in the largest valid TF is reserved for each transport channel in the radio frame. If there are not enough data bits to fill the frame, DTX indication bits are inserted. For example, as shown in Figure 7.18, a particular transport channel may have three valid transport formats, TF1, TF2, and TF3. When the transport format TF2 is chosen, DTX indication bits are added to fill up the difference between TF2 and TF3.

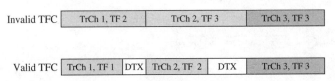

Figure 7.16 Fixed position channels with DTX.

Physical Layer   131

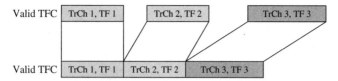

Figure 7.17 Variable position transport channels.

7.5.7 First Interleaving and Radio Frame Segmentation

The purpose of interleaving is to distribute the burst errors to multiple radio frames, reducing the BLER. First interleaving is performed right after 1st DTX insertion and before radio frame segmentation. The process of 1st interleaving is as follows:

1. Write the data within a TTI into a matrix row by row. The number of columns is equal to TTI divided by 10 ms, and the number of rows is equal to the number of bits per 10 ms in the transport channel.
2. Permute columns.
3. Read the data in the matrix column by column, starting from the far left column.

The column permutation for each TTI (10, 20, 40, 80 ms) is predefined [2]. It is <0, 1> for a 20-ms TTI, <0, 2, 1, 3> for a 40-ms TTI, and <0, 4, 2, 6, 1, 5, 3, 7> for an 80-ms TTI. The example in Figure 7.19 shows the 1st interleaving of a 40-ms TTI containing 20 bits. The 20 bits are written in rows to form a 4 × 5 matrix. Then, the second and third columns are permuted. After the permutation, the bits in the matrix are read column by column.

After 1st interleaving, a transport channel is broken up into 10 ms segments if its TTI is greater than 10 ms. This is called radio frame segmentation. Figure 7.20 illustrates that the 40-ms TTI in Figure 7.19 is broken into four 10-ms segments after 1st interleaving.

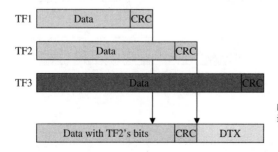

Figure 7.18 Insertion of DTX indication bits.

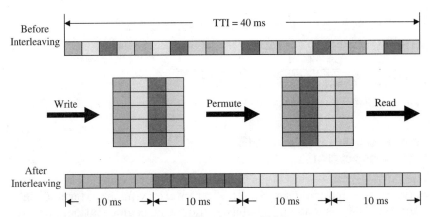

Figure 7.19 Interleaving for a 40-ms TTI containing 20 bits.

7.5.8 Transport Channel Multiplexing and Second DTX Insertion

Transport channel multiplexing arranges the transport channels to form a CCTrCh. After radio frame segmentation, one radio frame is extracted from each transport channel every 10 ms for transport channel multiplexing. These extracted radio frames are serially multiplexed into a CCTrCh. Thus, there are multiple transport channels within a 10-ms radio frame on the CCTrCh. Figure 7.21 illustrates transport channel multiplexing for five transport channels with different TTIs.

Second DTX insertion is carried out after transport channel multiplexing. The DTX indication bits are added to each radio frame at the end, as shown in the example in Figure 7.22. After 2nd insertion, the bit stream becomes a genuine CCTrCh.

7.5.9 Second Interleaving

The purpose of second interleaving is to protect the Viterbi or turbo decoder from bursts of errors. On the transmitting side, the bits are written into the matrix row by row from the top. The bits are read out of the matrix column by column from the left. On the receiving side, the

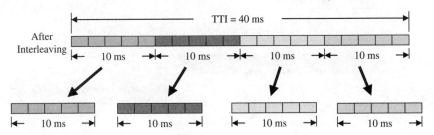

Figure 7.20 Example of radio frame segmentation.

Physical Layer 133

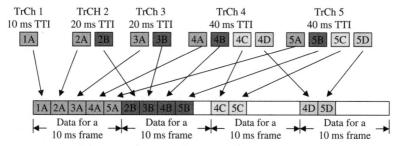

Figure 7.21 Multiplexing of transport channels.

receiver performs the operation in reverse to restore the bits to the original order. Second interleaving is performed on a per-physical-channel per-radio-frame basis. When more than one physical channel is used, the bits are divided among the different physical channels. Assuming the number of bits in a 10-ms radio frame in the CCTrCh is X and the number of physical channels used is P, the number of bits for one physical channel in one radio frame is $U = X/P$. This is called physical channel segmentation [2].

At the transmitter, second interleaving includes the following steps:

1. Write the U bits into a matrix, row by row, 30 columns wide and pad dummy bits in the last row, if it is not full.
2. Designate the columns from the left side of the matrix as column 0, column 1, and so forth, up to column 29 at the right side.
3. Permute columns such that the matrix reaches the following column pattern [2]:

   <0,20,10,5,15,25,3,13,23,8,18,28,1,11,21,6,16,26,4,14,24,19,9,29,1,2,
   7,33,27,17>.

4. Read the bits in the matrix column by column, starting from the far left column.
5. Prune the dummy bits away.

After 2nd interleaving, the DTX bits are distributed over all slots. Also, the burst errors are more uniformly distributed in the output data. This improves the decoder performance for channels experiencing fading in an RF environment.

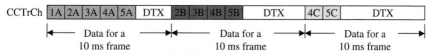

Figure 7.22 Illustration of 2nd DTX insertion.

7.5.10 Mapping to Physical Channel

After 2nd interleaving, each 10-ms radio frame in a CCTrCh is placed into the data fields of a dedicated physical channel (DPCH) radio frame. The physical layer control fields are multiplexed with the data fields in a DPCH. That is, the dedicated physical data channel (DPDCH) and dedicated physical control channel (DPCCH) are multiplexed to form a DPCH. The physical layer control fields include the fields for transport format combination indicator (TFCI), pilot bits, and transmit power control commands for the UE. Figure 7.23 shows the structure of a downlink DPCH [3]. The physical channel data rate ranges from 15 kbps to 1920 kbps.

7.5.11 Spreading and Scrambling

On the downlink, all physical channels, except synchronization channels, are spread by individual OVSF codes and then scrambled with the same scrambling code. Figure 7.24 shows the schematic of spreading and scrambling. The bits of a physical channel are serial-to-parallel converted. After serial-to-parallel conversion, the bits become symbols with two bits forming a symbol pair. The channel rate reduces to half of the original rate and ranges from 7.5 ksps to 960 ksps. For each pair of symbols, one symbol is sent to the I path and the other symbol is sent to the Q path. The symbols are then spread using the same OVSF code. After spreading, the symbols become chips with a chip rate of 3.84 Mcps. The chips in the Q path are phase shifted by 90 degrees and combined with the chips in the I path to become a complex-valued physical channel signal (chip stream). The signal is then scrambled using a scrambling code.

After scrambling, the physical channels are summed together with the synchronization channels, which are neither spread nor scrambled,

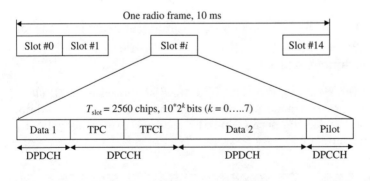

Figure 7.23 Downlink dedicated physical channel structure.

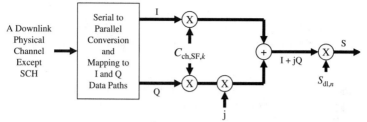

$C_{ch,SF,k}$ = OVSF code with the kth code of spreading factor SF

$S_{dl,n}$ = n^{th} downlink scrambling code

Figure 7.24 Schematic of spreading and scrambling.

and sent to the modulator for modulation. Figure 7.25 illustrates this process. In the figure, the downlink weight factors $G_1, G_2, ..., G_N$ are for the physical channels, and G_p and G_s are for the primary and secondary synchronization channels, respectively.

7.5.11.1 Assignment of Downlink OVSF and Scrambling Codes With the exception of synchronization channels, an OVSF code is assigned to each downlink physical channel. The OVSF codes for some channels are fixed. For example, CPICH and PCCPCH always use $C_{ch,256,0}$ and $C_{ch,256,1}$, respectively. The spreading factor for these two channels is fixed at 256. That means CPICH and PCCPCH are always operating at 30 kbps. The OVSF codes for all other physical channels, such as DPCH and SCCPCH, are assigned by UTRAN. For these channels, the spreading factor of the OVSF codes depends on the data rate. Channels at high data rates are assigned OVSF codes with low spreading factors, while channels at low data rates are assigned OVSF codes with high spreading factor. A spreading factor of 4 means 4 chips per symbol, while a spreading factor of 256 means 256 chips per symbol. The data rate of

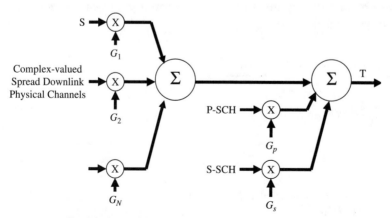

Figure 7.25 Combination of downlink physical channels.

the former is 64 times that of the latter. Due to the orthogonal property of OVSF codes, if an OVSF code in a given generation of the OVSF code tree family is used, its ancestral and descendant OVSF codes cannot be used.

A unique primary scrambling code is assigned to each cell (sector) for scrambling the physical channels. A secondary scrambling code may also be used if the OVSF codes scrambled by the PSC are to be reused in the same cell.

7.5.12 Modulation

When scrambling and summing are completed, the complex-valued chip stream is transmitted with QPSK modulation. As shown in Figure 7.26, the complex-valued chip stream is split into a real and an imaginary part. Each part then individually passes through a root-raised cosine filter for pulse shaping. After that, the real and imaginary parts are mixed with an in-phase and quadrature-phase signal, respectively. The outputs from the mixers are then combined and transmitted.

7.6 Uplink Procedure

The uplink procedure also starts from receiving the transport channel data from the MAC layer and ends with transmitting the physical channels into the air. As shown in Figure 7.27, it consists of many steps, including attachment of CRC, concatenation of transport block, segmentation of the data stream into code blocks, channel coding, radio frame equalization, 1st interleaving, segmentation of transport channel into 10-ms radio frames, rate matching, multiplexing of transport radio frames into a CCTrCh, 2nd interleaving, mapping to physical channels, spreading of physical channels using OVSF codes, scrambling of physical channels using uplink scrambling codes, and modulation. After the modulation step, the signal is transmitted into the air, completing the uplink physical procedure.

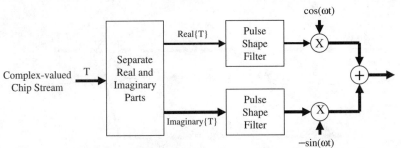

Figure 7.26 Modulation of downlink physical channels.

Physical Layer   137

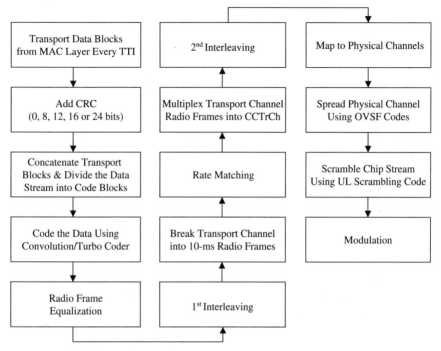

Figure 7.27 Uplink procedure.

The major differences between the uplink and the downlink procedures are in the steps of radio frame equalization, rate matching, mapping to physical channels, and spreading and scrambling. These steps are discussed in the following subsections.

7.6.1 Radio Frame Equalization

Radio frame equalization is adding 0 to 7 pad bits prior to the 1st interleaving. The purpose of radio frame equalization is to make sure that the data bits within any TTI length can be equally segmented into radio frames such that there are equal numbers of bits in the radio frame segmentations. Since the largest TTI is 80 ms, the maximum number of columns in the 1st interleaving matrix is 8. The pad bits are added to the bottom row if it is not full. As such, the possible pad bits are 0 to 7.

7.6.2 Rate Matching

Rate matching is required because bit rates in different transport channels may be different and the bit rate for a logical channel may also change in different TTIs. When the number of bits is changed for different TTIs on the uplink, bits are repeated or removed to ensure that the total bit rate after transport channel multiplexing is identical to the bit rate of the allocated dedicated physical channels. The key difference between the uplink and the downlink is that there is no DTX insertion on the uplink. Therefore, when the TFC changes, the number of bits for all transport channels may also change, which causes rate matching to change accordingly.

7.6.3 Mapping to Physical Channels

After 2nd interleaving, each 10-ms radio frame in a CCTrCh is placed into either a PRACH or a DPDCH radio frame. For the random access channel, the number of transport channels in a CCTrCh is 1. For the dedicated channel, it is derived from the UE capability class.

On the uplink, as will be seen in Section 7.7, the dedicated physical control channel (DPCCH) carries the physical layer control using a different OVSF code with a fixed spreading factor of 256. Physical layer control includes the TFCI, pilot bits, TPC commands for the downlink, and feedback indicator bits (FBI). The DPCCH and the DPDCH are two separate channels. The DPCCH is still active even when the DPDCH does not transmit data.

7.6.4 Spreading and Scrambling

Similar to the downlink physical channels, the uplink physical channels are spread to the chip rate with individual OVSF codes and then scrambled with the same scrambling code. On the uplink, the DPCCH is always in the Q path, while the DPDCHs can be in both the I and the Q paths. If there is only one DPDCH, it is in the I path. The spreading and scrambling of uplink channels is shown in Figure 7.28.

The G_d and G_c in Figure 7.28 are the uplink weight factors for data and control, respectively. $S_{ul,n}$ is the nth scrambling code for the uplink physical channels, where n is one of the 2^{24} codes.

7.6.4.1 Assignment of Uplink OVSF Codes and Scrambling Codes
On the uplink, the DPCCH and the DPDCH use separate OVSF codes with the same or different spreading factor(s). The DPCCH always uses $C_{ch,256,0}$. For the DPDCH, the code number is related to the spreading factor by SF/4. So, the OVSF code for uplink DPDCH is $C_{ch,SF,SF/4}$. The UE can use a spreading factor that is equal to or greater than the minimum spreading factor that is signaled by the UTRAN RRC in the uplink DPCH information IE.

The scrambling codes for the uplink are used to separate the UEs. Each UE uses one uplink scrambling code, which is also signaled by the

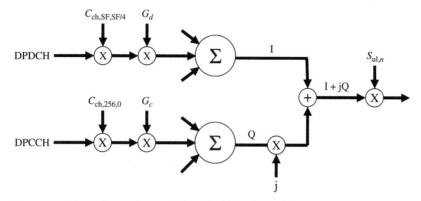

Figure 7.28 Spreading and scrambling of uplink physical channels.

UTRAN RRC in the uplink DPCH information IE. As with the downlink scrambling code, the uplink scrambling code repeats every 10 ms.

7.7 Physical Channel Structures and Channel Timing

On the downlink, physical channels include the primary common control physical channel (PCCPCH), secondary common control physical channel (SCCPCH), common pilot channel (CPICH), primary synchronization channel (P-SCH), secondary synchronization channel (S-SCH), paging indicator channel (PICH) and acquisition indicator channel (AICH), and downlink dedicated physical channel (DPCH), which consists of dedicated physical data channel (DPDCH) and dedicated physical control channel (DPCCH). On the uplink, physical channels include the physical random access channel (PRACH) and uplink dedicated physical data channel and uplink dedicated physical control channel. The structure of each of these physical channels and the channel timing are addressed in the following subsections.

7.7.1 PCCPCH

The primary common control physical channel (PCCPCH) is a downlink physical channel used to carry the BCH transport channel. The BCH carries the system information and is broadcast to the entire cell coverage area. There is one PCCPCH for each cell. The PCCPCH has a fixed rate of 30 kbps, and it uses the OVSF code, $C_{ch,256,1}$. Figure 7.29 shows the frame structure of the PCCPCH [3]. It contains a data field of 18 bits within each slot. However, during the first 256 chips of each time slot, there is nothing in the PCCPCH. That is, the PCCPCH is DTX for the first 256 chips of each slot. In other words, PCCPCH is not transmitted

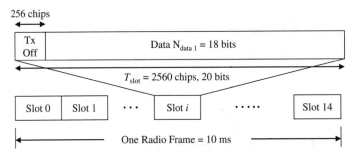

Figure 7.29 Frame structure for PCCPCH (courtesy of ETSI).

during the first 256 chips of each slot. During that period, the P-SCH and S-SCH are transmitted instead. As a matter of factor, PCCPCH and SCH are time multiplexed; they are not transmitted simultaneously.

The PCCPCH aligns with slot number 0 of S-SCH, as shown in Figure 7.30, in which the secondary synchronization code is denoted $C_s^{i,k}$, where $i = 0, 1, \ldots, 63$ is the number of the scrambling code group, and $k = 0, 1, \ldots, 14$ is the slot number.

7.7.2 SCCPCH

The secondary common control physical channel (SCCPCH) is used to carry the FACH and PCH transport channels. The FACH is used when the UE is in the Cell_FACH state, while the PCH transport channel is used when the UE is in idle mode, URA_PCH state, or Cell_PCH state. The FACH transport channel carries control signaling and short user data to the UE. It also carries the RRC Connection Setup message to the UE to set up a call. The PCH transport channel carries paging notifications to the UE.

Up to 16 SCCPCHs may be transmitted from a cell (sector). The FACH and the PCH may be carried by the same or different SCCPCH(s).

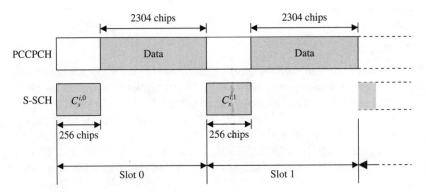

Figure 7.30 Timing alignment of PCCPCH and S-SCH.

There are two types of SCCPCH. One type contains TFCI and the other does not. The SCCPCH can support multiple transport format combinations by using TFCI. UTRAN determines whether a TFCI should be transmitted. However, all UEs must support the use of TFCI.

The frame structure and slot format for the SCCPCH are shown in Figure 7.31 [3]. There are three fields in the slot format: TFCI, data, and pilot. In Figure 7.31, the parameter k, ranging from 0 to 6, determines the total number of bits per SCCPCH slot. It is related to the spreading factor SF of the SCCPCH as SF = $256/2^k$. The spreading factor for the SCCPCH thus ranges from 256 down to 4. For example, when $k = 2$, there are 80 bits in a slot such that the bit rate is 120 kbps. After serial-to-parallel conversion, the symbol rate before spreading is 60 ksps. A spreading factor of $256/2^2 = 64$ spreads the symbols into chips at a chip rate of exactly 3.84 Mcps.

There are 18 different slot formats with different combinations of TFCI, data, and pilot bits for the SCCPCH. They are shown in Table 7.2, in which the number of bits per field is given. The channel bit rates and symbol rates given in Table 7.2 are respectively the rates before and after the serial-to-parallel conversion immediately preceding spreading.

The pilot field has 0, 8, or 16 bits. The pilot pattern can be found in [3]. Half of the slot formats do not have pilot bits transmitted.

As mentioned earlier, FACH and PCH can be mapped to the same or different SCCPCHs. If they are mapped to the same SCCPCH, they can be mapped to the same frame. The main difference between the PCCPCH and SCCPCH is that the transport channel (BCH) mapped to the PCCPCH has a fixed predefined transport format combination, while the SCCPCH supports multiple transport format combinations using TFCI. Also, there is no power control for SCCPCH; therefore, the messages on SCCPCH are often short.

For channel timing, each SCCPCH may be offset from the PCCPCH by an amount that is a multiple of 256 chips.

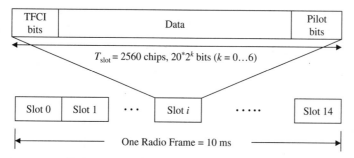

Figure 7.31 Frame structure and slot format for SCCPCH (courtesy of ETSI).

TABLE 7.2 Slot Format for SCCPCH (Courtesy of ETSI)

Slot Format Number	Channel Bit Rate (kbps)	Channel Symbol Rate (ksps)	Spreading Factor	Bits Per Frame	Bits Per Slot	N_{data1}	N_{pilot}	N_{TFCI}
0	30	15	256	300	20	20	0	0
1	30	15	256	300	20	12	8	0
2	30	15	256	300	20	18	0	2
3	30	15	256	300	20	10	8	2
4	60	30	128	600	40	40	0	0
5	60	30	128	600	40	32	8	0
6	60	30	128	600	40	38	0	2
7	60	30	128	600	40	30	8	2
8	120	60	64	1200	80	72	0	8*
9	120	60	64	1200	80	64	8	8*
10	240	120	32	2400	160	152	0	8*
11	240	120	32	2400	160	144	8	8*
12	480	240	16	4800	320	312	0	8*
13	480	240	16	4800	320	296	16	8*
14	960	480	8	9600	640	632	0	8*
15	960	480	8	9600	640	616	16	8*
16	1920	960	4	19200	1280	1272	0	8*
17	1920	960	4	19200	1280	1256	16	8*

* If TFCI bits are not used, then DTX shall be used in TFCI field.

7.7.3 SCH

The synchronization channel (SCH) consists of two subchannels: the primary synchronization channel (P-SCH) and the secondary synchronization channel (S-SCH). They are coded differently. There is only one primary synchronization code, while there are 16 secondary synchronization codes. The P-SCH is coded with the primary synchronization code, C_p, while the S-SCH is coded with the secondary synchronization code, $C_s^{i,k}$, where the subscript s stands for secondary, the superscript i stands for the ith scrambling code sequence ($i = 0, \ldots, 63$), and the superscript k stands for the kth slot ($k = 0,\ldots, 14$). Details of the secondary synchronization codes have been discussed in Section 7.3.

As shown in Figure 7.32, the P-SCH and the S-SCH are sent simultaneously at the beginning of each slot. The P-SCH code is the same for every slot and every cell in the system. For S-SCH, a sequence of 15 secondary synchronization codes is transmitted within a radio frame with one code for each slot. The primary and secondary synchronization codes are modulated by the symbol a shown in Figure 7.32, which indicates the presence/absence of STTD encoding on the PCCPCH. The value of a is equal to +1 if the PCCPCH is STTD encoded, while it is equal to −1 if the PCCPCH is not STTD encoded [3].

There are 64 different secondary synchronization code sequences, ranging from sequence 0 to sequence 63, with each sequence corresponding to a scrambling code group. Table 7.3 shows the allocation of secondary synchronization codes for SCH. As can be seen from the table,

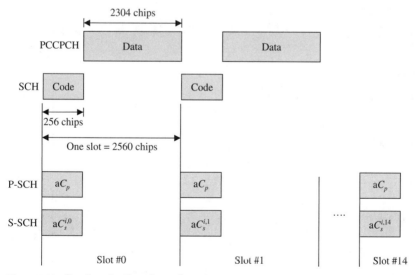

Figure 7.32 Synchronization channels.

TABLE 7.3 Allocation of Secondary Synchronization Codes for SCH (Courtesy of ETSI)

Scrambling Code Group	Slot Number														
	#0	#1	#2	#3	#4	#5	#6	#7	#8	#9	#10	#11	#12	#13	#14
0	1	1	2	8	9	10	15	8	10	16	2	7	15	7	16
1	1	1	5	16	7	3	14	16	3	10	5	12	14	12	10
2	1	2	1	15	5	5	12	16	6	11	2	16	11	15	12
3	1	2	3	1	8	6	5	2	5	8	4	4	6	3	7
4	1	2	16	6	6	11	15	5	12	1	15	12	16	11	2
5	1	3	4	7	4	1	5	5	3	6	2	8	7	6	8
6	1	4	11	3	4	10	9	2	11	2	10	12	12	9	3
7	1	5	6	6	14	9	10	2	13	9	2	5	14	1	13
8	1	6	10	10	4	11	7	13	16	11	13	6	4	1	16
9	1	6	13	2	14	2	6	5	5	13	10	9	1	14	10
10	1	7	8	5	7	2	4	3	8	3	2	6	6	4	5
11	1	7	10	9	16	7	9	15	1	8	16	8	15	2	2
12	1	8	12	9	9	4	13	16	5	1	13	5	12	4	6
13	1	8	14	10	14	1	15	15	8	5	11	4	10	5	4
14	1	9	2	15	15	16	10	7	8	1	10	8	2	16	9
15	1	9	15	6	16	2	13	14	10	11	7	4	5	12	3
16	1	10	9	11	15	7	6	4	16	5	2	12	13	3	14

(*Cont.*)

Physical Layer   145

TABLE 7.3 Allocation of Secondary Synchronization Codes for SCH (Courtesy of ETSI)

Scrambling Code Group	Slot Number														
	#0	#1	#2	#3	#4	#5	#6	#7	#8	#9	#10	#11	#12	#13	#14
17	1	11	4	4	13	2	9	10	12	16	8	5	3	15	6
18	1	12	12	13	14	7	2	8	14	2	1	13	11	8	11
19	1	12	15	5	4	14	3	16	7	8	6	2	10	11	13
20	1	15	4	3	7	6	10	13	12	5	14	16	8	2	11
21	1	16	3	12	11	9	13	5	8	2	14	7	4	10	15
22	2	2	5	10	16	11	3	10	11	8	5	13	3	13	8
23	2	2	12	3	15	5	8	3	5	14	12	9	8	9	14
24	2	3	6	16	12	16	3	13	13	6	7	9	2	12	7
25	2	3	8	2	9	15	14	3	14	9	5	5	15	8	12
26	2	4	7	9	5	4	9	11	2	14	5	14	11	16	16
27	2	4	13	12	12	7	15	10	5	2	15	5	13	7	4
28	2	5	9	9	3	12	8	14	15	12	14	5	3	2	15
29	2	5	11	7	2	11	9	4	16	7	16	9	14	4	4
30	2	6	2	13	3	3	12	9	7	16	6	9	16	13	12
31	2	6	9	7	7	16	13	3	12	2	13	12	9	16	6
32	2	7	12	15	2	12	4	10	13	15	13	4	5	5	10
33	2	7	14	16	5	9	2	9	16	11	11	5	7	4	14
34	2	8	5	12	5	2	14	14	8	15	3	9	12	15	9

(Cont.)

TABLE 7.3 Allocation of Cecondary Cynchronization Codes for SCH (Courtesy of ETSI)

Scrambling Code Group	Slot Number														
	#0	#1	#2	#3	#4	#5	#6	#7	#8	#9	#10	#11	#12	#13	#14
35	2	9	13	4	2	13	8	11	6	4	6	8	15	15	11
36	2	10	3	2	13	16	8	10	8	13	11	11	16	3	5
37	2	11	15	3	11	6	14	10	15	10	6	7	7	14	13
38	2	16	4	5	16	14	7	11	4	11	14	9	9	7	5
39	3	3	6	6	11	12	13	6	12	14	4	5	13	5	14
40	3	3	5	5	16	9	15	5	9	10	6	4	15	4	10
41	3	4	5	14	4	6	12	13	5	13	6	11	11	12	14
42	3	4	9	16	10	4	16	15	3	5	10	5	15	6	6
43	3	4	16	10	5	10	4	9	9	16	15	6	3	5	15
44	3	5	12	11	14	5	11	13	3	6	14	6	13	4	4
45	3	6	4	10	6	5	9	15	4	15	5	16	16	9	10
46	3	7	8	8	16	11	12	4	15	11	4	7	16	3	15
47	3	7	16	11	4	15	3	15	11	12	12	4	7	8	16
48	3	8	7	15	4	8	15	12	3	16	4	16	12	11	11
49	3	8	15	4	16	4	8	7	7	15	12	11	3	16	12
50	3	10	10	15	16	5	4	6	16	4	3	15	9	6	9
51	3	13	11	5	4	12	4	11	6	6	5	3	14	13	12
52	3	14	7	9	14	10	13	8	7	8	10	4	4	13	9
53	5	5	8	14	16	13	6	14	13	7	8	15	6	15	7

(Cont.)

TABLE 7.3   Allocation of Secondary Synchronization Codes for SCH (Courtesy of ETSI)

Scrambling Code Group	Slot Number														
	#0	#1	#2	#3	#4	#5	#6	#7	#8	#9	#10	#11	#12	#13	#14
54	5	6	11	7	10	8	5	8	7	12	12	10	6	9	11
55	5	6	13	6	13	5	7	7	6	16	14	15	8	16	15
56	5	7	9	10	7	11	6	12	9	12	11	8	8	6	10
57	5	9	6	8	10	9	8	12	5	11	10	11	12	7	7
58	5	10	10	12	8	11	9	7	8	9	5	12	6	7	6
59	5	10	12	6	5	12	8	9	7	6	7	8	11	11	9
60	5	13	15	15	14	8	6	7	16	8	7	13	14	5	16
61	9	10	13	10	11	15	15	9	16	12	14	13	16	14	11
62	9	11	12	15	12	9	13	13	11	14	10	16	15	14	16
63	9	12	10	15	13	14	9	14	15	11	11	13	12	16	10

secondary synchronization codes may be repeated within a code sequence. For example, in code sequence 0, there are 7 repeated codes, while there is no repeated code in code sequence 16.

The P-SCH provides slot timing. It is used to signify the start of a time slot in the process of UE's initial acquisition of the WCDMA system. The S-SCH provides frame timing. The scrambling code group identification reduces the primary scrambling code (PSC) search space from 512 to 8.

7.7.4 CPICH

The common pilot channel (CPICH) is a fixed rate downlink physical channel with a data rate of 30 kbps and a spreading factor of 256. It carries a pre-defined bit sequence. Figure 7.33 shows the frame structure of the CPICH [3].

The CPICH provides timing reference in a cell. There are two types of CPICHs: the primary CPICH (P-CPICH) and the secondary CPICH (S-CPICH). They are different in their usage and the limitations of their physical features.

The P-CPICH provides phase reference for the SCH, PCCPCH, AICH, PICH, SCCPCH, CPCH, DPCH, PDSCH, HS-PDSCH and HS-SCCH. Higher layer signaling will inform the UE if the P-CPICH is not a phase reference for a downlink DPCH and any associated PDSCH, HSPDSCH and HS-SCCH.

There is one and only one P-CPICH per cell (sector) and it is broadcast over the entire cell. The same OVSF code, $C_{ch,256,0}$, is always used for the P-CPICH [4]. The P-CPICH is scrambled by a primary scrambling code.

The S-CPICH can use arbitrary OVSF code with a spreading factor of 256, and it can be scrambled by either a primary scramble code or a secondary scramble code. A cell may have zero, one, or multiple S-CPICH. An S-CPICH may be broadcast over the whole cell or only over a portion of the cell. Therefore, it is anticipated to use S-CPICH in conjunction with smart antennas that can transmit energy to a specific spot or area

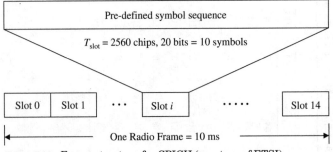

Figure 7.33 Frame structure for CPICH (courtesy of ETSI).

within a cell. When the S-CPICH is scrambled by a different scrambling code, all OVSF codes that were used can be reused.

7.7.5 PICH

The paging indicator channel (PICH) is a fixed rate downlink physical channel with a spreading factor of 256. It is used to carry the paging indicators (PI) and is always associated with an SCCPCH to which a PCH transport channel is mapped. The PICH is used to prolong the UE's battery life. In idle mode, URA_PCH, and Cell_PCH states, the UE sleeps between paging cycles. It wakes up briefly to listen to a designated PI on the PICH. The paging indicators inform the UE that there is a message for it on the transport PCH. That is, page indicators are associated with a paging occasion on the PCH.

Figure 7.34 illustrates the frame structure of the PICH [3]. There are 300 bits ($b_0, b_1, \ldots, b_{299}$) per radio frame in a PICH. However, only the first 288 bits ($b_0, b_1, \ldots, b_{287}$) are used to carry the paging indicators. The remaining 12 bits are not formally part of the PICH and are off (DTX). They are reserved for possible future use.

In each PICH frame, there are Np paging indicators $\{P_0, \ldots, P_{Np-1}\}$, where Np = 18, 36, 72, or 144. Each paging indicator may contain 16, 8, 4, or 2 bits. For example, if Np = 18, each paging indicator has 16 bits.

The PICH is always T_{pich} ahead of the SCCPCH, where T_{pich} is equal to 3 slots or 7680 chips.

The UE should monitor one of the paging indicators, say, P_q, where q is calculated by using SFN, Np, and the PI value calculated by RRC. In mathematical form, the q is given as [3]

$$q = \left(PI + \left\{ [18 \times (SFN + \lfloor SFN/8 \rfloor + \lfloor SFN/64 \rfloor + \lfloor SFN/512 \rfloor)] \bmod 144 \times \frac{Np}{144} \right\} \right) \bmod Np,$$

(7-4)

where PI is given by (4-4) and SFN is the Paging Occasion given by (4-3).

If a paging indicator in a certain PICH radio frame is set to −1, it is an indication that UE is associated with this paging indicator and the UE should read the corresponding frame of the associated SCCPCH,

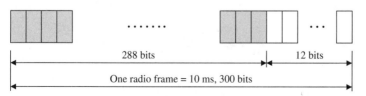

Figure 7.34 Frame structure for PICH.

which is T_{pich} behind the PICH. In other words, the PICH is sent T_{pich} before the SCCPCH, where T_{pich} is 3 slots or 7680 chips.

Use of PI and SFN to calculate the location of a set of paging indicator bits on the PICH causes the bit locations for a given IMSI to shift within the PICH frame each time the PICH is monitored. Therefore, all UEs would have same standby performance statistically, no matter whether their PICH bit locations are close to or far away from the frame boundary of the associated SCCPCH frame.

7.7.6 AICH

The acquisition indicator channel (AICH) is also a fixed rate downlink physical channel with a spreading factor of 256. It is used to carry the acquisition indicator (AI), which acknowledges (ACK) that the preamble on the physical random access channel (PRACH) has been detected by the UE, and that the PRACH message can now be transmitted. The network may also send a negative acknowledgement (negative ACK or NACK) to indicate that although the preamble has been detected, the UE is not allowed to send its PRACH message at this time. The AI uses 1 to represent a positive ACK, –1 to represent a negative ACK, and 0 to represent that no information is transmitted.

The AICH has 16 signatures, which correspond to the uplink signatures received on the PRACH. The codes for the signatures on the AICH are identical to that for the signatures on the PRACH, but each bit is repeated once.

Figure 7.35 illustrates the structure of the AICH [3]. An AICH frame consists of a repeated sequence of 15 consecutive access slots (AS) with each access slot having a length of 5120 chips. However, only the first 4096 chips, which is the real acquisition indicator part, are used to carry

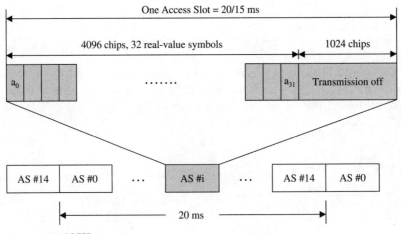

Figure 7.35 AICH structure.

the AIs; the remaining 1024 chips, which are not formally part of the AICH, are off (DTX). The acquisition indicator part consists of 32 real-valued symbols (a_0, ..., a_{31}), and the DTX part is reserved for possible future use by other physical channels. The first access slot of AICH is aligned with the first slot of S-SCH.

7.7.7 PRACH

The physical random access channel (PRACH) is an uplink physical channel used to carry RACH. The UE uses RACH to access the system. There are two parts in an RACH: RACH preamble and RACH message. The PRACH carries both parts. The RACH preamble is 4096 chips in length and consists of 256 repetitions of a 16-chip signature. There are a total of 16 available signatures. The RACH message may be 10 ms or 20 ms long. Each random access transmission consists of one or several preambles and a message of length 10 ms or 20 ms as shown in Figure 7.36 [3].

7.7.7.1 PRACH Preamble
The preamble part in a PRACH is a 4096-chip complex-valued sequence built from a preamble scrambling code $S_{\text{r-pre},n}$ and a preamble signature $C_{\text{sig},s}$ as follows:

$$P_{\text{pre},n,s}(k) = S_{\text{r-pre},n}(k) \times C_{\text{sig},s}(k)$$
$$\times \exp\{j[\pi/4 + (\pi/2)k]\}, k = 0, 1, ..., 4095. \quad (7\text{-}5)$$

The nth preamble scrambling code $S_{\text{r-pre},n}(i)$ is constructed from the long scrambling sequences and is defined as

$$S_{\text{r-pre},n}(i) = C_{\text{long},1,n}(i), i = 0, 1, ..., 4095, \quad (7\text{-}6)$$

where $C_{\text{long},1,n}$ is a real-valued sequence defined in Subsection 4.3.2.2 of [4].

The index n in (7-5) and (7-6) runs from 0 to 8191. That is, there are 8192 different PRACH-preamble-scrambling codes.

The 8192 PRACH-preamble-scrambling codes are divided into 512 groups with each group having 16 codes. The 512 groups are in a one-to-one correspondence with the cell's PSC. The scrambling code used on the PRACH is based on the primary scrambling code of the cell used

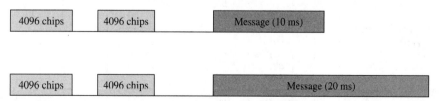

Figure 7.36 Random access transmission.

on the downlink. The kth PRACH preamble scrambling code within the cell with downlink primary scrambling code m is $S_{\text{r-pre},n}(i)$, as defined by (7-6) with

$$n = 16 \cdot m + k, k = 0, 1, 2, ..., 15 \quad \text{and} \quad m = 0, 1, 2, ..., 511.$$

The preamble signature $C_{\text{sig},s}$ is a signature consisting of 256 repetitions of a 16-chip signature $P_s(n)$, $n = 0...15$. It is defined as

$$C_{\text{sig},s}(i) = P_s(i \bmod 16), i = 0, 1, ..., 4095. \tag{7-7}$$

The signature $P_s(n)$ is a Hadamard sequence from the matrix H_4 that is generated from (7-1) and (7-2) in Section 7.3. There are a total of 16 Hadamard sequences. They are listed in Table 7.4.

7.7.7.2 PRACH Message An RACH message can be 10 ms or 20 ms long. Figure 7.37 shows the structure of a random access message part radio frame with a 10 ms length [3]. The 10-ms message radio frame is split into 15 slots, each with a length of 2560 chips. Each slot consists of a data part to which the RACH transport channel is mapped, and a control part that carries physical layer control information. The data and control parts are transmitted in parallel.

A 10-ms random access message part consists of one message radio frame, while a 20-ms message part consists of two consecutive message radio frames. The message part length is equal to the TTI of the RACH transport channel in use. The higher layers configure the TTI length.

TABLE 7.4 PRACH Preamble Signatures (Courtesy of ETSI)

Preamble signature	n															
	0	1	2	3	4	5	6	7	8	9	10	11	12	13	14	15
$P_0(n)$	1	1	1	1	1	1	1	1	1	1	1	1	1	1	1	1
$P_1(n)$	1	-1	1	-1	1	-1	1	-1	1	-1	1	-1	1	-1	1	-1
$P_2(n)$	1	1	-1	-1	1	1	-1	-1	1	1	-1	-1	1	1	-1	-1
$P_3(n)$	1	-1	-1	1	1	-1	-1	1	1	-1	-1	1	1	-1	-1	1
$P_4(n)$	1	1	1	1	-1	-1	-1	-1	1	1	1	1	-1	-1	-1	-1
$P_5(n)$	1	-1	1	-1	-1	1	-1	1	1	-1	1	-1	-1	1	-1	1
$P_6(n)$	1	1	-1	-1	-1	-1	1	1	1	1	-1	-1	-1	-1	1	1
$P_7(n)$	1	-1	-1	1	-1	1	1	-1	1	-1	-1	1	-1	1	1	-1
$P_8(n)$	1	1	1	1	1	1	1	1	-1	-1	-1	-1	-1	-1	-1	-1
$P_9(n)$	1	-1	1	-1	1	-1	1	-1	-1	1	-1	1	-1	1	-1	1
$P_{10}(n)$	1	1	-1	-1	1	1	-1	-1	-1	-1	1	1	-1	-1	1	1
$P_{11}(n)$	1	-1	-1	1	1	-1	-1	1	-1	1	1	-1	-1	1	1	-1
$P_{12}(n)$	1	1	1	1	-1	-1	-1	-1	-1	-1	-1	-1	1	1	1	1
$P_{13}(n)$	1	-1	1	-1	-1	1	-1	1	-1	1	-1	1	1	-1	1	-1
$P_{14}(n)$	1	1	-1	-1	-1	-1	1	1	-1	-1	1	1	1	1	-1	-1
$P_{15}(n)$	1	-1	-1	1	-1	1	1	-1	-1	1	1	-1	1	-1	-1	1

Physical Layer 153

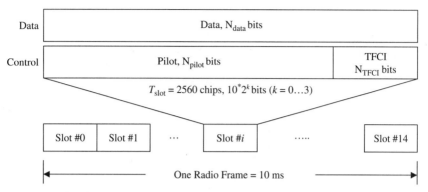

Figure 7.37 Random-access message part radio frame structure (courtesy of ETSI).

In each slot, the data part may have $10 \times 2k$ bits, where $k = 0,1,2,3$. This corresponds to a data rate of 15, 30, 60, and 120 kbps and a spreading factor of 256, 128, 64, and 32 respectively for the message data part. The control part is made up of 8 known pilot bits to support channel estimation for coherent detection and 2 TFCI bits. This corresponds to a spreading factor of 256 for the message control part. The pilot bit pattern may be different for different slots as listed in Table 8 of [3].

The total number of TFCI bits in the 10-ms message radio frame is $15 \times 2 = 30$. The TFCI of a radio frame indicates the transport format of the RACH transport channel mapped to the data part of the simultaneously transmitted message part radio frame. If the RACH message part is 20 ms long, the TFCI is repeated in the second 10-ms radio frame.

The PRACH message part uses a scrambling code of 10 ms in length. There are 8192 different scrambling codes in total for PRACH message. The PRACH message part scrambling code, denoted $S_{r\text{-msg},n}$, where $n = 0, 1, ..., 8191$, is based on the long scrambling sequence and is defined as:

$$S_{r\text{-msg},n}(i) = C_{\text{long},n}(i + 4096), \quad i = 0, 1, ..., 38399, \quad (7\text{-}8)$$

where the lowest index corresponds to the chip transmitted first in time, and $C_{\text{long},n}$ is defined in Subsection 4.3.2.2 of [4].

The message part scrambling code has a one-to-one correspondence to the preamble scrambling code. The same code number is used for both the PRACH preamble and the PRACH message. That is, if the PRACH preamble scrambling code used is $S_{r\text{-pre},j}$, then the PRACH message part scrambling code is $S_{r\text{-msg},j}$, where the number j is the same for both codes.

7.7.7.3 Timing Relation between PRACH Preamble and AICH
Based on a slotted ALOHA approach with fast acquisition indication, a UE can start random-access transmission at the beginning of a PRACH access slot.

There are 15 access slots per two 10-ms radio frames with a slot length of 5120 chips. In other words, there are 15 access slots for every 20 ms. As shown in Figure 7.38, these 15 PRACH access slots are grouped into two sets: access slot set 1 and access slot set 2. Access slot set 1 contains PRACH slots 0–7 and starts T_{p-a} chips ahead of the PCCPCH frame for which SFN mod 2 = 0. Access slot set 2 contains PRACH slots 8 – 14 and starts $(T_{p-a}-2560)$ chips ahead of the PCCPCH frame for which SFN mod 2 = 1. The parameter T_{p-a} could be 7680 or 12800 chips (2 or 3.33 ms). The timing relation of the PRACH access slots and the AICH access slots is also shown in Figure 7.38. AICH access slot #0 is aligned with the PCCPCH frames with SFN mod 2 = 0.

There are 12 PRACH subchannels with each subchannel equaling one access slot. The UE can transmit a preamble using the same subchannel per 12 access slots. For example, if the UE transmits in access slot #0 (called subchannel #0) and does not hear the associated AI on the AICH, it transmits again 12 access slots later in access slot #12 (which is also called subchannel #0). Since there are 15 access slots per two radio frames, the next time access slot #0 coincides with subchannel #0 is 8 radio frames later. That is, the UE would transmit on access slot #0 again 60 access slots (80 ms) after its first transmission on access slot #0.

7.7.8 Downlink DPCH

The downlink dedicated physical channel (DPCH) carries the bulk of user data. It transports the downlink DCH. Within one downlink DPCH, dedicated data generated in Layer 2 and above is transmitted in time-multiplex with control information generated in the physical layer, including pilot bits, transmit power control commands, and TFCI. As such, the downlink DPCH can be regarded as a time multiplex of a downlink DPDCH and a downlink DPCCH.

Figure 7.39 shows the frame structure of the downlink DPCH [3]. Each 10-ms radio frame is split into 15 slots with each slot corresponding to

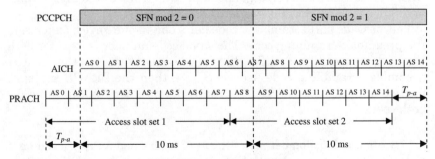

Figure 7.38 Relative timing of PRACH preamble and AICH.

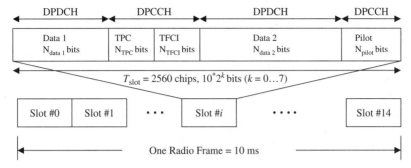

Figure 7.39 Frame structure for downlink DPCH (courtesy of ETSI).

one power-control period. The parameter k in Figure 7.39 determines the total number of bits per downlink DPCH slot. The spreading factor SF of the DPCH is given as SF = $512/2^k$. Therefore, the spreading factor may range from 512 down to 4.

On DPCCH, there are pilot bits used for synchronization, transmit power control (TPC) bits for closed loop power control, and optional TFCI bits for transport format specification for data in the 10-ms radio frame. TPC is sent 1500 times per second. TFCI, if any, is coded and then sent across all 15 slots. The DPCHs for several simultaneous services normally have TFCI, while those for fixed-rate services normally do not. UTRAN determines if a TFCI should be transmitted, and it is mandatory for all UE to support the use of TFCI on the downlink.

The exact numbers of bits for different downlink DPCH fields are given in [3], which lists 49 different slot formats. Different formats are used for different data rates, data types, and compressed mode. The slot format to be used is configured or reconfigured by higher layers.

In compressed mode, a different slot format is used compared to normal mode [3]. There are two possible compressed slot formats, one is used in frames compressed by spreading factor reduction and the other is used in frames compressed by puncturing or higher layer scheduling.

In the case of multi-code, where multiple DPCHs are supported, physical layer information is transmitted only on the first downlink DPCH; DTX (discontinuous transmission) is transmitted during the control period of other DPCHs.

Different DPCHs may transmit at a different time. The DPCH transmission timing is determined by DPCH offset (T_{DOFF}). The T_{DOFF}, ranging from 0 to 599 × 512 chips, offsets the DPCH transmission in units of 512 chips. Figure 7.40 illustrates a DPCH offset example in which one DPCH is aligned with the slot boundary while the other is offset by T_{DOFF}. Since pilot symbols always locate at the fixed position within a slot, DPCH offset can spread out the location of the pilot symbols reducing the peak downlink power.

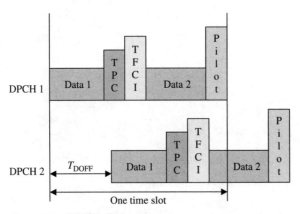

Figure 7.40 DPCH offset.

In addition to DPCH offset, there is another offset called DPCH frame offset [3]. This offset, ranging from 0 to 10 ms in units of 256 chips, allows the UE to soft combine multiple radio links during handover.

7.7.9 Uplink DPCH

Uplink dedicated physical channels include the uplink dedicated physical data channel (uplink DPDCH), the uplink dedicated physical control channel (uplink DPCCH), and the high-speed dedicated physical control channel (HS-DPCCH), which is associated with high-speed downlink shared channel transmission. Each radio link has one uplink DPCCH and zero, one, or several uplink DPDCHs. While the uplink DPDCH is used to carry the DCH transport channel, the uplink DPCCH is used to carry control information generated in the physical layer. The physical layer control information consists of pilot bits, transmit power control (TPC) bits, feedback information (FBI) bits, and an optional transport format combination indicator (TFCI). Pilot bits are used to support channel estimation for coherent detection. Transmit power control bits are for closed loop power control, and feedback information bits are for closed-loop transmit-diversity and site selection diversity transmission (SSDT). While closed-loop transmit-diversity uses multiple antennas at a single cell or sector, SSDT uses multiple cells. The optional transport-format combination indicator informs the receiver about the instantaneous transport format combination of the transport channels mapped to the simultaneously transmitted uplink DPDCH radio frame. UTRAN tells the UE whether it must transmit the TFCI. It is mandatory for all UE to support the use of TFCI on the uplink. Details on the bit patterns for pilot bits, TPC, and FBI are described in [3].

Figure 7.41 shows the frame structure of uplink DPDCH and DPCCH. Each 10-ms radio frame is split into 15 slots, with each slot corresponding

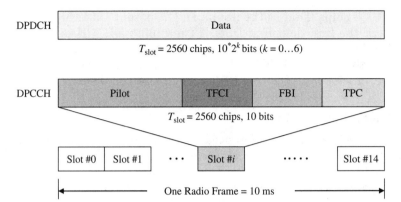

Figure 7.41 Frame structure for uplink DPDCH and DPCCH.

to one power-control period. The DPDCH and the DPCCH are always frame-aligned with each other.

The parameter k in Figure 7.41 determines the number of bits per uplink DPDCH slot. Its relation to the DPDCH spreading factor is given by $SF = 256/2^k$. The uplink DPDCH spreading factor ranges from 256 down to 4. The spreading factor for the uplink DPCCH is always equal to 256. That is, there are 10 bits per uplink DPCCH slot.

There are seven types of slot formats available for the uplink DPDCH [3] with different formats being used for different data rates. Similarly, there are six types of slot formats, ranging from slot format number 0 to slot format number 5, available for the uplink DPCCH with different formats being used for different data types and compressed mode. Each slot format number 0, 2, and 5 has two derivative formats, such as slot format number 0A and 0B.

Figure 7.42 shows the timing relation between uplink DPCH and downlink DPCH. The delay between reception of downlink DPCH and transmission of uplink DPDCH and DPCCH is specified by T_0, which is

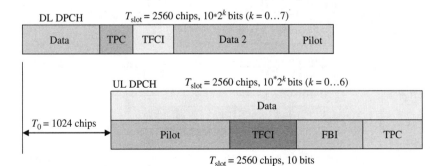

Figure 7.42 Relative timing of uplink DPCH and downlink DPCH.

equal to 1024 chips. That is, the uplink DPCH is sent 1024 chips after detection of the first downlink DPCH. The uplink DPCCH is always transmitted before the uplink DPDCH. The start of DPDCH is delayed relative to the start of DPCCH by using a power control preamble procedure, in which the UTRAN RRC gives the preamble length.

7.7.10 Physical Channel Timing

The CPICH provides timing reference in a cell. Figure 7.43 summarizes the timing relations among the physical channels. The common pilot channel (CPICH), primary synchronization channel (P-SCH), secondary synchronization channel (S-SCH), and primary common control physical channel (PCCPCH) all start at the beginning of a 10-ms radio frame boundary.

Secondary common control physical channel (SCCPCH) is offset from the radio frame boundary by T_{SCCPCH}, in units of 256 chips. In the case that there are multiple SCCPCHs, each may have its own offset. A paging indicator channel (PICH) precedes the SCCPCH carrying paging notification by T_{PICH}, which is equal to 3 time slots or 7680 chips. The acquisition indicator channel (AICH) starts at the 10-ms radio frame boundary and lasts 20 ms. The physical random access channel (PRACH) precedes AICH by T_{p-a}, which is equal to 7680 or 12800 chips. This channel also lasts 20 ms. The downlink DPCH is offset from the radio frame boundary by T_{DPCH}, in units of 256 chips. The uplink DPCH is offset from the downlink DPCH by T_0, which is equal to 1024 chips.

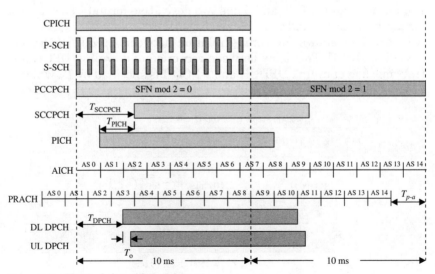

Figure 7.43 Physical channel timing.

7.8 Physical Layer Procedures

There are many physical layer procedures. Major physical procedures to be discussed in this chapter include initial acquisition procedure, PRACH procedure, paging procedure, DPDCH and DPCCH synchronization, radio link establishment and radio link failure/restore, and measurements.

7.8.1 Initial Acquisition Procedure

A UE uses the initial acquisition procedure to acquire the system when it is powered on or when it has lost service. Acquisition of the system means identifying the cell that the UE is trying to camp on and demodulating the system information on the broadcast channel. Therefore, in essence, initial acquisition determines the primary scrambling code (PSC) and frame timing of the cell so that the UE can demodulate the data on the PCCPCH that carries BCH. There are a total of 512 PSCs and each PSC has 38,400 possible chip offsets, making it a search space of $512 \times 38,400$ in size for UE to search. To simplify the search space, the initial acquisition procedure is carried out in three steps:

- **Step 1: Slot synchronization**—The first step of initial acquisition is slot synchronization. In this step, the UE uses the primary synchronization code of the synchronization channel to find the start of slots in the 10-ms radio frame. As described in Subsection 7.7.3, the primary synchronization code is the same for all cells in the system and transmitted with each slot. Therefore, the UE can find the start of slots by using a single matched filter to perform the autocorrelation of the primary synchronization code. A slot begins at the moment that a peak is detected at the matched filter output. The search space for this step is 2560.

- **Step 2: Frame synchronization and code-group identification**—In this step, UE finds the start of the 10-ms radio frame and identifies the scrambling code group of the cell by using the secondary synchronization code of the synchronization channel. As described in Subsection 7.7.3, a sequence of 15 secondary synchronization codes are transmitted within a radio frame; with one code for each slot. There are a total of 64 different secondary synchronization code sequences, ranging from sequence 0 to sequence 63, with each sequence corresponding to a primary scrambling code group. In the search process, the UE correlates the received signal with all 64 secondary synchronization code sequences, and identifies the maximum correlation value. The maximum correlation value not only signifies the start of a radio frame, but also points out the correct secondary synchronization code sequence and thus the corresponding primary scrambling group, because the cyclic shifts of the secondary synchronization code sequences are unique. The search space for this step is 64 × 15 = 960.

- **Step 3: Scrambling-code identification**—In this step, the UE determines the exact primary scrambling code of the cell. There are eight primary scrambling codes in the scrambling code group identified in the second step. The PSC is identified through symbol-by-symbol correlation over the CPICH with all eight PSCs within the code group. The CPICH carries a known symbol sequence that is repeated every slot and scrambled with the PSC. The search space for this step is 8. Note that there are a total of 512 different PSCs. Reducing the search space from 512 to 8 makes this initial acquisition procedure very effective.

Once the PSC is identified, the UE needs to resolve a 10-ms ambiguity on the BCH because BCH has a 20-ms TTI. That is, if the first radio frame is not the start of the transport channel data, the next radio frame will be. After resolving the ambiguity, the UE can detect the PCCPCH and read the system specific and cell specific BCH information.

7.8.2 Physical Random Access Procedure

To gain access to the system, a UE must perform the physical random access procedure at the physical layer, which receives the following access information from the RRC before the procedure is initiated [5]:

- The PRACH preamble scrambling code
- The message length in time, either 10 or 20 ms
- The AICH transmission timing parameter (0 or 1)
- The set of available signatures and the set of available RACH subchannels for each access service class (ASC)
- The power ramp step
- The number of preamble cycles and random back-off parameters
- The initial preamble power
- The power offset $P_{p-m} = P_{\text{message-control}} - P_{\text{preamble}}$, measured in dB, between the power of the last transmitted preamble and the control part of the random-access message.
- Transport format set parameters including the power offset between the data part and the control part of the random-access message for each transport format.

The RRC may update the access information before each physical random access procedure is initiated.

In addition to the information received from the RRC, the physical layer also receives the following information from the MAC at each initiation:

- The transport format to be used for the PRACH message part
- The ASC of the PRACH transmission
- The data to be transmitted

It should be noted that the ASC is initially set by the RRC when sending the RRC Connection Request message. For all other RACH transmissions, MAC sets the ASC.

The PRACH procedure is illustrated in Figure 7.44. The parameters T_{p-a}, T_{p-m}, and T_{p-p} depend on the AICH transmission timing parameter (0 or 1): T_{p-a} is 1.5 or 2.5 access slots (2 or 3.33 ms), T_{p-m} is 3 or 4 access slots (4 or 5.33 ms), and T_{p-p} is equal to or greater than T_{p-m}.

Basically, the UE performs the physical random access procedure by taking the following sequential steps:

1. The UE determines the available PRACH access slots based on the access information provided by the RRC and MAC, and then randomly selects one access slot in the set of available RACH subchannels within the given ASC. The random function must ensure each of the allowed selections is chosen with equal probability.

2. It then randomly selects a signature from the set of available signatures within the given ASC.

3. The UE sets the maximum number of retransmissions.

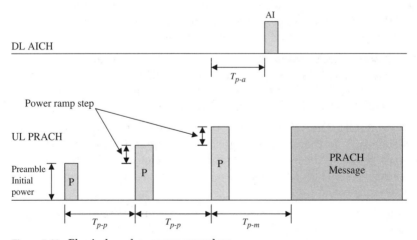

Figure 7.44 Physical random access procedure.

4. It then sets the preamble initial power according to the following formula:

$$\text{Preamble\_Initial\_Power} = \text{Primary CPICH TX power} - \text{CPICH\_RSCP} \\ + \text{UL interference} + \text{Constant Value}, \quad (7\text{-}9)$$

where CPICH_RSCP (CPICH received signal code power) is measured, and the other three parameters are signaled. Note that Preamble_Initial_Power is not calculated again until the next PRACH procedure begins.

5. The UE transmits a preamble using the selected PRACH access slot, signature, and preamble transmission power.

6. If, on AICH, the UE does not detect a positive or a negative acquisition indicator corresponding to the selected signature in the AICH access slot that corresponds to the selected PRACH access slot, it performs the following substeps:

   a. The UE selects the next available access slot in the set of available RACH subchannels within the given ASC, since the ASC determines which subchannels the UE can use. However, the next access slot used must be $T_{p\text{-}p}$ access slots later on the same or a different subchannel, as determined by the ASC. If the preamble is on the same subchannel, it is 12 access slots later.

   b. The UE then randomly selects a new signature from the set of available signatures within the given ASC.

   c. It increases the preamble power by a power ramp step. If the preamble power exceeds the maximum allowed power by 6 dB, the UE may pass the No-Ack-on-AICH to the MAC layer and exit the physical random access procedure.

   d. It then checks whether the number of preamble retransmissions has reached its maximum. If not, the UE restarts the procedure from Step 5. Otherwise, it passes the No-Ack-on-AICH to the MAC layer and exits the physical random access procedure.

7. If, on AICH, the UE detects a negative acquisition indicator corresponding to the selected signature in the AICH access slot that corresponds to the selected PRACH access slot, it passes the received Nack-on-AICH to the MAC layer and exits the physical random access procedure.

8. The UE transmits the random access message three or four access slots after the access slot of the last transmitted preamble, depending on the AICH transmission timing parameter. Transmission power of the control part of the random access message should be

P_{p-m} dB higher than the power of the last transmitted preamble. Transmission power of the data part of the random access message is set using a power gain factor [5].

9. The UE passes RACH-message-transmitted to the MAC layer, completing the physical random access procedure.

In the physical random access procedure, many UEs can transmit on the same access slot. They randomly choose among 16 possible signatures. The 16 signatures do not interfere with each other because they are formed by different Hadamard codes. However, collisions could still occur if two or more UEs pick the same signature.

7.8.3 Page Procedure

When UTRAN wants to communicate with UE, it must page the UE first. In idle mode, URA_PCH state, or Cell_PCH state, the UE sleeps between paging cycles to prolong the battery life using discontinuous reception (DRX). It wakes up periodically to read the designated paging indicator bits on the PICH. As described in Section 4.3, the DRX cycle length determines the interval between wake-ups. When the UE wakes up, it monitors the designated paging indicator on the PICH in a paging occasion, which is calculated by using (4-3). The location of the designated paging indicator on the PICH that the UE will monitor can be calculated by using (7-4) and (4-4).

As described in Subsection 7.7.5, there is a total of Np paging indicators in a paging occasion, where Np could be 18, 36, 72 or 144. Each paging indicator on the PICH is a sequence of repeated bits. The number of repeated bits is equal to 288 divided by the number of paging indicators, Np. If all bits are −1, the paging indicator is on. On the contrary, if they all are 1, the paging indicator is not on.

Figure 7.45 shows an example of a PICH in which Np = 18 and the paging indicators are designated as P_0, P_1, ..., P_{17}, with each paging

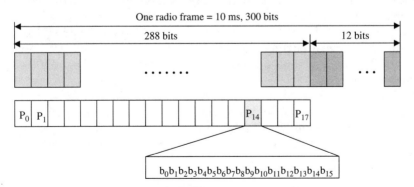

Figure 7.45 Paging indicators in a PICH.

indicator having 16 repeated bits. If PI = 5, SFN = 256, then, according to (7-4), the paging indicator that the UE will monitor is P_{14}. If the bits, $b_0....b_{15}$, of P_{14} are all set to −1, the UE must listen to the transport paging channel (PCH) on the associated SCCPCH, which is 3 radio slots behind the PICH frame. If there are no other pages, all the bits of all other paging indicators are set to 1.

As mentioned earlier in Chapter 4, the BCCH information IE in the Paging Type 1 message contains system information changes such as value tag and modification time. Therefore, if there is a change of system information when the UE is in idle mode, URA_PCH state, or Cell_PCH state, all bits in all paging indicators are set to −1. This will force all UE to read the corresponding Paging Type 1 message on the SCCPCH.

Sometimes, two UEs with different IMSIs may be hashed to the same paging indicator. This is called a *hashing collision*. When this happens, a UE will find that the designated paging indicator on the PICH requests it to read the PCH on the associated SCCPCH. But, it turns out that the paging message is not intended for that UE.

7.8.4 DPDCH/DPCCH Synchronization

As described in [5], synchronization primitives are used to indicate the synchronization status of radio links for dedicated channels both in uplink and downlink. For uplink, at the physical layer, Node B checks the synchronization status of all radio link sets at every radio frame. Synchronization status is indicated to the radio link failure/restored triggering function using either the CPHY-Sync-IND or CPHY-Out-of-Sync-IND primitive. Only one synchronization status indication is given per radio link set. However, the exact criteria for indicating in-sync/out-of-sync are not specified in the standard.

For downlink, in the physical layer, the UE checks the synchronization status of the downlink dedicated channels at every radio frame and uses the CPHY-Sync-IND and CPHY-Out-of-Sync-IND primitives to report the synchronization status to higher layers. The criteria for reporting synchronization status are defined in two different phases: Phase 1 starts at the beginning of radio link establishment as described in [7] and lasts until 160 ms after the radio link is established, and Phase 2 starts 160 ms after the downlink dedicated channel is considered established by higher layers.

In Phase 1, there is no out-of-sync reporting and in-sync is reported if the DPCCH quality measured by the UE over the previous 40 ms is better than a threshold Q_{in}, which is implicitly defined in [8]. No report is required before DPCCH quality measurements have been collected for a 40-ms period.

In Phase 2, out-of-sync and in-sync are reported based on different criteria. Out-of-sync is reported using the CPHY-Out-of-Sync-IND primitive if the one of the following two criteria is fulfilled:

- The DPCCH quality measured by the UE over the previous 160 ms period is worse than a threshold, Q_{out}, which is implicitly defined in [8].
- All of the last 20 transport blocks received and all transport blocks received in the last 160 ms have CRC failures. Transport blocks without CRCs are ignored.

In-sync is reported using the CPHY-Sync-IND primitive if both of the following criteria are fulfilled:

- The DPCCH quality measured by the UE over the previous 160 ms period is better than Q_{in}.
- At least one transport block passes the CRC test within the previous 160 ms. If none of the transport blocks received over the previous 160 ms has CRC attached, the in-sync condition is still fulfilled.

7.8.5 Radio Link Establishment and Radio Link Failure

In Node B, each radio link set can be in three different states: initial state, out-of-sync state, and in-sync state. Radio link establishment involves the transitions between the initial state and in-sync state, while radio link failure involves the transitions between in-sync state and out-of-sync state.

7.8.5.1 Radio Link Establishment In the radio link establishment, two synchronization procedures are defined [5] for physical layer synchronization of dedicated channels between UE and UTRAN:

- Synchronization procedure A—This procedure is used when at least one downlink and one uplink dedicated physical channel is to be set up on a frequency and none of the radio links existed prior to the establishment/reconfiguration.
- Synchronization procedure B—This procedure is used when one or more radio links are added to the active set and at least one of the radio links which existed prior to the establishment/reconfiguration still exists after the establishment/reconfiguration.

7.8.5.1.1 Synchronization Procedure A The synchronization procedure A starts at the time indicated by higher layers. This could be immediately at receipt of upper layer signaling or at an indicated activation time. The procedure is as follows:

1. First, each Node B involved sets all radio links to be set up for this UE in the initial state.

2. Next, UTRAN starts to transmit the downlink DPCCH and may start to transmit the downlink DPDCH if there is any data is to be transmitted. The initial transmit power of the downlink DPCCH is set by higher layers [9]. The TPC commands sent in the downlink DPCCH follow the pattern [n (0, 1), 1] where n is specified by UTRAN to the Node B, but is not signaled to the UE. That is, the TPC pattern consists of n instances of the pair of TPC commands (0, 1), followed by one instance of TPC command 1, where (0, 1) means that the TPC commands are transmitted in 2 consecutive slots. The TPC pattern restarts at the beginning of each frame where CFN mod 4 = 0, and terminates once uplink synchronization is achieved.

3. Then UE establishes downlink chip and frame synchronization of DPCCH using the PCCPCH timing and timing offset information given by UTRAN. Downlink synchronization status is reported to higher layers every radio frame as described in Subsection 7.8.4.

4. The UE then decodes the TPC bits for a minimum of 40 ms. The RRC layer determines whether the downlink DCH is established. The criterion is N312 (default = 1) in-sync indicators within T312 (default = 1) seconds. When the downlink DCH is established, the UE sends its power control preamble (PCP) on the uplink DPCCH for N_{pcp} frames; N_{pcp} is specified in the RRC Connection Setup message. After N_{pcp} frames have been sent, the UE begins transmission on the DPDCH.

5. Finally, UTRAN establishes uplink chip and frame synchronization. Radio links remain in the initial state until N_INSYNC_IND successive in-sync indications are received from the physical layer. Then, Node B triggers the radio link restore procedure indicating which radio link set has obtained synchronization and the radio link set is considered to be in the in-sync state. The radio link restore procedure may be triggered several times, indicating the instances synchronization is obtained for different radio link sets.

7.8.5.1.2 Synchronization Procedure B The synchronization procedure B starts at the time indicated by higher layers. This could be immediately at receipt of upper layer signaling or at an indicated activation time. The procedure is as follows:

1. UTRAN transmits the downlink DPCCH/DPDCH for each new radio link at frame timing such that the frame timing received at the

UE is within $T_0 \pm 148$ chips prior to the frame timing of the uplink DPCCH/DPDCH at the UE. Simultaneously, UTRAN establishes uplink chip and frame synchronization of each new radio link. Radio link sets considered to be in the initial state must remain in the initial state until N_INSYNC_IND successive in-sync indications are received from the physical layer. Then, Node B triggers the radio link restore procedure, indicating which radio link set has obtained synchronization. At this point the radio link set is considered to be in the in-sync state. The radio link restore procedure may be triggered several times, indicating the instances synchronization is obtained for different radio link sets.

2. The UE establishes downlink chip and frame synchronization for each new radio link. The physical layer in the UE keeps reporting the downlink synchronization status to higher layers every radio frame.

7.8.5.2 Radio Link Failure and Restore

Node B monitors the uplink radio link sets to trigger radio failure/restore procedures. The radio link sets are in either in-sync state or out-of-sync state once they are established. Radio link failure is based on physical layer reporting of the out-of-sync condition.

For the uplink, when the radio link set is in the in-sync state, Node B starts timer T_RLFAILURE after receiving N_OUTSYNC_IND consecutive out-of-sync indications. If Node B receives N_INSYNC_IND successive in-sync indications before T_RLFAILURE expires, it stops and resets timer T_RLFAILURE, otherwise it triggers the reverse link failure procedure and indicates which radio link set is out-of-sync. When the radio link failure procedure is triggered, the state of the radio link set changes to the out-of-sync state.

When the radio link set is in the out-of-sync state, Node B triggers the radio link restore procedure after receiving N_INSYNC_IND successive in-sync indications and indicates which radio link set has re-established synchronization. When the radio link restore procedure is triggered, the state of the radio link set changes to the in-sync state.

The parameters (T_RLFAILURE, N_OUTSYNC_IND, and N_INSYNC_IND) are configurable and signaled to Node B over the Iub interface [9].

For downlink, there is a similar procedure, which is handled by RRC. The radio links are monitored by the UE to trigger radio link failure procedures. A timer T313 (default = 3 seconds) is triggered after receiving N313 (default = 20) consecutive out-of-sync indications. If N315 (default = 1) consecutive in-sync indications are received before the timer expires, the timer is stopped and reset, otherwise, the radio link failure procedure starts.

7.8.6 Measurements

The physical layer in UE and UTRAN provides measurements of various quantities to the upper layers to perform a number of functions including random-access, handover, power control, positioning, and UE maximum transmit power. Each of these functions may require one or more measurements. The standard does not specify the method to perform these measurements.

Measurements could be made periodically and reported to the upper layers or could be triggered by events. Measurements may also be a combination of both periodical approach and the event-triggered approach. They are tightly related to the service primitives in that the primitives' parameters may constitute some of the measurements.

In addition, the same measurement may have two different types of measurement: Type 1 and Type 2. For example, both handover and location services require the SFN-SFN observed time-difference measurement. Handover uses SFN-SFN observed time difference Type 1 while location services use Type 2, which is more accurate.

The detailed definitions of measurement control and abilities can be found in [6] for FDD and in [13] for TDD. The measurement performance requirements together with accuracy, range, and mapping are specified in [6] for FDD and in [12] for TDD. The physical layer does the actual measurements and filtering. The results are passed to the RRC. The period for this delivery is signaled by the RRC. The physical layer needs to deliver measurements to the RRC within required accuracies. The performance requirements are described in [11].

7.8.6.1 UE Measurements The measurements that UE performs include CPICH Ec/Io, receive power, CPICH RSCP, SFN-CFN observed time difference, SFN-SFN observed time difference, BLER, Rx-Tx time difference, and Tx power.

7.8.6.1.1 CPICH Ec/Io CPICH Ec/Io is the ratio of received chip energy to the power density in the band. It is equivalent to CPICH RSCP divided by the received power. CPICH Ec/Io measurement is performed on the primary CPICH at the UE antenna connector. If transmit diversity is applied on the primary CPICH, the received chip energy from each antenna must be individually measured and summed together to a total received chip energy before calculating the Ec/Io.

7.8.6.1.2 Receive Power The receive power is the received wideband power (including thermal noise and noise generated in the receiver) within the pulse shaping filter's bandwidth. It is also known as RSSI. The reference point for the measurement is the UE antenna connector. It is used for WCDMA inter-frequency cell reporting.

7.8.6.1.3 CPICH RSCP CPICH RSCP is the received power on one code measured on the primary CPICH. The reference point for the measurement is the UE antenna connector. If transmit diversity is used for the primary CPICH, the received code power from each antenna must be individually measured and summed together to a total received code power on the primary CPICH. The CPICH RSCP is used to report cells for possible handover and to set both the initial random access power and the initial DPCCH power.

7.8.6.1.4 SFN-CFN Observed Time Difference The SFN-CFN observed time difference is an important measurement for handovers. This measurement enables UTRAN to order handovers to other asynchronous Node Bs. The UE reports the time difference between the serving cell and the potential handover candidates. When UTRAN adds a new radio link to the active set, the new cell must send its frames so that they make it to the UE at nearly the same time as other radio links' frames. This is required by Synchronization procedure B, as described in Subsection 7.8.5.1.2. The reference point for the SFN-CFN observed time difference is the UE antenna connector.

The UE measures the time difference using OFF and T_m. T_m reports the chip level difference between the UE transmission and the UE reception of the neighboring cell's signal, and OFF reports the difference between system frame number and the UE's current connection frame number. More specifically, the SFN-CFN observed time difference is defined as: OFF $\times$ 38400 + T_m, where $T_m = (T_{UETx}-T_0) - T_{RxSFN}$, given in chip units ranging from 0 to 38,399 chips. T_{UETx} is the time when the UE transmits an uplink DPCCH/DPDCH frame. T_0 is the uplink DPCH offset from the downlink DPCH, which is equal to 1024 chips. T_{RxSFN} is the time at the beginning of the neighboring PCCPCH frame received most recently before the time instant $T_{UETx}-T_0$ in the UE. OFF = (SFN-CFN$_{Tx}$) mod 256, given in frame units ranging from 0 to 255. CFN$_{Tx}$ is the connection frame number of an uplink DPCCH/DPDCH frame that UE transmits at the time T_{UETx}. SFN is the system frame number of the neighboring cell PCCPCH frame received by the UE at the time T$_{RxSFN}$. OFF may be reported as 0, if the inter-frequency measurement is made during compressed mode.

7.8.6.1.5 SFN-SFN Observed Time Difference There are two types of SFN-SFN observed time difference: Type 1 and Type 2. For Type 1, the SFN-SFN observed time difference to cell is defined as: OFF $\times$ 38,400 + T_m, where $T_m = T_{RxSFNj} - T_{RxSFNi}$, given in chip units ranging from 0 to 38,399. T_{RxSFNj} is the time at the beginning of a PCCPCH frame received from the neighboring cell j. T_{RxSFNi} is the time at the beginning of the PCCPCH frame received most recently from the neighboring cell i

before the time instant $T_{\text{RxSFN}j}$ in the UE. OFF = $(\text{SFN}_i - \text{SFN}_j)$ mod 256, given in frame units ranging from 0 to 255, where SFN_j is the system frame number for the downlink PCCPCH frame from cell j received in the UE at the time $T_{\text{RxSFN}j}$, and SFN_i is the system frame number for the PCCPCH frame from cell i received in the UE at the time $T_{\text{RxSFN}i}$. The reference point for the SFN-SFN observed time difference Type 1 is the UE antenna connector.

Type 2 is the relative timing difference between cell j and cell i, defined as $T_{\text{CPICHR}xj} - T_{\text{CPICHR}xi}$, where $T_{\text{CPICHR}xj}$ is the time when the UE receives one primary CPICH slot from cell j, and $T_{\text{CPICHR}xi}$ is the time when the UE receives the primary CPICH slot from cell i that is closest in time to the primary CPICH slot received from cell j. The reference point for the SFN-SFN observed time difference Type 2 is again the UE antenna connector. SFN observed time difference Type 1 is used for handovers. Type 2, which is more precise, is used for location services (LCS).

7.8.6.1.6 Block Error Rate Block error rate (BLER) is used for transport channels. Non-zero CRC is required for a BLER measurement. The BLER is computed over the measurement period as the ratio of received transport blocks with a CRC error to the total received transport blocks.

When the CRC size is non-zero, the transport channel BLER measurement may be requested for a transport channel only if at least one transport format in the associated transport format set contains at least one transport block when TFCI or guided detection is used. If neither TFCI nor guided detection is used, the measurement may be requested only if all transport formats in the associated transport format set contain at least one transport block. In addition, the measurement does not apply to transport channels mapped to a PCCPCH and an SCCPCH.

7.8.6.1.7 Rx-Tx Time Difference Rx-Tx time difference is the difference in time between UE transmission of the uplink DPCCH/DPDCH frame and the first detected path downlink DPCH frame of the measured radio link. There are two types of measurements defined: Type 1 and Type 2. Type 1 is used for synchronization procedures when starting an uplink DPCH. Type 2 can be used for LCS. For Type 1, the reference Rx path is the first detected path in time among the paths used in the demodulation process. For Type 2, the reference Rx path is the first detected path in time among all paths detected by the UE. Therefore, the reference path used for the measurement may be different for Type 1 and Type 2. The reference point for the UE Rx-Tx time difference is the UE antenna connector. Measurement must be made for each cell included in the active set.

7.8.6.1.8 UE Tx Power The UE transmit power is measured at the UE antenna connector. It is the UE transmit power on one carrier. The UE may be instructed not to exceed a maximum transmit power.

7.8.6.2 UTRAN Measurements The measurements UTRAN performs include receive power, transmit carrier power, transmit code power, SIR, BER, and RTT.

7.8.6.2.1 Receive Power Receive power is the received wideband power, including thermal noise and noise generated in the receiver, within the pulse shaping filter's bandwidth. The reference point for the measurement is the Rx antenna connector. If receiver diversity exists, the reported power must be the linear average of the power measured for the individual branches.

7.8.6.2.2 Transmit Carrier Power Transmit carrier power is the ratio of average total transmit cell power in one carrier to the average maximum transmit power on that carrier. This power measurement is used for keeping track of available cell power. The reference point for the power measurement is the Tx antenna connector. If transmit diversity exists, the transmit carrier power is the ratio of the sum of the transmit power of all branches to the maximum transmission power.

7.8.6.2.3 Transmit Code Power Transmit code power is the power per OVSF code per scrambling code per carrier. Measurement is possible on the DPCCH field of any downlink dedicated channel and reflects the power on the pilot bits of the DPCCH field. When measuring the transmitted code power in compressed mode, all slots, including the slots in the transmission gap, must be included in the measurement. The reference point for the transmit code power measurement is the Tx antenna connector. If transmit diversity exists, the transmit code power for each branch must be measured and summed together.

7.8.6.2.4 Signal to Interference Ratio Signal to interference ratio (SIR) is defined as (RSCP/ISCP) × SF, where ISCP is the interference code power, RSCP is the power on one OVSF code, and SF is the spreading factor on the DPCCH. This measurement is used for power control and is performed on the DPCCH of a radio link set. In compressed mode, the SIR is not measured in the transmission gap. The reference point for the SIR measurements is the Rx antenna connector. If the radio link set consists of more than one radio link, the reported SIR is the sum of the SIR from each individual radio link. If Rx diversity is used in a cell, the SIR for a radio link is the sum of the SIR from each Rx antenna for that radio link.

7.8.6.2.5 Bit Error Rate Bit error rate (BER) measurement applies to transport channels and physical channels. The BER measurement is used for outer loop power control. The transport channel BER is an estimation of the average BER of the DPDCH data of a radio link set. It is measured from the data consisting of only non-punctured bits at the input of the channel decoder in Node B. The physical channel BER is an estimation of the BER on the DPCCH of a radio link set.

7.8.6.2.6 Round-Trip Time Round-trip time (RTT) is the time difference between the beginning of a downlink DPCH transmission and the first arriving path of the corresponding uplink DPCH reception. Specifically, it is defined as $R_{TT} = T_{RX} - T_{TX}$, where T_{TX} = the time of transmission of the beginning of a downlink DPCH frame to the UE, T_{RX} = the time of reception of the beginning of the corresponding uplink DPCCH/DPDCH frame from the UE. The reference point for T_{TX} is the Tx antenna connector, while the reference point for T_{RX} is the Rx antenna connector.

7.9 Summary

The physical layer is the lowest layer of the protocol stack. It provides services to the higher layers. A number of physical channels are defined in the physical layer. The physical layer performs many functions, which include error detection on transport channels, FEC encoding/decoding of transport channels, multiplexing of transport channels and de-multiplexing of coded composite transport channels (CCTrCh), rate matching of coded transport channels to physical channels, mapping of coded composite transport channels on physical channels, power weighting and combining of physical channels, modulation and spreading/demodulation and de-spreading of physical channels, frequency and time (chip, bit, slot, frame) synchronization, radio characteristics measurements, macro-diversity distribution/combining and soft handover execution, inner-loop power control, RF processing, synchronization shift control, and beam-forming.

WCDMA uses three types of codes: OVSF codes, scrambling codes, and synchronization codes. OVSF codes spread the data symbols to chips. OVSF gives the number of chips per symbol. Spreading factors can range from 4 to 512 for the downlink and 4 to 256 for the uplink. Codes with different spreading factors have different lengths. On the downlink, different OVSF codes separate UEs within a cell. On the uplink, different OVSF codes separate dedicated physical channels in UE.

Scrambling codes are truncated Gold codes, which are pseudorandom noise sequences simulating a random noise process with good cross-correlation properties. Scrambling codes are used to scramble the data, separate the cells (sectors) on the downlink, and differentiate the users on the uplink. Each scrambling code has a length of 38,400 chips. In the normal mode, there are 8192 scrambling codes for the downlink. These 8192 codes are divided into 512 sets with each set containing a primary scrambling code (PSC) and 15 secondary scrambling codes (SSC). The 512 PSCs are further divided into 64 scrambling code group with each group having 8 PSCs. On the uplink, there are 2^{24} scrambling codes. The UE is informed by UTRAN of which uplink scrambling code to use when a dedicated physical channel is assigned. Each UE is assigned a unique scrambling code, thus differentiating the users.

The synchronization codes are complex-valued sequences with a length of 256 chips. There is one primary synchronization code and 16 secondary synchronization codes. The primary synchronization code is used to identify the time slot boundaries, while the secondary synchronization codes are used to identify the frame boundaries.

The physical layer uses downlink procedure as well as uplink procedure to perform many of its functions. The downlink procedure starts from receiving the transport channel data from the MAC layer and ends with transmitting the physical channels into the radio link. It consists of many sequential steps, including attachment of cyclic redundancy-check (CRC), concatenation of transport block, segmentation of the data stream into code blocks, channel coding, rate matching, 1st DTX, 1st interleaving, segmentation of transport channel into 10-ms radio frames, multiplexing of transport radio frames into coded composite transport channel (CCTrCh), 2nd DTX insertion, 2nd interleaving, mapping to physical channels, spreading of physical channels using OVSF codes, scrambling of physical channels using downlink scrambling codes, and modulation. After the modulation step, the signal is transmitted into the radio link, completing the downlink physical procedure.

The uplink procedure also starts from receiving the transport channel data from the MAC layer and ends with transmitting the physical channels into the radio link. The sequential steps in the uplink procedure include attachment of cyclic redundancy-check (CRC), concatenation of transport block, segmentation of the data stream into code blocks, channel coding, radio frame equalization, 1st interleaving, segmentation of transport channel into 10-ms radio frames, rate matching, multiplexing of transport radio frames into coded composite transport channel (CCTrCh), 2nd interleaving, mapping to physical channels, spreading of physical channels using OVSF codes, scrambling of physical channels using uplink scrambling codes, and modulation.

On the downlink, physical channels include the primary common control physical channel (PCCPCH), secondary common control physical channel (SCCPCH), common pilot channel (CPICH), primary synchronization channel (P-SCH), secondary synchronization channel

(S-SCH), paging indicator channel (PICH) and acquisition indicator channel (AICH), and the downlink dedicated physical channel (DPCH), which consists of a dedicated physical data channel (DPDCH) and a dedicated physical control channel (DPCCH). On the uplink, physical channels include the physical random access channel (PRACH), and an uplink dedicated physical data channel and uplink dedicated physical control channel. The key physical layer procedures include initial acquisition procedure, physical random access procedure, and page procedure among others. A UE performs the initial acquisition procedure to acquire the system when it is powered on or when it has lost service. Basically, initial acquisition determines the primary scrambling code (PSC) and frame timing of the cell so that the UE can demodulate the data on the PCCPCH that carries BCH. The initial acquisition procedure is carried out in three steps: 1) slot synchronization using the primary synchronization code, 2) frame synchronization and code-group identification using secondary synchronization codes, and 3) scrambling-code identification through symbol-by-symbol correlation over the CPICH.

A UE gains access to the system by performing the physical random access procedure at the physical layer. The sequential steps that a UE must take in the physical random access procedure include: 1) determine the available PRACH access slots based on the information provided by the RRC and MAC, and randomly select one access slot in the set of available RACH subchannels within the given ASC; 2) randomly select a signature from the set of available signatures within the given ASC; 3) set the maximum number of retransmissions; 4) set the preamble initial power; 5) transmit a preamble using the selected PRACH access slot, signature, and preamble transmission power; 6) transmit the random access message three or four access slots after the uplink access slot of the last transmitted preamble; and 7) pass RACH-message-transmitted to the MAC layer, completing the physical random access procedure.

UTRAN initiates a communication with UE by first paging the UE. The UE sleeps between paging cycles to prolong the battery life when it is in idle mode, URA_PCH state, or Cell_PCH state. It wakes up periodically to read the designated paging indicator bits on the PICH. If the paging indicator bits are set to -1, the UE will listen to the transport page channel (PCH) on the associated SCCPCH, which is 3 radio slots behind the PICH frame.

In DPDCH/DPCCH synchronization, synchronization primitives are used to indicate the synchronization status of radio links for dedicated channels both in uplink and downlink. For uplink, Node B checks the synchronization status of all radio link sets at every radio frame. For downlink, the UE checks the synchronization status of the downlink dedicated channels at every radio frame.

In Node B, each radio link set can be in the initial state, out-of-sync state, or in-sync state. Radio link establishment deals with the transitions between the initial state and the in-sync state, while radio link failure involves the transitions between the in-sync state and the out-of-sync state.

In radio link establishment, two synchronization procedures called synchronization procedures A and B are defined for physical layer synchronization of dedicated channels between UE and UTRAN. Synchronization procedure A is used when at least one downlink and one uplink dedicated physical channel is to be set up on a frequency and none of the radio links existed before the establishment/reconfiguration. Synchronization procedure B is used when one or more radio links are added to the active set and at least one of the radio links which existed before still exists after the establishment/reconfiguration.

In radio failure and restore, Node B monitors the uplink radio sets so that it can trigger radio failure/restore procedures. Once established, the radio link sets are in either the in-sync state or out-of-sync state. Radio link failure is based on physical layer reporting of the out-of-sync condition. For downlink, the radio links are monitored by the UE to trigger radio link failure procedures.

The physical layer in UE and UTRAN provides measurements of a variety of quantities to the upper layers to perform a number of functions, including random-access, handover, power control, positioning, and UE maximum transmit power. Each of these functions may require one or more measurements. The measurements that a UE performs include CPICH Ec/Io, receive power, CPICH RSCP, SFN-CFN observed time difference, SFN-SFN observed time difference, BLER, Rx-Tx time difference, and Tx power. The measurements UTRAN performs include receive power, transmit carrier power, transmit code power, SIR, BER, and RTT.

References

[1] 3GPP TS 25.201, v5.2.0, "Physical layer—general description," (Release 5).

[2] 3GPP TS 25.212, v5.4.0, "Multiplexing and channel coding (FDD)," (Release 5).

[3] 3GPP TS 25.211, v5.3.0, "Physical channels and mapping of transport channels onto physical channels (FDD)," (Release 5).

[4] 3GPP TS 25.213, v5.3.0, "Spreading and modulation (FDD)," (Release 5).

[5] 3GPP TS 25.214, v5.6.0, "Physical layer procedures (FDD)," (Release 5).

[6] 3GPP TS 25.215, v5.5.0, "Physical layer measurements (FDD)," (Release 5).

[7] 3GPP TS 25.331, v5.7.1, "Radio resource Control (RRC); protocol specification," (Release 5).

[8] 3GPP TS 25.101, v5.6.0, "UE radio transmission and reception (FDD)," (Release 5).

[9] 3GPP TS 25.433, v5.6.0, "UTRAN Iub Interface NBAP Signaling," (Release 5).

[10] 3GPP TS 25.302, v5.8.0, "Services provided by the physical layer," (Release 5).

[11] 3GPP TS 25.133, v5.8.0, "Requirements for support of radio resource management (FDD)," (Release 5).

[12] 3GPP TS 25.123, v5.10.0, "Requirements for support of radio resource management (TDD)," (Release 5).

[13] 3GPP TS 25.225, v5.7.0, "Physical layer; Measurements (TDD)," (Release 5).

Chapter

8

Cell Reselection

The purpose of cell reselection is to continuously reselect the most suitable cell in the network when the UE is not operating on a dedicated channel. The number of cells that are available for the UE to reselect depends on the capabilities of the UE and the restrictions to access certain networks. Based on RF conditions and priorities, the UE has to decide which cell to reselect. This chapter discusses the cell reselection process.

8.1 Types of Cell Reselection

There are three types of cell reselection: intra-frequency, inter-frequency, and inter-RAT cell reselection. Intra-frequency cell reselection occurs between cells on the same radio frequency. In intra-frequency cell reselection, the UE measures the signal strength of other cells, with no interruption of its connectivity with the current cell. Inter-frequency cell reselection happens between cells on different radio frequencies. In this case, the UE must tune to the neighbor cell's frequency to measure the signal strength of that cell. Inter-RAT cell reselection takes place between cells on different radio access technologies. Similar to inter-frequency cell reselection, the UE must tune to another RAT's frequency to measure the signal strength of a cell. A typical inter-RAT cell reselection example is cell reselection between UMTS and GSM. Cell reselection between UMTS FDD and UMTS TDD is also considered as inter-RAT cell reselection.

To determine whether and when UE should make a cell reselection, the UE must perform measurements on the cells neighboring to the cell that it is camping on. The occasions that UE may perform measurements are determined by the UE state, the types of measurements, and the capability of the UE receiver. UE may take intra-frequency measurements on other cells while continuing to receive the information from the current serving cell. However, for inter-frequency and inter-RAT measurements, the UE must tune to the other frequencies for measurements if it has only one receiver.

Figure 8.1 shows the occasions that UE may take measurements. As shown in part (a) of Figure 8.1, when UE is in idle mode, the Cell_PCH state, or URA_PCH state, it sleeps most of the time and only wakes up to monitor the paging indicator channel at its assigned paging occasion. It may also perform intra-frequency, inter-frequency, and inter-RAT measurements during the paging occasion. How often the UE takes these measurements depends on the DRX cycle length coefficient received.

When UE is in the Cell_FACH state, it may perform intra-frequency measurements at any time. However, it may take inter-frequency and inter-RAT measurements only during the time interval determined by the FACH measurement occasion cycle length coefficient. The value of the FACH measurement occasion cycle length coefficient is read from the FACH measurement occasion information IE in SIB Type 11 or SIB Type 12. During an FACH measurement occasion, UTRAN does not transmit any data to the UE on FACH. As such, the UE can tune to another frequency or radio access technology without missing any data from the current serving cell.

When UE is in the Cell_DCH state, it may take intra-frequency measurement at any time. But, it may take inter-frequency and inter-RAT measurements only during the compressed mode gaps. Compressed mode will be discussed in Chapter 9.

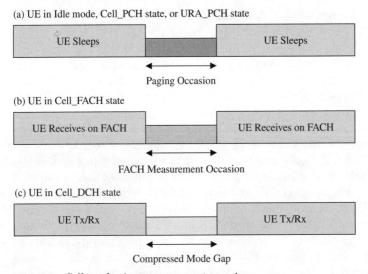

Figure 8.1 Cell reselection measurement occasions.

8.2 Cell Reselection Fundamentals

Cell reselection fundaments involve some UE procedures. UE procedures are described in the 3GPP specification TS25.304 [1] for the access stratum part. The non-access stratum part is described in 3GPP specification TS23.122 [2].

When UE is powered on, it identifies a public land mobile network (PLMN) and searches for a suitable cell of this PLMN to camp on using the initial acquisition procedure described in Subsection 7.8.1. The UE continuously reselects the most suitable cell in the network even after it has camped on a cell. The reselection of a new cell is carried out autonomously by the UE without requiring intervention from UTRAN. However, UTRAN provides parameters in the system information messages that affect the cell reselection decision of the UE.

When UE camps on a cell, it uses that cell to provide services and tunes to its BCH and receives system information from the PLMN. The UE registers in the registration area of the chosen cell using a NAS registration procedure and can establish an RRC connection, if it wishes, by initially accessing the network on the RACH of the cell it camped on. If the PLMN receives a call for the registered UE, it generally knows the registration area of the UE-camped cell and can send a paging message for the UE on PCHs of all the cells in the registration area. The UE will then receive the paging message, since it is tuned to the control channel of a cell in that registration area and the UE can respond on the RACH.

If UE is unable to find a suitable cell to camp on regardless of the cause, it will still attempt to camp on a cell and enters a limited-service state in which the UE can only attempt to make emergency calls.

If the UE finds a more suitable cell, it will reselect the cell and camp on it. The UE will perform location registration if the new cell belongs to a different registration area.

8.2.1 Cell Reselection Criteria

When camped on a cell, the UE regularly searches for a better cell according to the cell reselection criteria. During cell reselection, the UE takes measurements of the serving cell, as well as the neighbor cells provided in SIB Type 11 or 12. The S criteria (suitable criteria) and R criteria (ranking criteria) are calculated for each measured cell. The cells that meet the S criteria are suitable cells. All suitable cells are ranked based on R criteria. If the highest ranked cell is a cell other than the current serving cell, then that cell is chosen for cell reselection.

8.2.1.1 S Criteria

S criteria determine whether a cell is suitable. For a UMTS cell to be suitable, the necessary condition is that the cell must meet the following S criteria:

$$\text{Srxlev} > 0 \tag{8-1}$$

and

$$\text{Squal} > 0 \tag{8-2}$$

where

$$\text{Srxlev} = Q_{\text{rxlevmeas}} - \text{Qrxlevmin} - \text{Pcompensation} \tag{8-3}$$

$$\text{Squal} = Q_{\text{qualmeas}} - \text{Qqualmin} \tag{8-4}$$

$$\text{Pcompensation} = \max(\text{UE\_TX\_PWR\_MAX\_RACH} - \text{P\_MAX}, 0), \tag{8-5}$$

where UE_TX_PWR_MAX_RACH is the maximum allowed transmission power on the PRACH, P_MAX is the maximum transmission power of the UE, $Q_{\text{rxlevmeas}}$ is the measured CPICH RSCP, Q_{qualmeas} is the measured CPICH Ec/Io, Qrxlevmin is the minimum CPICH RSCP for a cell to be suitable, and Qqualmin is the minimum CPICH Ec/Io for the cell to be suitable. For the current serving cell and each neighbor cell, the UE obtains from SIB Type 11 or SIB Type 12 the following parameters: Qrxlevmin, Qqualmin, and UE_TX_PWR_MAX_RACH.

For GSM cells to be suitable, only (8-1) needs to be fulfilled, as (8-2) does not apply to GSM cells.

In addition to meeting the S criteria, a suitable cell must belong to the currently registered PLMN, must not be barred, and must not be in a forbidden location area.

8.2.1.2 R criteria

R criteria determine the ranking of the serving cell and other suitable neighbor cells. The highest ranked cell is chosen for cell reselection. The definition of R criterion for a serving cell is given as

$$R_s = Q_{\text{meas},s} + \text{Qhyst}_s, \tag{8-6}$$

where $Q_{\text{meas},s}$ is the measured quality of the serving cell and Qhyst_s is the hysteresis applied to the serving cell. $Q_{\text{meas},s}$ stands for both measured CPICH RSCP and CPICH Ec/Io. When $Q_{\text{meas},s}$ stands for the measured CPICH RSCP, Qhyst_s is equal to Qhyst1_s, which is a hysteresis parameter applied on CPICH RSCP measurements. Similarly, when $Q_{\text{meas},s}$ stands for the measured CPICH Ec/Io, Qhyst_s is equal to Qhyst2_s, which is a hysteresis parameter applied on CPICH Ec/Io measurements.

The definition of R criterion for neighbor cells is given as

$$R_n = Q_{meas,n} - Qoffset_{s,n}, \qquad (8\text{-}7)$$

where $Q_{meas,n}$ is the measured quality of the neighboring cell and $Qoffset_{s,n}$ is the offset applied to the neighbor cell. $Q_{meas,n}$ stands for both measured CPICH RSCP and CPICH Ec/Io. When $Q_{meas,n}$ stands for the measured CPICH RSCP, $Qoffset_{s,n}$ is equal to $Qoffset1_{s,n}$, which is an offset applied to the neighbor cell on CPICH RSCP measurements. Similarly, when $Q_{meas,n}$ stands for the measured CPICH Ec/Io, $Qoffset_{s,n}$ is equal to $Qoffset2_{s,n}$, which is an offset applied to the neighbor cell CPICH Ec/Io measurements.

8.2.2 Cell Reselection Ranking Process

The cell reselection ranking process is illustrated by the flow chart in Figure 8.2. The first step is to rank all the suitable cells by using $Q_{rxlevmeas}$ (CPICH RSCP for UMTS and Rxlev for GSM), $Qhyst1_s$ and $Qoffset1_{s,n}$. If a GSM cell is the highest ranked cell, then the ranking is done. If a UMTS cell is the highest ranked cell, then check the information element,

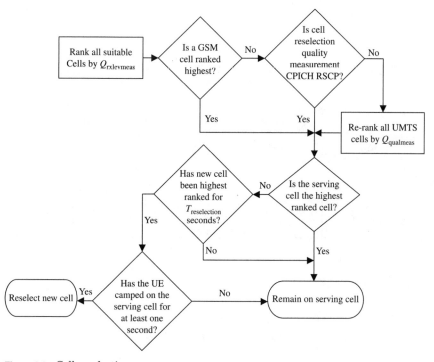

Figure 8.2 Cell reselection process.

cell-selection and reselection-quality-measure, received from SIB Type 3 or SIB Type 4 to see what it contains [3]. If it contains CPICH RSCP, then the ranking is done. Otherwise, rank the suitable UMTS cells again by using $Q_{qualmeas}$ (CPICH Ec/Io), Qhyst2_s and Qoffset$2_{s,n}$. Then the UE chooses the highest-ranking cell to camp on if the cell ranks higher than the current serving cell for $T_{reselection}$ seconds and the UE has camped on the current serving cell for at least one second. The parameter $T_{reselection}$ is specified by UTRAN in SIB Type 3 or SIB Type 4 [3].

To further illustrate the cell reselection process, we use the example in Figure 8.3 to demonstrate in the time domain how a new cell is selected. In this example, Cell 1 is the initial serving cell. The vertical axis represents the ranking, while the horizontal axis represents time. At point A, Cell 2 starts to rank higher than Cell 1 until point B. However, the time interval between points A and B is shorter than $T_{reselection}$. Therefore, Cell 2 cannot replace Cell 1 as the serving cell, even though it is ranked higher than Cell 1 during that interval. At point B, Cell 1 becomes the highest ranked cell again and it remains as the serving cell. At point C, Cell 2 once again surpasses Cell 1 as the highest ranked cell. And at point D, Cell 3 becomes the highest ranked cell.

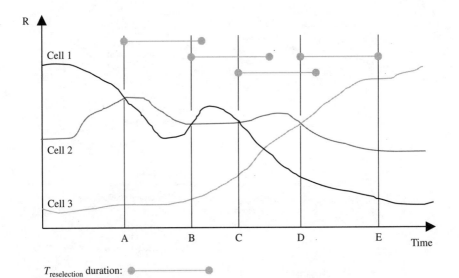

$T_{reselection}$ duration:

- Cell 1 is the initial serving cell.
- At points A, B, C, D, Cell 1 remains the serving cell.
- At point E, Cell 3 becomes the new serving cell.

Figure 8.3 Cell reselection example.

The time interval between points C and D is still shorter than $T_{reselection}$. Therefore, Cell 1 still remains as the serving cell between points C and D. Between points D and E, Cell 1 remains as the serving cell because the duration in which Cell 3 is ranked the highest has not yet exceeded $T_{reselection}$ seconds. At point E, Cell 3 has been highest ranked for $T_{reselection}$ seconds, and it becomes the serving cell from that point on.

8.2.3 Inter-Frequency Cell Reselection

The inter-frequency cell reselection procedure is the same as that of intra-frequency cell reselection. In the system information messages, UTRAN may include up to 32 cells on other frequencies for inter-frequency cell reselection. The system information messages may also include up to 2 additional frequencies.

UTRAN controls whether the UE takes measurements of inter-frequency cells by setting the $S_{intersearch}$ threshold. When the serving cell quality measurement Squal is greater than this threshold, the UE does not need to measure inter-frequency cells. However, if the $S_{intersearch}$ is not set for the serving cell, the UE must perform inter-frequency measurements. Additional details on inter-frequency cell reselection can be found in [1].

8.2.4 Inter-RAT Cell Reselection

The cell reselection procedure for Inter-RAT (GSM) cells is the same as for intra-frequency cells. As for inter-frequency cell reselection, UTRAN may include up to 32 GSM cells in the system information messages. It may also include up to 32 GSM carrier frequencies in the system information messages. Again, UTRAN controls whether the UE takes measurements of inter-RAT cells by setting the $Ssearch_{RATGSM}$ threshold. If the serving cell quality measurement Squal is greater than this threshold, the UE does not need to measure GSM cells. If this threshold is not set for the serving cell, the UE has to perform measurements on GSM cells. Details on inter-RAT cell reselection are also given in [1].

When measurements are performed on GSM cells, the measurements are based only on average received signal levels. That is why the ranking procedure starts by ranking all cells based on $Q_{rxlevmeas}$. When taking measurements on GSM cells, the UE has to periodically decode the SCH of the four best GSM cells to determine their base station identification codes (BSIC). For cell reselection, the UE may consider only those GSM cells with their BSIC matching that of a GSM cell listed in the system information messages.

8.3 Summary

UE continuously reselects the most suitable cell in the network when the UE is not operating on a dedicated channel. The number of cells that are available for the UE to reselect is determined by the capabilities of the UE and network access restrictions. Based on RF conditions and priorities, the UE decides which cell to reselect.

Cell reselection can occur on the same radio frequency (intra-frequency), between different radio frequencies (inter-frequency), or between different RATs (inter-RAT). Cell reselections between UMTS FDD and UMTS TDD are regarded as inter-RAT cell reselections.

UE performs measurements on the neighbor cells to determine if and when it should make a cell reselection. The occasions that UE may perform measurements are determined by the UE state, the types of measurements, and the capability of the UE receiver. UE may take intra-frequency measurements on other cells and receive the information from the current serving cell at the same time. But, for inter-frequency and inter-RAT measurements, the UE must stop receiving information from the current serving cell and tune to the other frequencies for measurements if it has only one receiver.

When camped on a cell, the UE constantly searches for a better cell according to the cell reselection criteria. During cell reselection, the UE takes measurements of the serving cell and the neighbor cells provided in SIB Type 11 or 12. The S criteria and R criteria are calculated for each measured cell. The cells meeting the S criteria are called suitable cells. All suitable cells are ranked against one another. The top ranked cell is chosen for cell reselection.

The inter-frequency and inter-RAT cell reselection processes are essentially the same as intra-frequency cell reselection process. For inter-frequency cell reselection, UTRAN may include up to 32 other-frequency cells and 2 additional frequencies in the system information messages. UTRAN uses the $S_{intersearch}$ threshold to decide whether the UE should perform inter-frequency measurements or not. Similarly, for inter-RAT cell reselection, UTRAN may include up to 32 GSM cells and 32 GSM frequencies in the system information messages. UTRAN uses the $Ssearch_{RATGSM}$ threshold to decide whether the UE should perform inter-RAT measurements or not.

References

[1] 3GPP TS 25.304, v5.3.0, "UE procedures in idle mode and procedures for cell reselection in connected mode,"(Release 5).

[2] 3GPP TS 23.122, v5.3.0, "NAS functions related to Mobile Station (MS) in idle mode,"(Release 5).

[3] 3GPP TS 25.331, v5.7.1, "Radio Resource Control (RRC) protocol specification,"(Release 5).

Chapter 9

Handover

The purposes of handovers are to maintain the continuity of a connection as the UE moves around in the network and also to maintain the quality of the radio links. There are a variety of handovers defined in the 3GPP specification [1], including soft handover, softer handover, hard handover, and inter-radio access technology handover (inter-RAT handover).

Soft handover establishes the connection with a new Node B before breaking the connection with an old one. During a soft handover, the UE maintains connections with at least two Node Bs. On the uplink, the RNC selects the best frame received from these radio links.

Softer handover occurs when radio link connections exist between the UE and the cells belonging to the same Node B. It is an intra Node B handover. During a softer handover, the Node B combines radio signals received from these radio links, demodulates the combined signal, and passes the information to the RNC.

Hard handovers break the old connection before establishing the new one. So, it follows that there is a brief interruption in communication during a hard handover. A hard handover occurs when the radio link needs to transition to another carrier frequency or another RAT. It also occurs during a certain physical channel reconfiguration, such as a timing change. If the Iur interface is not supported, a hard handover is required for radio link transitions from one Node B to another Node B belonging to a different RNC.

For all handovers, it is always necessary to evaluate the radio links. There are two types of handover evaluation processes: network evaluated handovers, in which a handover is triggered by evaluations made in the network, and UE evaluated handovers, in which a handover is triggered by evaluations made in the UE.

In a UE evaluated handover, the UE reports measurements to UTRAN when the reporting conditions set by UTRAN are met. UTRAN executes the relevant procedures to add, drop, or replace the radio links based on the UE reports. The following sections will address measurement and reporting, soft/softer handover, inter-frequency handover, inter-RAT handover, and handover in compressed mode.

9.1 UE Measurements and Reporting

UE measurements and reporting are a part of handover procedure, though not all measurements and reporting are for handover only. UTRAN controls a measurement in the UE, either by using the measurement information broadcast on SIB Type 11, or by transmitting a Measurement Control message. The information for UE measurements and reporting of measurement results includes: measurement identity, measurement command, measurement type, measurement objects, measurement quantity, reporting quantities, measurement reporting criteria, measurement validity, measurement reporting mode, and additional measurement identities [2]. These measurement parameters have been briefly described in Subsection 4.8.1. Next is a short description of each of these parameters.

Measurement identity is a reference number that is used by UTRAN when setting up, modifying, or releasing the measurement. It is also used by UE in the measurement report.

Measurement command can be a setup, modification or release command. The setup command orders the UE to set up a new measurement. The modification command tells the UE to modify a previously defined measurement, such as changing the reporting criteria. The release command instructs the UE to stop a measurement and clear all information in the UE relating to that measurement.

Measurement object lists the objects on which the UE measures, including the corresponding object information.

Measurement quantity specifies the quantity (CPICH Ec/Io, CPICH RSCP, or path loss) the UE measures on the measurement object, including the filtering of the measurements.

Reporting quantity indicates the reporting quantities that the UE must include in the report, in addition to the mandatory report quantities for the specific event. These include cell identity, CPICH Ec/Io, CPICH RSCP, path loss, observed time difference, and cell synchronization.

Measurement reporting criteria define triggering of the measurement report, such as periodical or event-triggered reporting. Periodic reporting may be used with intra-frequency, inter-frequency, or inter-RAT measurements. If a measurement is configured to use periodic reporting, then a report is sent at the specified time interval, which ranges from 250 ms to 64 seconds. If a measurement is configured as event-triggered, then a report is sent after the event occurs and the time-to-trigger period expires.

Measurement validity defines the UE states in which the measurements are to be performed. It could be Cell_DCH, Idle, Cell_FACH, Cell_PCH, or URA_PCH. Measurements that the UE takes in all states except Cell_DCH are used for cell reselection decisions. However, measurements reported in RACH messages may be used by UTRAN for soft/softer handover decisions when the UE transitions to the Cell_DCH state.

Measurement reporting mode specifies whether the UE must transmit the measurement report using AM or UM transfer mode.

Additional measurement identity is a list of reference numbers for previously configured measurements that should be reported whenever a measurement report for the current measurement is triggered.

Measurement type defines the types of measurements, including intra-frequency measurements, inter-frequency measurements, inter-RAT measurements, traffic volume measurements, quality measurements, UE internal measurements, and UE positioning measurements.

9.1.1 UE Measurements for Handover

For handover purposes, UE measurements take place when the UE is at call setup or in the Cell_DCH state.

When the UE is at call setup, it uses the relevant measurement and reporting parameters broadcast by UTRAN through SIB Type 11 for measurements. The UE just performs the measurements of the cells without any filtering and reports the results to UTRAN on the RACH. UTRAN then evaluates the results and decides whether the UE goes into soft/softer handover or just sets up a single link before the call becomes active. UTRAN identifies the cells to be included in the active set, prepares the radio links on the UTRAN side, and sends the corresponding commands to the UE. As such, handover at call setup is considered as a network evaluated handover.

For Cell_DCH, the initial information for measurements and reports are also broadcast on SIB Type 11 or SIB Type 12. Upon entering the Cell_DCH state, the UE starts to perform measurements and also evaluates the measurement results. If the reporting criteria defined by the measurement information are met, then the UE sends an event measurement report to UTRAN.

When a UE is in the Cell_DCH state, UTRAN may request the UE to set up, modify or release a measurement by sending a Measurement Control message to the UE on a downlink DCCH using AM RLC. The message tells the UE what measurements to perform and on which cells the measurements should be performed. The Measurement Control message also defines how and when the UE should send a Measurement Report message to UTRAN. The measurements pertaining to handovers are intra-frequency measurements for soft and softer handovers, and inter-frequency measurements for hard handover and inter_RAT measurements.

When UTRAN receives the UE measurement report, it will add and/or remove the radio link(s) for the UE. No additional evaluation process in UTRAN is needed to decide on handover actions, because this decision is already made by the UE. That is, the actual handover algorithm sits in the UE. UTRAN only defines the handover rules by setting the parameters for the measurement and reporting procedures. As mentioned earlier, the information on these parameters are either broadcast on SIB Type 11 or sent to the UE by using a Measurement Control message. In general, whenever the UE moves through the network, the handover parameters may need to be changed. It is the Measurement Control message that carries the new parameter settings to the UE.

9.2 Categories of Cells

Cells that the UE keeps track of are grouped into three mutually exclusive categories: active set, monitored set, and detected set.

- **Active set:** Cells in the active set are in soft or softer handover with the UE. The signals from the cells in the active set are simultaneously demodulated and coherently combined by the UE. The UE measures and reports on cells in the active set to detect conditions in which the signal quality of an active set cell becomes too poor to be usable. The measurements and reports on active set cells allow UTRAN to remove a cell from the active set through the RRC Active Set Update procedure. The active set size depends on services. Normally it is limited to 3 or 4 cells. The maximum number of reporting cells for an event is set relative to the active set size, as it would make no sense to ask for a large number of cells to be reported when actually the active set size is very small. Also, active set applies to intra-frequency handover only. For inter-frequency handover, a virtual active set is defined.

- **Monitored set:** Cells belonging to this category are not part of the active set, but are also measured and reported by the UE if certain reporting criteria are met. The monitored set is normally a subset of the neighbor list and is updated by UTRAN with a neighbor list selection algorithm. The cells in the monitored set are the most likely targets for handover. When the signal strength of a cell in the monitored set rises to a certain level, measurement reports from the UE trigger UTRAN to add that cell to the active set. When a cell is added to the active set, another cell in the active set will be removed if the number of cells in the active set exceeds the active set size.

- **Detected set:** When measuring the cells in an active set and a monitored set, the UE may detect a cell that is in neither of these two sets. This cell becomes a member of the detected set. In other words, the cells included in the detected set are the unlisted ones, which are not included in the active set or monitored set provided by UTRAN, but are detected by the UE. The UE may report on a detected set cell, triggering UTRAN to add that cell to either the monitored set or the active set. However, assuming that a detected cell can be measured and reported by the UE, it will be added to the active set only if it actually belongs to the neighbor list of one of the current active set cells. If it is not registered in any neighbor list of the active set cells, it cannot be added. In this case, the neighbor list needs to be re-optimized, or other RF parameters, such as the antenna azimuth and down tilt of this cell, need to be adjusted so that it will not be detected by the UE.

If a Measurement Control message contains the cells-for-measurement IE, only the monitored set cells explicitly indicated by the IE for a given intra-frequency (respectively inter-frequency, inter-RAT) measurement need to be considered for measurement. If a Measurement Control message does not include the cells-for-measurement IE, all of the intra-frequency (respectively inter-frequency, inter RAT) cells stored in the variable CELL_INFO_LIST must be considered for measurement. The cells-for-measurement IE is not applicable to active set cells or virtual active set cells. For example, when the triggering condition refers to active set cells, the UE must consider all active set cells in the CELL_INFO_LIST for measurement regardless of whether these cells are explicitly indicated by the cells-for-measurement IE.

9.3 Soft and Softer Handover

As mentioned before, soft and softer handover can happen at call setup and when UE is in the Cell_DCH state.

9.3.1 Soft and Softer Handovers at Call Setup

At call setup, it is possible for the UE to go directly into soft or softer handover when the UE is transitioning to the Cell_DCH state from the idle mode, Cell_PCH state, or the URA_PCH state. The UE performs measurements at call setup based on the measurement information broadcast in SIB Type 11 and reports the results on the RACH along with the RRC Connection Request message. The broadcast measurement information includes the measurement and reporting quantities, the neighboring cell information, and the maximum number of cells to be reported on the RACH.

The neighboring cell information contains the primary downlink scrambling codes used in the neighbor cells of the current cell that the UE camps on. The maximum number of cells to be reported is defined by UTRAN and can be up to 6 neighbor cells plus the current cell. It is normally recommended that the UE report two best neighbor cells in addition to the current cell. However, in most practical cases, only the current cell is reported on the RACH. As mentioned before, filtering is not applied to the measurements to be reported on the RACH.

Once UTRAN receives the measurement report from the UE, it will check whether the reported cells are good enough to be included in the active set at call setup using a certain criterion. For example, if a reported cell, in addition to the current cell, meets the criterion given by (9-1) below, it will be included in the active set.

$$(\text{CPICH Ec}/\text{Io})_i > (\text{CPICH Ec}/\text{Io})_{\text{current}} - (\text{CPICH Ec}/\text{Io})_{\text{SHOthreshold}}, \qquad (9\text{-}1)$$

where $(\text{CPICH Ec}/\text{Io})_i$ is the measured pilot Ec/Io of the ith reported cell, $(\text{CPICH Ec}/\text{Io})_{\text{current}}$ is the pilot Ec/Io of the current cell that the UE camps on, and $(\text{CPICH Ec}/\text{Io})_{\text{SHOthreshold}}$ is a hysteresis.

If one or more cells, in addition to the current cell, are to be included in the active set at call setup, radio links between the UE and all of these cells are established before the UE sends the RRC Connection Setup Complete message that signifies the transition to the Cell_DCH state. If for some reason, at least one radio link cannot be set up, the call is not set up and all the relevant resources are released.

If no cell besides the current cell is to be included in the active set, only one link to the current cell is established. Similarly, if for some reason, the radio link cannot be established, the call is not set up and the relevant resources are released.

9.3.2 Soft and Softer Handovers in Cell_DCH State

When UE enters the Cell_DCH state, it stores all the relevant information received from the SIB Type 11 message in order to continue performing measurements and sending report. In the Cell_DCH state, the UE samples the intra-frequency measurements once every measurement period, which is 200 ms [3]. After measurement sampling, the UE performs filtering of the measured quantity according to the following formula.

$$F_n = (1-\psi)F_{n-1} + \psi M_n, \qquad (9\text{-}2)$$

where F_n is the updated filtered measurement result, F_{n-1} is the old filtered measurement result, M_n the latest received measurement result from physical layer measurements, and ψ is the filter factor given by $\psi = 1/2^{(k/2)}$, where k is the parameter received in the filter coefficient IE. The unit used for M_n is the same as the reported unit in the Measurement Report message or the unit used in the event evaluation. The filtering must be performed by the UE before UE event evaluation.

In order to initialize the averaging filter, F_0 is set to M_1 when the first measurement result from the physical layer measurement is received. The filtering guarantees that large fluctuations of the UE measurements are prevented. The filter coefficient k affects the weight given to previous measurements. The larger the filter coefficient value, the more influence the previous measurements have on the updated filtered measurement. If k is equal to zero, there is no filtering. In general, a low filter coefficient value results in a fast soft handover decision for high-speed UEs, and may also result in a ping-pong effect for low-speed UEs.

9.3.2.1 Intra-Frequency Reporting Events
After filtering, the filtered measurement results are compared with the thresholds defined for the event triggered reporting. The UMTS standard specifies six main reporting events for intra-frequency FDD measurements numbered 1A, 1B, 1C, 1D, 1E, and 1F. Most of these six reporting events are important to UMTS soft/softer handovers. Events 1A and 1E are used to indicate to UTRAN when a new cell should be added to the active set. Events 1B and 1F are used to indicate to UTRAN when a cell should be removed from the active set. Event 1C is used to indicate to UTRAN when a cell in the active set should be replaced with a different cell. Event 1D is not important for handovers and will not be discussed here.

For each of the previously mentioned events, there are specific parameters that define the triggering conditions and the reporting rules that apply to different sets of cells. These parameters include reporting set cells, maximum number of event-triggered reports sent by the UE upon event triggering, and measurement report periodicity. They are broadcast on SIB Type 11 or sent on a Measurement Control message and are individually set for each reporting event.

During the initial phase of network optimization, the reporting set cells include the cells in both the monitored set and detected set for the purpose of neighbor list optimization. However, when the network is in normal operation, only the cells in the monitored set can be added to the active set in order to minimize the pilot search processing time and thus speed up handover reporting.

In the following we will describe reporting events 1A, 1B, 1C, 1E, and 1F using, for example, CPICH Ec/Io as the measurement and reporting quantity. Most network equipment vendors currently support only events 1A, 1B, and 1C.

9.3.2.1.1 Reporting Event 1A
Reporting event 1A is triggered when a primary CPICH enters the reporting range. In this case the cell is added to the active set. A reporting range is calculated based on the measurements of the cell and the current active set cells and the measurement information received from UTRAN. The triggering condition for a cell entering the reporting range is given as follows [2]:

$$10\log(M_{\text{New}}) + \text{CIO}_{\text{New}} \geq W_{1A}10\log\left\{\sum_{i=1}^{N_A} M_i\right\} + (1-W_{1A})10\log(M_{\text{Best}}) - (R_{1A} - H_{1A}/2) \quad (9\text{-}3)$$

where

- M_{New} = Measurement result of the cell entering the reporting range
- M_i = Measurement result that affects the reporting range in the active set
- M_{Best} = Measurement result of the strongest cell in the active set
- N_A = Number of cells that affect the reporting range in the current active set
- W_{1A} = Weighting constant valued 0.0 to 2.0 in steps of 0.1
- R_{1A} = Reporting range constant in dB for event 1A
- H_{1A} = Hysteresis value in dB for event 1A
- CIO_{New} = Cell individual offset for the cell entering the reporting range

If (9-3) has been fulfilled for a time period indicated by Time-to-trigger, event 1A is triggered and reported to UTRAN.

Figure 9.1 shows an example of event 1A occurrence assuming the active set size is equal to 3 and the weighting constant is equal to 0. Initially, P-CPICH 1 and P-CPICH 2 are in the active set and the UE is in two-way soft handover. As soon as P-CPICH 3 enters the reporting range (stronger than P-CPICH 1 minus the reporting range value), event 1A is detected. After a time-to-trigger period, event 1A is reported to UTRAN.

Normally, after receiving the first Measurement Report message sent from the UE, UTRAN should execute the soft/softer handover via RRC active set update procedure and adds P-CPICH 3 to the active set and thus the UE is now in a three-way soft handover. However, in some cases, the UTRAN may not be able to successfully initiate the soft handover because the Measurement Report message is not successfully decoded

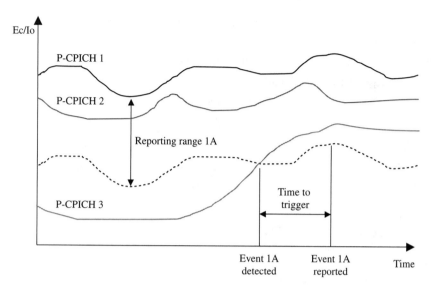

Figure 9.1 Example of event 1A with an active set size of 3.

or there is not enough resource in Node B. If this happens, the UE continues to report after the first report by using periodic reporting at a predefined time interval until the maximum number of event-triggered reports is reached. It should be noted that this event-triggered periodic reporting is configured separately from the global periodic reporting and may be used only for events 1A and 1C.

9.3.2.1.2 Reporting Event 1B Reporting event 1B is triggered when a primary CPICH leaves the reporting range. Event 1B occurs only when the current active set size is greater than one. In this case a cell is removed from the active set. The triggering condition for a cell leaving the reporting range is given as follows [2]:

$$10 \log (M_{Old}) + CIO_{Old}$$
$$\leq W_{1B} 10 \log \left\{ \sum_{i=1}^{N_A} M_i \right\} + (1 - W_{1B}) 10 \log (M_{Best}) - (R_{1B} + H_{1B} / 2) \quad (9\text{-}4)$$

where

- M_{Old} = Measurement result of the cell leaving the reporting range
- M_i = Measurement result that affects the reporting range in the active set
- M_{Best} = Measurement result of the strongest cell in the active set

- N_A = Number of cells that affect the reporting range in the current active set
- W_{1B} = Weighting constant valued 0.0 to 2.0 in steps of 0.1
- R_{1B} = Reporting range constant in dB for event 1B
- H_{1B} = Hysteresis value in dB for event 1B
- CIO_{Old} = Cell individual offset for the cell leaving the reporting range.

If (9-4) has been fulfilled for a time period indicated by Time-to-trigger, event 1B is triggered and reported to UTRAN.

Figure 9.2 shows an example of event 1B occurrence assuming the active set size is equal to 2 and the weighting constant is equal to 0. Initially, P-CPICH 1 and P-CPICH 2 are in the active set and the UE is in a two-way soft handover. As soon as P-CPICH 2 leaves the reporting range (weaker than P-CPICH 1 minus the reporting range value), event 1B is detected. After a time-to-trigger period, event 1B is reported to the UTRAN. After receiving the measurement report from the UE, UTRAN removes P-CPICH 2 from the active set and the UE is no longer in soft handover.

For event 1B, the maximum number of event-triggered reports and the report periodicity are not specified. As such, the UE sends one and only one measurement report when event 1B is triggered.

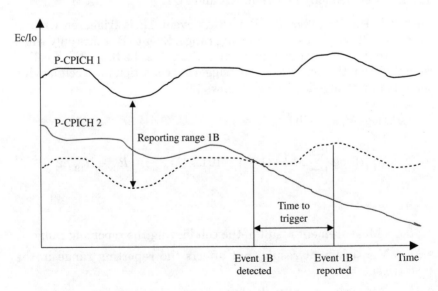

Figure 9.2 Example of event 1B with an active set size of 2.

9.3.2.1.3 Reporting Event 1C If the active set is full, reporting event 1C is triggered when a primary CPICH in the monitored set becomes stronger than the weakest primary CPICH in the active set. In this case, the weakest primary CPICH in the active set is removed and the stronger primary CPICH is added to the active set. The triggering condition for event 1C is given as follows [2]:

$$10\log(M_{\text{New}}) + \text{CIO}_{\text{New}} \geq 10\log(M_{\text{InAS}}) + \text{CIO}_{\text{InAS}} + H_{1C}/2, \quad (9\text{-}5)$$

where

- M_{New} = Measurement result of the best candidate cell to be included in the active set
- M_{InAS} = Measurement result of the worst cell in the active set
- CIO_{InAS} = Cell individual offset for the worst cell in the active set
- H_{1C} = Hysteresis value for event 1C

CIO_{New} = Cell individual offset for the cell becoming better than the cells in the active set. If (9-5) has been fulfilled for a time period indicated by Time-to-trigger, event 1C is triggered and reported to UTRAN.

Figure 9.3 shows an example of event 1C occurrence assuming the active set size is equal to 3. Initially, P-CPICH 1, P-CPICH 2, and P-CPICH 3 are in the active set and the UE is in a three-way soft handover. When P-CPICH 4 becomes better than P-CPICH 3, event 1C is detected.

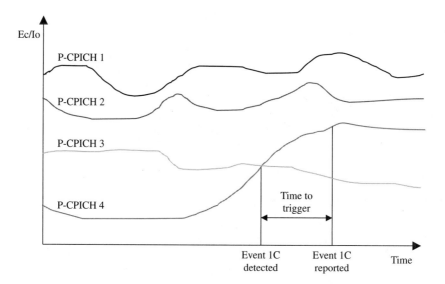

Figure 9.3 Example of event 1C with an active set size of 3.

After a time-to-trigger period, event 1C is reported to UTRAN. After receiving the measurement report from the UE, UTRAN replaces P-CPICH 3 with P-CPICH 4 in the active set and the UE is still in a three-way handover.

Similar to event 1A, in some abnormal cases, if UTRAN fails to react to the Measurement Report message for event 1C by sending an Active Set Update message to the UE, the UE begins using periodic reporting.

9.3.2.1.4 Reporting Event 1E Reporting event 1E is triggered when a primary CPICH becomes better than an absolute threshold. The triggering condition for event 1E is given as follows [2]:

$$10 \log (M_{\text{New}}) + \text{CIO}_{\text{New}} \geq T_{1E} + H_{1E} / 2, \qquad (9\text{-}6)$$

where M_{New} is the measurement result of a cell that becomes better than an absolute threshold, CIO_{New} is the individual cell offset for the cell becoming better than the absolute threshold, T_{1E} is an absolute threshold, and H_{1E} is the hysteresis parameter for event 1E. If (9-6) has been fulfilled for a time period indicated by Time-to-trigger, event 1E is triggered and reported to UTRAN. After receiving a Measurement Report message from the UE for reporting event 1E, UTRAN adds the reported cell to the active set.

Figure 9.4 illustrates an example of event 1E occurrence. The active set originally has two pilots, P-CPICH 1 and P-CPICH 2. The UE is in a two-way handover. When the pilot P-CPICH 3 grows stronger than the

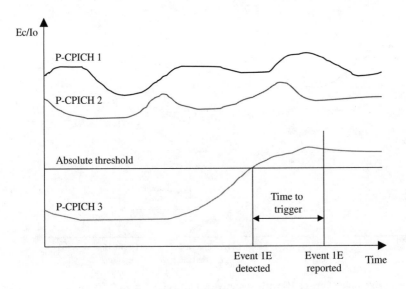

Figure 9.4 Example of event 1E occurrence.

absolute threshold, event 1E is detected. After a time-to-trigger period, event 1E is reported to UTRAN. After receiving the measurement report from the UE, UTRAN adds P-CPICH 3 to the active set and the UE is now in a three-way handover.

9.3.2.1.5 Reporting Event 1F Reporting event 1F is triggered when a primary CPICH in the active set becomes worse than an absolute threshold. The triggering condition for event 1F is given as follows [2]:

$$10\log(M_{Old}) + CIO_{Old} \leq T_{1F} - H_{1F}/2, \qquad (9\text{-}7)$$

where M_{Old} is the measurement result of a cell that becomes worse than an absolute threshold, CIO_{Old} is the individual cell offset for the cell becoming worse than the absolute threshold, T_{1F} is an absolute threshold, and H_{1F} is the hysteresis parameter for event 1F. If (9-7) has been fulfilled for a time period indicated by Time-to-trigger, event 1F is triggered and reported to UTRAN. After receiving a Measurement Report message from the UE for event 1F, UTRAN removes the reported cell from the active set.

Figure 9.5 illustrates an example of event 1F occurrence. The active set originally has two pilots, P-CPICH 1 and P-CPICH 2. The UE is in a two-way handover. When the pilot P-CPICH 2 becomes worse than the absolute threshold, event 1F is detected. After a time-to-trigger period, event 1F is reported to UTRAN. After receiving the measurement report from the UE, UTRAN removes P-CPICH 2 from the active set and the UE is now no longer in handover.

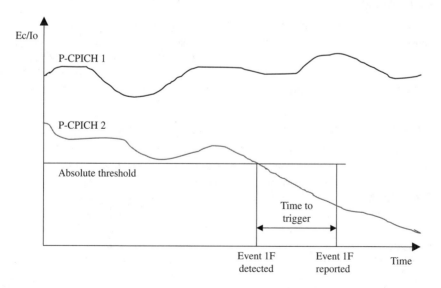

Figure 9.5 Example of event 1F occurrence.

It should be noted that the triggering conditions given by (9-3), (9-4), (9-5), (9-6), and (9-7), respectively for reporting events 1A, 1B, 1C, 1E, and 1F, are valid for the measured and reported quantity CPICH Ec/Io. Other triggering conditions are defined in [2] for other measured and reported quantities.

9.4 Inter-Frequency Handover

For inter-frequency handover, UTRAN sends an inter-frequency measurement and reporting information in a Measurement Control message. This message contains information about inter-frequency measurements. Much of the information is the same as that for intra-frequency measurements. However, there are some differences as discussed below.

The cell information list contains similar information as for intra-frequency cells but also includes the frequency of each cell. The reporting quantity includes the reporting quantity for both intra-frequency and inter-frequency cells. There is an inter-frequency set update, which defines changes to the virtual active set. In addition to the intra-frequency measurement reporting criteria that define the intra-frequency reporting events for the inter-frequency cells, there are inter-frequency measurement reporting criteria that define reporting events 2A to 2F and the corresponding inter-frequency measurement parameters, including threshold values and time-to-trigger.

9.4.1 Virtual Active Set

All inter-frequency measurements adopt the concept of virtual active set. A virtual active set is a set of the best cells of a non-used frequency. There is one virtual active set for each non-used frequency. The virtual active set can be updated either autonomously by the UE or by UTRAN. A detailed discussion on virtual active set updating is given in [2].

9.4.2 Inter-Frequency Handover Procedure

The inter-frequency handover procedure involves three steps. The first step is that the UE has to detect and report that there is a new frequency that is better than the frequency currently being used. UTRAN may decide to hand over the UE to that new frequency based on cell load and other criteria. The second step is to identify the best cell on that new frequency. This means that the UE has to take some cell-specific measurements on that new frequency and report the measurement results. The last step is that UTRAN instructs the UE to transition to that new frequency.

The first step is dictated by the inter-frequency reporting events 2A through 2F. These reporting events are based on calculations of a frequency quality estimate. In the frequency quality estimate calculations, cells in the active set are used for the frequency currently being used, while cells in a virtual active set are used for a new frequency. The frequency quality estimate used in events 2A through 2F is given as [2]:

$$Q_{\text{frequency }j} = 10\log(M_{\text{frequency }j}) = W_j \cdot 10\log\left\{\sum_{i=1}^{N_{Aj}} M_{ij}\right\} + (1-W_j) \cdot 10\log(M_{\text{Best }j}),$$

(9-8)

where

- $Q_{\text{frequency }j}$ = estimated quality of the virtual active set on frequency j in dB
- $M_{\text{frequency }j}$ = estimated quality of the virtual active set on frequency j in decimal value
- M_{ij} = a measurement result of cell i in the virtual active set on frequency j
- N_{Aj} = number of cells in the virtual active set on frequency j
- $M_{\text{Best }j}$ = measurement result of the cell in the virtual active set on frequency j with the highest measurement result, and W_j is a parameter sent from UTRAN to UE and used for frequency j.

In the second step, intra-frequency reporting events 1A through 1C are configured for the cells on the new frequency. The virtual active set is now regarded as an active set as far as the new frequency is concerned. The cells in the virtual active set define the reporting range for events 1A and 1B. When a cell not in the virtual active set becomes better than a cell in the virtual active set, reporting event 1C is triggered.

9.4.3 Inter-Frequency Reporting Events

As mentioned before, the inter-frequency reporting events include reporting events 2A through 2F. These reporting events are briefly described in the next subsections.

9.4.3.1 Reporting Event 2A
Reporting event 2A is for change of best frequency. If event 2A is configured in the UE within a measurement, the UE regards the frequency currently being used as the best frequency and stores it in the variable BEST_FREQUENCY_2A_EVENT when the measurement is initiated or resumed.

Reporting event 2A is triggered when the quality estimate of another frequency is better than the current one. The triggering condition is given as [2]:

$$Q_{\text{NotBest}} \geq Q_{\text{Best}} + H_{2A} / 2, \qquad (9\text{-}9)$$

where Q_{NotBest} is the quality estimate of the new frequency, Q_{Best} is the quality estimate of the current frequency, and H_{2A} is the hysteresis parameter for event 2A.

If (9-9) has been fulfilled for a time period indicated by Time-to-trigger, event 2A is triggered. The UE will send a measurement report to UTRAN and update the variable BEST_FREQUENCY_2A_EVENT with the new frequency. UTRAN may decide to hand over the UE to the new frequency based on cell loading and other criteria.

9.4.3.2 Reporting Event 2B Reporting event 2B is triggered when the quality estimate of the currently used frequency is below a certain threshold and the quality estimate of a non-used frequency is above a certain threshold. When an inter-frequency measurement configuring event 2B is set up, the UE creates a variable TRIGGERED_2B_EVENT related to that measurement. This variable is empty initially. The UE deletes this variable when the measurement is released.

When event 2B is configured in the UE within a measurement, it is triggered when the following two conditions are fulfilled for a non-used frequency for a period of time indicated by Time-to-trigger.

$$Q_{\text{Non used}} \geq T_{\text{Non used 2B}} + H_{2B} / 2 \qquad (9\text{-}10)$$

$$Q_{\text{Used}} \leq T_{\text{Used 2B}} - H_{2B} / 2, \qquad (9\text{-}11)$$

where

- $Q_{\text{Non used}}$ = quality estimate of a non-used frequency that becomes better than an absolute threshold
- $T_{\text{Non used 2B}}$ = absolute threshold that applies to this non-used frequency in that measurement
- H_{2B} = hysteresis parameter for event 2B
- Q_{Used} = quality estimate of the used frequency
- $T_{\text{Used 2B}}$ = absolute threshold that applies to the used frequency in that measurement

The UE will then store the non-used frequency that triggered the event into the variable TRIGGERED_2B_EVENT and send a measurement report to UTRAN.

There may be more than one non-used frequency that triggered event 2B. In this case, the UE must store these non-used frequencies into the TRIGGERED_2B_EVENT if they are not already there. The measurement report sent to UTRAN must include the frequency information and measured results of these non-used frequencies beginning with the best frequency. UTRAN may decide to hand over the UE to the best frequency in the variable TRIGGERED_2B_EVENT based on cell loading and other criteria.

The non-used frequencies stored in the variable TRIGGERED_2B_EVENT, and even the variable itself, may not stay there forever. If the condition given by (9-12) is fulfilled for a stored non-used frequency, that non-used frequency is removed from the variable TRIGGERED_2B_EVENT.

$$Q_{\text{Non used}} < T_{\text{Non used 2B}} - H_{2B} / 2 \qquad (9\text{-}12)$$

Furthermore, if (9-13) is fulfilled for the used frequency, the UE will clear the variable TRIGGERED_2B_EVENT.

$$Q_{\text{Used}} > T_{\text{Used 2B}} + H_{2B} / 2 \qquad (9\text{-}13)$$

9.4.3.3 Reporting Event 2C Reporting event 2C is triggered when the quality estimate of a non-used frequency is above a certain threshold. The UE operation for event 2C triggering is similar to that for event 2B. When an inter-frequency measurement configuring event 2C is set up, the UE must create a variable TRIGGERED_2C_EVENT related to that measurement. Initially, this variable is empty and it is deleted when the measurement is released.

When event 2C is configured in the UE within a measurement, it is triggered when the condition given by (9.14) is fulfilled for one or more non-used frequencies for a period of time indicated by Time-to-trigger.

$$Q_{\text{Non used}} \geq T_{\text{Non used 2C}} + H_{2C} / 2, \qquad (9\text{-}14)$$

where $T_{\text{Non used 2C}}$ is the absolute threshold that applies to this non-used frequency in that measurement and H_{2C} is the hysteresis parameter for event 2C.

The UE then stores the non-used frequencies that triggered the event into the variable TRIGGERED_2C_EVENT and sends a measurement report to UTRAN. The measurement report includes the frequency information and measured results of these non-used frequencies beginning with the best frequency. The UTRAN may decide to hand over the UE to the best frequency in the variable based on cell loading and other criteria.

Again, the non-used frequencies stored in the variable TRIGGERED_2C_EVENT may not stay in there forever. If (9-15) is fulfilled for a stored non-used frequency, that non-used frequency is removed from the variable TRIGGERED_2C_EVENT.

$$Q_{\text{Non used}} < T_{\text{Non used 2C}} - H_{2C} / 2 \qquad (9\text{-}15)$$

9.4.3.4 Reporting Event 2D
Reporting event 2D is triggered when the quality estimate of the currently used frequency is below a certain threshold. It is required that a UE must be able to perform this measurement and the corresponding event reporting without requiring compressed mode.

When an inter-frequency measurement configuring event 2D is set up, the UE creates a variable TRIGGERED_2D_EVENT related to that measurement. Initially, the variable is set to FALSE and it is deleted when the measurement is released.

When event 2D is configured in the UE within a measurement and the condition given by (9-16) has been fulfilled for the used frequency for a time period indicated by Time-to-trigger, the UE will set the variable TRIGGERED_2D_EVENT to TRUE if it is not already set to TRUE. At the same time the UE sends a measurement report to UTRAN.

$$Q_{\text{Used}} \leq T_{\text{Used 2D}} - H_{2D} / 2, \qquad (9\text{-}16)$$

where $T_{\text{Used 2D}}$ is the absolute threshold that applies to the used frequency for event 2D, and $H_{2D}/2$ is the hysteresis parameter for event 2D.

On the contrary, if the variable TRIGGERED_2D_EVENT is set to true and the condition given by (9-17) is fulfilled for the used frequency, the UE will set the variable TRIGGERED_2D_EVENT to FALSE.

$$Q_{\text{Used}} > T_{\text{Used 2D}} + H_{2D} / 2 \qquad (9\text{-}17)$$

9.4.3.5 Reporting Event 2E
Reporting event 2E is triggered when the quality estimate of a non-used frequency is below a certain threshold. When an inter-frequency measurement configuring event 2E is set up, the UE creates a variable TRIGGERED_2E_EVENT related to that

measurement. Initially, this variable is empty and it is deleted when the measurement is released.

When event 2E is configured in the UE within a measurement, it is triggered when the condition given by (9-18) is fulfilled for one or more non-used frequencies for a period of time indicated by Time-to-trigger.

$$Q_{\text{Non used}} \leq T_{\text{Non used 2E}} - H_{2E}/2, \qquad (9\text{-}18)$$

where $T_{\text{Non used 2E}}$ is the absolute threshold that applies to this non-used frequency in that measurement, and H_{2E} is the hysteresis parameter for the event 2E.

The UE then stores the non-used frequencies that triggered the event into the variable TRIGGERED_2E_EVENT and sends a measurement report to UTRAN. The measurement report includes the frequency information and measured results of these non-used frequencies beginning with the best frequency.

Again, the non-used frequencies stored in the variable TRIGGERED_2E_EVENT may not stay in there forever. If (9-19) is fulfilled for a stored non-used frequency, that non-used frequency is removed from the variable TRIGGERED_2E_EVENT.

$$Q_{\text{Non used}} > T_{\text{Non used 2E}} - H_{2E}/2 \qquad (9\text{-}19)$$

9.4.3.6 Reporting Event 2F Reporting event 2F is triggered when the quality estimate of the currently used frequency is above a certain threshold. It is required that a UE must be able to perform this measurement and the corresponding event reporting without requiring compressed mode.

When an inter-frequency measurement configuring event 2F is set up, the UE creates a variable TRIGGERED_2F_EVENT related to that measurement. Initially, the variable is set to FALSE and it is deleted when the measurement is released.

When event 2F is configured in the UE within a measurement and the condition given by (9-20) has been fulfilled for the used frequency for a time period indicated by Time-to-trigger, the UE will set the variable TRIGGERED_2F_EVENT to TRUE if it is not already set to TRUE. At the same time the UE sends a measurement report to UTRAN.

$$Q_{\text{Used}} \geq T_{\text{Used 2F}} + H_{2F}/2, \qquad (9\text{-}20)$$

where $T_{\text{Used 2F}}$ is the absolute threshold that applies to the used frequency for event 2F, and $H_{2F}/2$ is the hysteresis parameter for event 2F.

On the contrary, if the variable TRIGGERED_2F_EVENT is set to true and the condition given by (9-21) is fulfilled for the used frequency, the UE will set the variable TRIGGERED_2F_EVENT to FALSE.

$$Q_{\text{Used}} < T_{\text{Used 2F}} - H_{2F} / 2 \qquad (9\text{-}21)$$

9.5 Inter-RAT Handover

The inter-RAT handover transfers a connection between the UE and UTRAN to another radio access technology, such as GSM, or a connection between the UE and other radio technology to UTRAN. This section focuses only on the inter-RAT handover that transfers a UE connection from UTRAN to another radio access technology, say GERAN.

For inter-RAT handover, UTRAN sends an information element in a Measurement Control message containing information about inter-RAT measurements. Much of the information is the same as that for intra-frequency measurements. The differences are:

1. The cell information list also includes GSM-specific information such as frequency, band indicator, and BSIC.
2. The measurement quantity defines the filter coefficient and the quantity that the UE should measure for the UMTS cells and for the GSM cells.
3. The reporting quantity defines the observed time difference and the GSM carrier RSSI in addition to those quantities that are mandatory for the event.
4. The inter-RAT reporting criteria define the reporting events 3A to 3D and the corresponding parameters for inter-RAT measurements including thresholds, hysteresis, and time-to-trigger values.

9.5.1 Inter-RAT Handover Triggering Events

The inter-RAT handover triggering events are defined as follows:

- **Event 3A**—The estimated quality of the currently used UTRAN frequency is below a certain threshold and the estimated quality of the other system is above a certain threshold.
- **Event 3B**—The estimated quality of the other system is below a certain threshold.
- **Event 3C**—The estimated quality of the other system is above a certain threshold.
- **Event 3D**—Change of best cell in other system

The frequency quality estimate for the serving UTRAN frequency used in event 3A is the same as that used for inter-frequency events. More specifically it is given as [2]:

$$Q_{\text{UTRAN}} = 10\log(M_{\text{UTRAN}}) = W \cdot 10\log\left\{\sum_{i=1}^{N_A} M_i\right\} + (1-W) \cdot 10\log(M_{\text{Best}}),$$

(9-22)

where

- Q_{UTRAN} = estimated quality of the active set on the currently used UTRAN frequency in dB
- M_{UTRAN} = estimated quality of the active set on currently used UTRAN frequency expressed in decimal value
- M_i = measurement result of cell i in the active set, according to what is indicated in the measurement-quantity-for-UTRAN-quality-estimate IE
- N_A = number of cells in the active set
- M_{Best} = the measurement result of the cell in the active set with the highest measurement result, and W is a parameter sent from UTRAN to the UE.

The triggering conditions for event 3A are defined as follows:

$$Q_{\text{Used}} \leq T_{\text{Used}} - H_{3A}/2 \quad (9\text{-}23)$$

$$M_{\text{Other RAT}} + \text{CIO}_{\text{Other RAT}} \geq T_{\text{Other RAT}} + H_{3A}/2, \quad (9\text{-}24)$$

where

- Q_{Used} = quality measurement of the serving UTRAN frequency
- T_{Used} = absolute threshold that applies to the UTRAN system in that measurement
- $M_{\text{Other RAT}}$ = measurement quantity for the cell of the other system. For GSM, this is RSSI.
- $\text{CIO}_{\text{Other RAT}}$ = cell individual offset for the cell of the other system
- $T_{\text{Other RAT}}$ = absolute threshold that applies to the other system in that measurement
- H_{3A} = hysteresis parameter for event 3A

Similar triggering conditions are defined for events 3B, 3C, and 3D. For details, interested readers may refer to [2].

To hand over the UE to a GSM cell, it needs to identify the GSM cell for handover. A GSM cell is identified by its ARFCN and its base station identification code (BSIC). GSM cell broadcasts its BSIC on the synchronization channel. For Inter-RAT measurements and reporting, the UE may or may not need to identify the BSIC.

If the UE does not need to identify the BSIC, the UE may send an inter-RAT measurement report on any GSM cell whose ARFCN is listed in the inter-RAT cell information list sent by UTRAN. UTRAN may configure inter-RAT measurement reporting that does not need BSIC identification.

If the UE needs to identify the BSIC, it must read the synchronization channel of the GSM cell and match the ARFCN and BSIC to the cells listed in the inter-RAT cell information list before making an inter-RAT measurement report. Before an inter-RAT handover, UTRAN may require BSIC identification to determine the identity of the target GSM cell.

9.6 Compressed Mode

Compressed mode plays an important role in inter-frequency and inter-RAT measurements. When instructed by UTRAN to do inter-frequency or inter-RAT measurements, the UE needs a second receiver or needs to operate in compressed mode to perform the measurements. With compressed mode, the UE is able to perform measurements on non-serving UMTS frequencies or on other RAT frequencies without the need of a full dual receiver. Specifically, for inter-RAT measurements, compressed mode may be used to perform GSM carrier RSSI measurement, initial BSIC identification, and BSIC reconfirmation.

Compressed mode must be used for inter-RAT measurements when a UE with a single receiver moves from UMTS coverage areas to GSM-only areas. It is also used when a UE moves in or out of an area where multiple UMTS frequencies are deployed. With compressed mode operation, the UE can perform measurements on another frequency without missing any of the data being sent on the dedicated channels by the serving UMTS cells.

9.6.1 Compressed Mode Basics

Since compressed mode must be used for inter-frequency and inter-RAT measurements if the UE has only one receiver, the logical questions are what the compressed mode is and how it works. By definition, compressed mode means a transmission mode in which the information is compressed in the time domain to create gaps in the transmission. In other words, it is a compression or reduction of transmission time and creation of transmission gaps during which the UE receiver can switch to other frequencies and take measurements.

The transmission gaps are always counted in terms of radio frame slots. Figure 9.6 shows an example of compressed mode transmission. The instantaneous power is increased in the compressed mode frame to keep the quality unaffected by the reduced processing gain (spreading factor). The amount of power increase is determined by the methods used for creating the transmission gaps.

The methods for creating the gaps include puncturing, spreading factor reduction, and higher layer scheduling. In puncturing, transmission gaps are created by additional puncturing or fewer repetitions relative to normal mode in rate matching. Sufficient physical channel bits must be punctured in order create the gaps. On the other hand, too many physical channel bits punctured results in loss of information. As such, normally it is not recommended that this method be used for creating long transmission gaps. In addition, puncturing applies only to the downlink.

For spreading factor reduction, the spreading factor of the compressed radio frame is reduced by a factor of two, allowing the same number of bits to be sent in a shorter time interval. In other words, the bit rate in the compressed radio frame increases so that less time is required to transmit the same number of bits, creating a time gap that allows no transmission. This method is used when the spreading factor is greater than four. It is applicable to both the uplink and downlink.

In the higher layer signaling method, the upper layers can virtually create a transmission gap by restricting the permitted transport

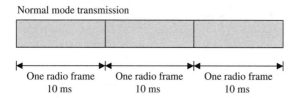

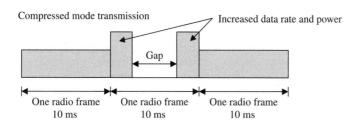

Figure 9.6 Illustration of compressed mode transmission.

format combinations. This method is applicable to both uplink and downlink. It works for packet data because they are burst data in nature. However, it is not desirable for circuit-switched data or voice.

9.6.2 Transmission Gap Pattern Sequence

If a UE uses compressed mode for inter-frequency or inter-RAT measurements, UTRAN must provide transmission gap pattern sequence (TGPS) [3]. A transmission gap pattern sequence contains alternating transmission gap patterns 1 and 2 [6]. Each of these two patterns contains one or two transmission gaps within a transmission gap pattern length (TGPL). A unique transmission gap pattern sequence identifier (TGPSI) is used to identify a transmission gap pattern sequence. Figure 9.7 shows an example of a transmission gap pattern sequence.

The compressed mode parameters associated with each TGPS include transmission gap connection frame number (TGCFN), transmission gap pattern repetition count (TGPRC), uplink/downlink selection, uplink compressed mode method, downlink compressed method, downlink frame type, and scrambling code change. These parameters are decided by UTRAN based on the measurement purpose.

TGCFN is the CFN of the first radio frame in the first transmission gap pattern 1 within the TGPS. TGPRC is the number of transmission gap patterns within a transmission gap pattern sequence. The uplink/downlink selection specifies whether the compressed mode is used in uplink only, downlink only, or both uplink and downlink. The uplink compressed mode method specifies the method for creating uplink compressed mode gaps, while the downlink compressed mode method specifies the method for creating downlink compressed mode gaps. Downlink frame type specifies whether frame structure type A or B (to be discussed in Subsection 9.6.3.3) must be used on downlink compressed mode frames. Scrambling code change indicates whether

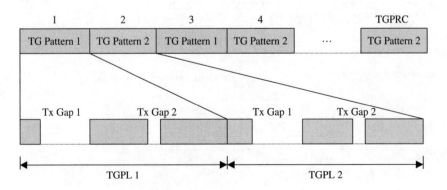

Figure 9.7 Illustration of a transmission gap pattern sequence (courtesy of ETSI).

an alternative scrambling code is used for the downlink compressed mode method in which the spreading factor is reduced by a factor of 2.

9.6.3 Transmission Gap Pattern

A transmission gap pattern sequence contains alternating transmission patterns 1 and 2. Figure 9.8 shows the details of a pair of transmission gap patterns, TGP 1 and TGP 2. The parameters shown in the figure define a transmission gap pattern.

Transmission gap starting slot number (TGSN) is the slot number of the first transmission gap slot within the first radio frame of the transmission gap pattern. Transmission gap length 1 (TGL1) is the length of the first transmission gap within the transmission gap pattern, expressed in number of slots. Similarly, transmission gap length 2 (TGL2) is the length of the second transmission gap within the transmission gap pattern, expressed in number of slots. If higher layers did not set a value for TGL2, then TGL2 is equal to TGL1. Transmission gap start distance (TGD) is the time difference between the starting slots of two consecutive transmission gaps within a transmission gap pattern, expressed in number of slots. Again, if higher layers did not set a value for this parameter, there is only one transmission gap in the transmission gap pattern. Transmission gap pattern length 1 (TGPL1) is the length of transmission gap pattern 1, expressed in number of radio frames. Similarly, transmission gap pattern length 2 (TGPL2) is the length of transmission gap pattern 2, expressed in number of radio frames. If higher layers did not set a value for this parameter, then TGPL2 is equal to TGPL1.

9.6.3.1 Transmission Gap
The length of a transmission gap cannot be greater than 14 time slots. That is, the maximum length of a transmission

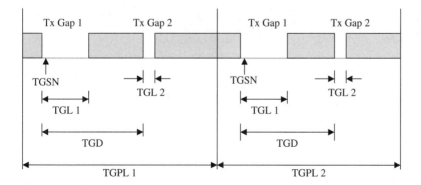

Figure 9.8 Illustration of compressed mode pattern parameters (courtesy of ETSI).

gap is 9.33 milliseconds. In addition, the maximum number of possible transmission gap slots per radio frame is 7. Therefore, if a transmission gap is longer than 7 slots, it must span two consecutive radio frames. Under the above conditions, a transmission gap can be created at any location depending on the measurement purpose.

The transmission gap locations are calculated as follows: Let N_{first} specify the starting slot of the consecutive idle slots with $N_{first} = 0,1,2,\ldots,14$. Also, let N_{last} specify the final idle slot of the gap. If $N_{first} + $ TGL ≤ 15, then $N_{last} = N_{first} + $ TGL -1. This is called a single-frame transmission gap. If $N_{first} + $ TGL > 15, then $N_{last} = (N_{first} + $ TGL $- 1) \bmod 15$ (in the next radio frame). This is called a double-frame transmission gap. For double-frame transmission gap, N_{first} and TGL must be chosen in such a way that at least 8 slots in each radio frame are not idle. Figure 9.9 shows an example of single-frame transmission gaps, while Figure 9.10 shows an example of double-frame transmission gaps.

9.6.3.2 Simultaneous Transmission Gap Pattern Sequence There are cases that simultaneous transmission gap pattern sequences exist because each measurement type requires one transmission gap pattern sequence. In these cases, UTRAN must ensure that compressed mode patterns do not overlap or fall on the same frame. If the compressed mode transmission gaps exceed the maximum gap length, the UE rejects the message as an invalid configuration. If overlapping transmission gap patterns are detected, the UE selects the pattern with the lowest transmission gap pattern sequence identifier (TGPSI) and drops the measurements associated with the other pattern(s).

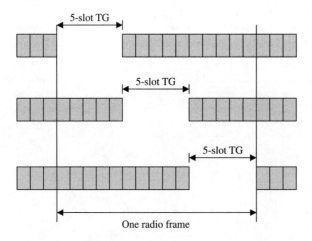

Figure 9.9 Single-frame transmission gaps.

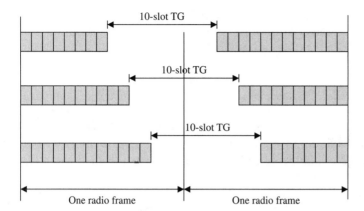

Figure 9.10 Double-frame transmission gaps.

9.6.3.3 Downlink Compressed Mode Frame Structure As mentioned in Subsection 9.6.2, the downlink frame type parameter specifies whether frame structure type A or B must be used on downlink compressed mode frames. The higher layers should set this parameter. If type A frame structure is used, the pilot field of the last slot in the transmission gap is transmitted. If type B frame structure is used, the TPC field of the first slot and the pilot field of the last slot in the transmission gap are transmitted. This structure improves power control performance because it includes some TPC bits during the transmission gap. Without TPC bits transmitted in the transmission gap, power control may not work properly in compressed mode because of lacking TPC bits.

9.6.3.4 Alternative Downlink Scrambling Codes The downlink DPCH channel is normally transmitted with either a primary scrambling code or a secondary scrambling code from the scrambling code group associated with the primary scrambling code of the cell. If the spreading factor is reduced by a factor of 2 for compressed mode operation, the SF/2 OVSF code could be in use by another user. In this case, an alternative scrambling code can be used to scramble the DPCH channel.

Each primary and secondary scrambling code has a left and a right alternative scrambling code. The left alternative scrambling code corresponding to a scrambling code c is $c + 8192$, while the corresponding right alternative scrambling code is $c + 16,384$ [7].

Higher layers will signal the use of alternative scrambling for each physical channel individually. Assume $C_{ch,SF,k}$ is the OVSF code used for non-compressed frames, then $C_{ch,SF/2,k \bmod(SF/2)}$ is used for the alternative scrambling code. Left alternative scrambling code is used when $k < SF/2$, while right alternative scrambling code is used when $k \geq SF/2$.

9.7 Summary

Handovers include soft handover, softer handover, hard handover, and inter-radio access technology handover. Soft handovers establish new connections before breaking the old ones. During a soft handover, the UE maintains radio link connections with two or more Node Bs simultaneously. The RNC selects the best frame received from these radio links. During a softer handover, the UE maintains radio link connections with the cells belonging to the same Node B. It is an intra Node B handover. The Node B combines radio signals received from these radio links, demodulates the combined signal and passes the information to the RNC.

Hard handovers breaks the old connection before establishing the new one. So, there is a brief interruption in communication during a hard handover. A hard handover occurs when the radio link needs to move to another carrier frequency or another RAT. It also occurs during a certain physical channel reconfiguration, such as a timing change. In addition, if the Iur interface is not supported, a hard handover is required when a radio link needs to transition from a Node B to another one belonging to a different RNC.

UE measurements and reporting are a part of the handover procedure. UTRAN controls UE measurements and reporting either by using the measurement information broadcast on SIB Type 11, or by transmitting a Measurement Control message. For handover purposes, UE measurements take place when the UE is at call setup or in the Cell_DCH state. When the UE is at call setup, it uses the relevant measurement and reporting parameters broadcast by UTRAN on SIB Type 11. When the UE is in the Cell_DCH state, UTRAN may request the UE to set up, modify or release a measurement by sending a Measurement Control message to the UE. The message instructs the UE regarding the types of measurements to perform and the cells on which the measurements should be performed. The Measurement Control message also specifies how and when the UE should send a Measurement Report message to UTRAN. After receiving the UE measurement report, UTRAN will add and/or remove the radio link(s) for the UE.

Cells that the UE keeps track of are grouped into three mutually exclusive categories: active set, monitored set, and detected set. All of the cells in the active set are in soft or softer handover with the UE. The signals from these cells are simultaneously demodulated and coherently combined by the UE. Cells in the monitored set are also measured and reported by the UE if certain reporting criteria are met. A monitored set is a subset of the neighbor list and is updated by UTRAN. A cell that is neither in the active set nor in the monitored set but is detected by the UE becomes a member of the detected set.

The UMTS standard specifies six main reporting events for intra-frequency FDD measurements called events 1A, 1B, 1C, 1D, 1E, and 1F. With the exception of event 1D, all other reporting events are related to UMTS soft/softer handovers. Events 1A and 1E are used to indicate to UTRAN when a new cell should be added to the active set. Events 1B and 1F are used to indicate to UTRAN when a cell should be removed from the active set. Event 1C is used to indicate to UTRAN when a cell in the active set should be replaced with a different cell.

For inter-frequency handover, UTRAN sends an inter-frequency measurement and reporting information to the UE in a Measurement Control message. The UE acts by following the instructions in the Measurement Control message. Instead of using the active set concept, inter-frequency measurements adopt the concept of virtual active set. A virtual active set is a set of the best cells of a non-used frequency. The virtual active set can be updated either autonomously by the UE or by UTRAN.

The inter-frequency handover procedure includes three steps. The first step is that the UE has to detect and report that there is a new frequency that is better than the current frequency. UTRAN may decide to hand over the UE to that new frequency based on cell load and other criteria. The second step is to identify the best cell on that new frequency. That is, the UE has to take some cell-specific measurements on that new frequency and report the measurement results. The last step is that UTRAN instructs the UE to transition to that new frequency.

The inter-frequency reporting events include reporting events 2A through 2F. Reporting event 2A is triggered when the quality estimate of another frequency is better than the current one. Reporting event 2B is triggered when the quality estimate of the currently used frequency is below a certain threshold and the quality estimate of a non-used frequency is above a certain threshold. Reporting event 2C is triggered when the quality estimate of a non-used frequency is above a certain threshold. Reporting event 2D is triggered when the quality estimate of the currently used frequency is below a certain threshold. Reporting event 2E is triggered when the quality estimate of a non-used frequency is below a certain threshold. Reporting event 2F is triggered when the quality estimate of the currently used frequency is above a certain threshold.

In an inter-RAT handover, a connection between the UE and UTRAN is transferred to another RAT, such as GSM. For inter-RAT measurements, UTRAN sends the UE an information element in a Measurement Control message containing information about the inter-RAT measurements. The inter-RAT handover triggering events include events 3A through 3D.

Event 3A is triggered when estimated quality of the currently used UTRAN frequency is below a certain threshold and the estimated quality of the other system's frequency is above a certain threshold. Event 3B is triggered when the estimated quality of the other system's frequency is below a certain threshold. Event 3C is triggered when the estimated quality of the other system's frequency is above a certain threshold. Event 3D is triggered when change of best cell in the other system occurs.

Compressed mode may be used in inter-frequency and inter-RAT measurements. When instructed by UTRAN to perform inter-frequency or inter-RAT measurements, the UE needs a second receiver to perform the measurements. However, with compressed mode, the UE is able to perform measurements on non-serving UMTS frequencies or on other RAT frequencies without the need of dual receivers. For example, for inter-RAT measurements, UE may perform a GSM carrier RSSI measurement, initial BSIC identification, and BSIC reconfirmation. Moreover, the UE will not miss any of the data being sent on the dedicated channels by the serving UMTS cells when it is performing measurements on another frequency.

Compressed mode means a transmission mode in which the information is compressed in the time domain to create gaps in the transmission. The methods for creating the gaps include puncturing, spreading factor reduction, and higher layer scheduling. If a UE uses compressed mode for inter-frequency or inter-RAT measurements, UTRAN must provide a transmission gap pattern sequence. The length of a transmission gap cannot be longer than 14 time slots. In addition, the maximum number of possible transmission gap slots per radio frame is 7. If a transmission gap is longer than 7 slots, it must span two consecutive radio frames.

References

[1] 3GPP TS 25.922, v5.2.0, "Radio resource management strategies," (Release 5).

[2] 3GPP TS 25.331, v5.7.1, "Radio Resource Control (RRC); protocol specification," (Release 5).

[3] 3GPP TS 25.133, v5.8.0, "Requirements for support of radio resource management (FDD)," (Release 5).

[4] 3GPP TS 25.433, v5.6.0, "UTRAN Iub Interface NBAP Signaling," (Release 5).

[5] 3GPP TS 25.423, v3.14.2, "UTRAN Iur Interface RNSAP Signaling," (Release 99).

[6] 3GPP TS 25.215, v5.5.0, "Physical layer – measurements (FDD)," (Release 5).

[7] 3GPP TS 25.213, v5.3.0, "Spreading and modulation (FDD)," (Release 5).

Chapter

10

Power Control

Power control is needed in robust WCDMA systems to minimize the transmit power of both UE and Node B, and at the same time maintain quality of service. Reducing UE power and Node B power increases the system capacity because WCDMA systems are interference-limited. Poor power control will cause a loss of system capacity. The basic problem in power control is the near-far problem. Signals from the transmitters near the receiver are more easily received than those that are from transmitters that are far away. Power control makes these transmitters transmit at the appropriate power levels, such that the signals received by the receiver are nearly the same. Efficient power control requires fast feedback so that it can adjust the transmit power in time to minimize the interference—and hence avoid system capacity loss. Fast power control in WCDMA runs at 1500 Hz and is called inner loop power control. It applies to both uplink and downlink.

To maintain quality of service, a certain block error rate (BLER), and thus a certain signal to interference ratio (SIR) target is required. A user in a poor radio channel condition needs a higher SIR target than a user in a better radio channel condition. It is power control that must find the right SIR target so that the quality of service required for every radio link is achieved, regardless of the radio channel conditions. The mechanism of finding the right SIR target is called outer loop power control. It also applies to both uplink and downlink. Outer loop power control runs at a much slower rate than inner loop power control. It may run at a maximum of 100 times per second, depending on the transmission time interval (TTI).

Power control applies to both common and dedicated channels. There is open loop power control and closed loop power control. Open loop power control is to set the initial transmit power of the channels. Closed loop power control consists of two parts: outer loop power control and inner loop power control. The outer loop sets the target SIR and delivers it to the inner loop, while the inner loop adjusts the code channel power such that the measured SIR meets the target SIR provided by the outer loop power control. Both open loop and closed loop power controls apply to dedicated channels. However, for common channels, there is no closed loop power control. The power control function for common channels is to set the power levels of these channels.

The outer loop power control autonomously runs at the RNC for the uplink and at the UE for the downlink. The 3GPP standards do not specify how outer loop power control should be done. It is vendor-specific. UE vendors may have their own outer loop power control mechanisms in their UE, while network vendors may have their own outer loop power control mechanisms in their RNC.

The inner loop power control is carried out between UE and Node B. It does not involve RNC. On the downlink, UE measures the downlink power, calculates the SIR, compares it with the target SIR, and sends TPC bits on the uplink DPCCH channel to Node B. Node B reads the received TPC bits and adjusts the DCH channel power accordingly. On the uplink, Node B measures the uplink power, calculates the SIR, compares it with the target SIR, and sends TPC bits in the downlink DPCH channel to UE. UE reads the received TPC bits and adjusts the power of its DCH channels accordingly.

Power controls for the downlink dedicated channels, downlink common channels, uplink dedicated channels, and uplink common channels, and power control in compressed mode are discussed in the following sections.

10.1 Power Control for Downlink Dedicated Channels

Power control for the downlink dedicated channels sets the initial power during call setup and when new legs are added using open loop power control. It also adjusts the dedicated channel power using closed loop power control.

10.1.1 Open Loop Power Control

Open loop power control is used at two occasions: at call setup and when a new radio link is added to existing ones.

At call setup, the RNC creates a new radio link for UE that does not have an existing one. The downlink transmit power is determined by using the measurements reported by the UE, in combination with the RNC parameters. The RRC Connection Request message sent by the UE to the RNC contains the UE measurement results. The RNC extracts the measurement results and then calculates the initial downlink power according to the following formula:

Initial_DL_Tx_Power

$$= \text{Min}\{P_{\text{Max}}, \text{Max}[P_{\text{Min}}, (\text{SIR\_target} - \text{CPICH\_Ec}/\text{Io} + \text{DL\_Power\_Margin})]\}. \quad (10\text{-}1)$$

In (10-1), Initial_DL_Tx_power is expressed in dB relative to the P-CPICH power. P_{Max} and P_{Min} are the maximum and minimum allowed downlink transmit power for the considered radio bearer configuration and are also expressed in dB relative to the P-CPICH power. SIR_target is the signal to interference ratio target for the considered radio bearer configuration expressed in dB. CPICH_Ec/Io is measured by UE and is reported in the RRC Connection Request message. It is also expressed in dB after being converted from the reported value to dB using the mapping defined in [1]. DL_Power_Margin is a margin allowing some link budget variation between the time the measurement is performed and the time the radio link is actually established. It is also expressed in dB.

When the UE is in the Cell_DCH state, at least one radio link has been established. However, when the UE moves into soft and/or softer handover, the result is that more radio links are established. Now the issue is, when the RNC adds a new radio link to the existing ones, what the Node B transmit power should be for the new link. The answer is that the RNC calculates the initial power according to the following formula:

$$\text{Initial\_DL\_Tx\_Power} = P_{\text{Min}} + P_{\text{New\_leg\_offset}}, \quad (10\text{-}2)$$

where P_{Min} is the minimum allowable Node B transmit power for the UE in question, and $P_{\text{New\_leg\_offset}}$ is an offset value.

Whether at call setup or when adding a new radio link, the RNC provides the initial downlink transmit power information to the Node B through NBAP signaling during radio setup or radio link addition procedures [2]. Upon reception of the NBAP message, the Node B applies the initial power level to each downlink DCH of the radio link when transmission starts. In the case of adding a new radio link, if the initial downlink transmission power is not specified in the Radio Link Addition Request message, the Node B uses the current power level for the UE in question.

10.1.2 Closed Loop Power Control

Closed loop power control contains two parts: outer loop power control and inner loop power control. The outer loop power control is responsible for setting the SIR_target, and inner loop power control is responsible for adjusting the transmit power to make sure that the received SIR will match the SIR_target.

10.1.2.1 Downlink Dedicated Channel Outer Loop Power Control The downlink dedicated channel outer loop power control runs autonomously in the UE. The UE is in charge of updating the SIR_target to provide the required BLER performance for meeting the requested quality of service (QoS). Since the minimum TTI of a transport block is 10 ms, the maximum speed of the outer loop power control is 100 Hz or 100 times per second.

The basic principle of the downlink dedicated channel outer loop power control is discussed next.

The RNC sends a target BLER value to the UE through RRC signaling, using the downlink dedicated control channel (DCCH). The DCCH is mapped to a downlink dedicated channel (DCH) and then to a downlink dedicated physical data channel (DPDCH). From this target BLER value the UE derives the initial SIR_target value. The UE continuously monitors (measures) the instant BLER value and changes the SIR_target accordingly, to ensure that the requested quality is achieved for any radio channel scenario. (This is a function of speed and RF channel conditions.) The 3GPP specification does not specify the algorithm for deriving the SIR_target from the BLER or how fast the algorithm should run. UE manufacturers have their own specific algorithms for carrying out this derivation in the UE. For example, in one algorithm the UE may monitor and set the SIR_target for every frame, while in another algorithm the UE may monitor the BLER over five frames and then set the SIR_target. The former runs at a rate of 100 times per second, while the latter runs at rate of 20 times per second. However, 3GPP specification does specify the minimum UE performance in given RF conditions [3].

If the outer loop power control is not working properly (the SIR_target is not updated according to the BLER measured in the downlink DCH), the following abnormal phenomena will occur:

- UE may request higher downlink DCH power than is actually required in good RF conditions.

- The SIR_target may not be increased to maintain the target BLER in poor RF conditions. As such, the downlink BLER rises sharply even if downlink DCH power is available.

- The UE may request higher downlink power than required, regardless of the measured BLER, resulting in higher downlink DCH power consumption and hence lower downlink capacity.

10.1.2.2 Downlink Dedicated Channel Inner Loop Power Control The purpose of inner loop power control is to change the Node B transmit power in order to achieve a target SIR of the DCH channel. Since the control loop can run as high as 1500 times per second, it involves only the UE and the Node B. The inner loop power control procedure includes four steps:

1. First the Node B transmits user data in the DCH to the UE.
2. Then the UE performs the SIR estimation and compares the estimation SIR_{est} with the SIR_target.
3. Next, depending on the results of the comparison, the UE sends the transmit power control commands to the Node B, using a dedicated uplink control channel (TPC in DPCCH).
4. Based on the information in the TPC command, the Node B decides to either increase or decrease the transmit power. However, the Node B transmit power must remain within the power range between P_{Min} and P_{Max}.

10.1.2.2.1 SIR Estimation and TPC Command Generation The UE performs SIR estimation and obtains the SIR_{est} value for each radio frame slot. The SIR_{est} is the ratio of the pilot bits, power in the downlink DPCCH to noise. 3GPP specification does not specify how the SIR_{est} is measured. It is UE-vendor-specific. However, some performance requirements are defined in [1].

Based on the SIR_{est}, the UE generates the TPC commands according to the following rule:

If $SIR_{est} \geq$ SIR_target, the TPC command is 0, requesting the Node B to decrease the transmit power. If $SIR_{est} <$ SIR_target, the TPC command is 1, requesting the Node B to increase the transmit power.

A TPC command is represented by a TPC bit pattern. How soon the Node B should apply the command to the downlink channel is governed by the downlink inner loop timing; and how fast the power control should run is determined by the power control mode used. Details on TPC bit pattern, downlink inner loop timing, and downlink inner loop power control modes are discussed below.

- **TPC bit patterns**—As specified in [4], depending on the slot format, the TPC field in the uplink DPCCH channel may contain one bit or two bits. The slot format is also defined in [4]. There are six types of slot formats ranging from slot format 0 to slot format 5. From slot format 0 up to slot format 4, the TPC field in the uplink DPCCH channel contains two bits. For slot format 5, the TPC field contains only one bit. If the TPC field in the uplink DPCCH channel has only one bit, the TPC command 1 is represented by the TPC bit pattern 1, while the TPC command 0 is represented by the TPC bit pattern 0. If the TPC field in the uplink DPCCH channel has two bits, the TPC command 1 is represented by the TPC bit pattern 11, while the TPC command 0 is represented by the TPC bit pattern 00. The slot format information can be read from the uplink DPCCH slot format IE in the Radio Link Setup Request and Radio Link Setup Reconfiguration NBAP messages. Normally, only slot format 0 is used and therefore the TPC bit pattern is always either 11 or 00.

- **Downlink inner loop timing**—When a Node B receives the TPC commands from the UE, it must apply the command to the downlink channel. Although the 3GPP specification does not specify the timing for the application to the downlink channel, indicative information is provided in [5]. Basically, the Node B manufacturer can decide the timing for updating the DPCH output power. Normally, it is recommended that the downlink inner loop power control should update the downlink DPCH power in the Node B at the start of the next available downlink pilot field. Depending on the propagation delay between the Node B and the UE, the pilot field where the update starts may belong to a slot that is a few slots later than the slot that the SIR_{est} calculated at the UE.

- **Downlink inner loop power control modes**—According to [5], two downlink inner loop power control modes are available: the single TPC (DPC_Mode = 0) and the TPC triplet (DPC_Mode = 1). In single TPC, the UE sends a unique TPC command in each slot and the TPC command is transmitted in the first available TPC field in the uplink DPCCH. In this mode, the power control speed is 1500 times per second. In TPC triplet, the UE repeats the same TPC command over 3 slots and there are five triplets of TPC commands in a radio frame. The TPC commands are transmitted in such a way that the beginning of a triplet is aligned with the frame boundary. In other words, there is a new TPC command at the beginning of the frame. In this mode, the power control speed is reduced to 500 times per second. The power control mode is configured in the UE through Radio Bearer Reconfiguration and Radio Bearer Setup messages. For more detailed information, the readers may refer to [6].

10.1.2.2.2 Downlink Limited Transmit Power Increase When the RLC mode is set to acknowledge mode (AM), it is possible to achieve a capacity gain by limiting the number of inner loop power control adjustments and hence the peak power. The purpose of this limited power increase is to have a more efficient use of the inner loop DL power control for non-real-time data. Limiting the transmit power increase may result in errors, which are recovered by RLC retransmissions.

The UTRAN Iub interface NBAP signaling determines whether to use the limited power increase feature [2]. If the received DL DPCH information IE includes the limited power increase IE that is set to *used*, the Node B will activate the limited power increase feature for the inner loop downlink power control in the new configuration. In case of softer handover, all the softer legs associated with the UE are activated.

If the received DL DPCH information IE includes the limited power increase IE that is set to *not used*, the Node B will not use limited power increase for the inner loop downlink power control in the new configuration. Instead, the Node B will deactivate, for the related radio link(s), the limited power increase feature, if it is active. In case of softer handover, all the softer legs associated with the UE are deactivated.

The downlink limited power increase feature can be implemented on a per cell basis. The parameters that regulate the downlink transmit power increase limits are Power_Raise_Limit and DL_power_averaging_window_size. Power_Raise_Limit ranges from 0 to 10 dB with a step size of 1 dB. DL_power_averaging_window_size represents the inner loop power adjustments and ranges from 1 to 60 with a step of 1.

These two parameters are contained in the Cell Setup Request message. Node B stores their values during the cell setup procedure. They are used each time a request for the activation of the downlink limited power increase feature on one of the radio links supported by the cell is received.

10.1.2.2.3 Downlink Inner Loop Power Control Algorithm The downlink inner loop power control algorithm is located in Node B and is used to update the transmitter power according to the TPC command received by the Node B. After reading the kth TPC command, the Node B adjusts the current downlink power $P(k-1)$ to a new power $P(k)$ according to the following formula [5]:

$$P(k) = P(k-1) + P_{TPC}(k) + P_{bal}(k), \qquad (10\text{-}3)$$

where $P_{TPC}(k)$ is the kth power adjustment due to the inner loop power control, and $P_{bal}(k)$ is a correction in dB, according to the downlink power control procedure for balancing radio link powers towards a common reference power in soft/softer handover scenarios.

$P_{TPC}(k)$ is set to power adjustment step $-\Delta_{TPC}$ if the kth TPC command, $TPC_{est}(k)$, received by the Node B is 0, regardless of whether the limited power increase feature is used or not. When the limited power increase feature is not used, $P_{TPC}(k)$ is set to $+\Delta_{TPC}$ if $TPC_{est}(k)$ received by the Node B is 1. When the limited power increase feature is used and the $TPC_{est}(k)$ received by the Node B is 1, $P_{TPC}(k)$ is set to +Δ_{TPC} if cumulative power adjustment in dB during the last DL_power_averaging_window_size inner loop power adjustments is less than Power_Raise_Limit; otherwise, $P_{TPC}(k)$ is set to 0.

10.1.2.2.4 Downlink Inner Loop Transmit Power Step Size The RNC sets the power step size through an NBAP primitive. The value contained in the TPC DL step size IE is found only in the Radio Link Setup Request primitive. After receiving a TPC downlink step size IE, the Node B will store and apply the new step size value when the message becomes effective.

If the TPC downlink step size IE is not found during creation of a radio link, the Node B will use the default value.

The 3GPP specification [5] requires that the UTRAN must support TPC command adjustment steps Δ_{TPC} of 1dB, while support of 0.5 dB is optional. The Node B transmitter must have the capability of setting the inner loop code domain power with a step size of 1dB mandatory and 0.5 dB optional.

The step size accuracy for the inner loop power control must comply with the 3GPP requirements as given in [7]. The power control step due to inner loop power control must be within the range shown in Table 10.1, and the aggregated output power change due to inner loop power control must be within the range shown in Table 10.2.

10.1.2.2.5 Downlink Power Dynamic Range The downlink power dynamic range is defined as the difference between the maximum and minimum transmit power of a downlink channelization code. The downlink power control dynamic range requirements for Node B are specified in [7]. The dynamic range must be at least 18 dB. As such, the minimum transmit power should be less than 25 dBm for a Node B with a maximum output power of 43 dBm, and 28 dBm for a Node B with a maximum output power of 46 dBm.

To limit the downlink interference generated on one sector by one call, the RNC defines a maximum transmit power level per radio link, P_{Max}. Similarly, to maintain the call quality, a minimum transmit power level P_{Min} is also defined. In other words, a minimum transmit power level is required so that the call can still be maintained in an area where coverage variations are large.

TABLE 10.1 Transmitter Power Control Step Tolerance (Courtesy of ETSI)

Power Control Commands in the Down Link	Transmitter Power Control Step Tolerance			
	1 dB Step Size		0.5 dB Step Size	
	Lower	Upper	Lower	Upper
Up (TPC Command 1)	+0.5 dB	+1.5 dB	+0.25 dB	+0.75 dB
Down (TPC Command 0)	−0.5 dB	−1.5 dB	−0.25 dB	−0.75 dB

TABLE 10.2 Transmitter Aggregated Power Control Step Range (Courtesy of ETSI)

Power Control Commands in the Down Link	Transmitter Aggregated Output Power Change After 10 Consecutive Equal Commands (Up or Down)			
	1 dB Step Size		0.5 dB Step Size	
	Lower	Upper	Lower	Upper
Up (TPC Command 1)	+8 dB	+12 dB	+4 dB	+6 dB
Down (TPC Command 0)	−8 dB	−12 dB	−4 dB	−6 dB

When Node B receives these power limits through a Radio Link Setup Request or Radio Link Addition Request message, it will not transmit the power less than P_{Min} or more than P_{Max}. Please note that this applies to the average power of the data part only, that is, the DPDCH symbols, which are complex QPSK symbols before spreading. For more details, the readers may refer to [2].

In the case where two or more sectors are in softer handover, the allowed dynamic range [P_{Min}, P_{Max}] for each sector is defined independently from one softer sector to the other.

To avoid downlink imbalance within a set of softer handover radio links, the RNC requires that the initial downlink power level of any added radio link be the same as that of the existing links.

Upon creation of a new radio link, the RNC must populate the maximum downlink power IE and the minimum downlink power IE in the Radio Link Setup Request message in NBAP signaling. If no maximum downlink power IE is included in the Radio Link Setup Request message, the maximum downlink power value stored for the existing radio links for this UE must be used. The same procedure is used to determine the value of the minimum downlink power if it is not included in the Radio Link Setup Request message.

10.1.2.2.6 Downlink Power Imbalance   During soft handover, the UE sends the same UL signal to all of the Node Bs involved. The individual signals to the Node Bs may experience different radio conditions. The TPC command detection error rate becomes high on the weakest link of a connection. If each active-set Node B keeps transmitting at power levels calculated from its TPC bits, then due to different TPC error rates the transmit powers of Node Bs may drift apart. For example, when a TPC error occurs, the TPC command interpreted as a *power up* on a leg may be interpreted as a *power down* on other legs. This results in downlink DCH power levels diverging by twice the power adjustment step size. Transmit power drift may lead to a random walk effect for the transmit power of the link, and the downlink transmission power may reach a much higher level than required. This effect is known as power imbalance.

Power imbalance can also happen when one of the transmit legs has a maximum transmit power value lower than that of other legs. In such cases, the power up TPC commands may be ignored on this leg while the TPC down commands would result in power decrease. Power imbalance, if resulting in an excessive transmitted power in a link, may pollute other radio links in the vicinity, leading to a lower radio network capacity. The power imbalance can be corrected by power balance procedures, which are Node B manufacturer specific.

In general, from the RNC perspective, a power imbalance correction procedure for a given UE can be activated or deactivated at the serving RNC by setting the imbalance correction parameter in the RNC MIB either to *active* or *inactive*. There is one imbalance correction MIB parameter per downlink radio access bearer combination. The activation of the imbalance correction for one particular UE can occur either at call setup or upon modification of the number of radio bearer services multiplexed onto the physical channel.

If power balancing is active on a radio link, the power balancing adjustments superimposing on the inner loop power control adjustment as shown in (10-3), is given as follows [2]:

$$P_{\text{bal}} = (1-r)(P_{\text{ref}} + P_{\text{P-CPICH}} - P_{\text{init}}), \qquad (10\text{-}4)$$

where P_{ref} is the value of the DL reference power IE, $P_{\text{P-CPICH}}$ is the power used on the primary CPICH, P_{init} is the code power of the last slot of the previous adjustment period, and r is the scale adjustment ratio given by the scale adjustment ratio IE.

10.1.2.2.7 Downlink Power Offset Between Downlink DPCCH and DPDCH
The downlink transmit power control procedure simultaneously controls the power of a DPCCH and its associated DPDCHs. The power of the different channels is adjusted in such a way that the relative power difference between the DPCCH and DPDCH is not changed. The S-RNC sets the relative transmit power offset between the DPCCH fields and DPDCHs by reading the values from the OA&M managed tables. As shown in Figure 10.1, the TFCI, TPC, and pilot fields of the DPCCH are offset relative to the DPDCH power by PO1, PO2, and PO3 dB, respectively.

Note that PO1 and PO3 can be signaled to the Node B only at Radio Link Setup [2]. In other words, they cannot be modified once the radio link is set up. PO2 can be modified only through DCH Framing Protocol (FP) signaling as defined in [8]. Based on the BLER_target sent by the RNC for the downlink DCH, the UE can deduce, via the outer loop procedure, a SIR_target for the DPDCH, but not for the pilot. To calculate the SIR_target for the pilot, the UE must add up the power offset PO3 between pilot field and data field. This is achieved via RRC signaling on the FACH as described in detail in [6].

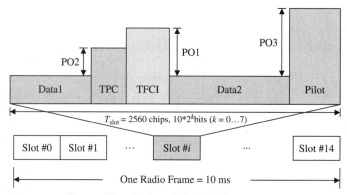

Figure 10.1 Power offsets in downlink DPCH.

Use of a larger PO1 value may help improve the BLER performance by reducing the TFCI decoding error probability. Since only one TFCI per radio frame has to be decoded, the probability of TFCI decoding failure is relatively small.

Increasing PO3 may help the UE achieve synchronization and consequently help its receiver perform better at a given SIR set point, resulting in downlink capacity gain.

Radio link quality may be improved when PO2 value increases with the number of radio link sets connected to the UE. Therefore, different PO2 values should be defined for different active set sizes and these PO2 values should be used at call setup and each time a new service is added or an old service is removed on the downlink. This is done through DCH FP signaling using the Radio Interface Parameter Update message as described in [8].

10.1.2.2.8 Deactivation of the Downlink Inner Loop Power Control By using the inner loop power control IE in the NBAP/RNSAP downlink Power Control Request message, it is possible to deactivate the downlink inner loop power control at the Node B for a group of radio links belonging to the same Node B communication context. However, it cannot deactivate the inner loop power control on one radio link and keep the other radio links of the same Node B–UE context active. If the inner loop power control is deactivated, the new power $P(k)$ can still be calculated by using (10-3) with the component $P_{TPC}(k)$ being excluded.

10.2 Power Control for Downlink Common Channels

Power control for downlink common channels basically deals with power settings of downlink common physical channels. Downlink common control physical channels include: common pilot channel (CPICH) with subchannels called primary CPICH (P-CPICH) and secondary CPICH (S-CPICH), primary common control physical channel (PCCPCH), secondary common control physical channel (SCCPCH), synchronization channel (SCH) with two subchannels called primary SCH (P-SCH) and secondary SCH (S-SCH), paging indicator channel (PICH), and acquisition indicator channel (AICH).

The transmit power of all downlink common channels are controlled by the RNC. There is no control loop for any of these channels. As such, the transmit power of these channels stays constant until the relevant RNC parameters are modified and a different transmit power is required.

The P-CPICH power level is expressed in terms of dBm. Power levels of the other downlink common physical channels are defined in dB relative to the P-CPICH power level. Therefore, when the P-CPICH power level is changed, the power level of other downlink common physical channels is also changed accordingly.

The configuration and reconfiguration of the transmit power of these downlink common physical channels is carried out by the RNC, which sends a Cell Setup Request NBAP message to the Node Bs to set the power of the P-SCH, S-SCH, S-CPICH, and PCCPCH channels. For the PICH and AICH, the transmit power level is set using a Common Transport Channel Setup Request NBAP message.

The following subsections will cover the power settings of individual downlink common physical channels.

10.2.1 Common Pilot Channel and Synchronization Channel Power Levels

The P-CPICH is transmitted continuously. Its transmit power level can be defined either as a fraction (percentage) of the total output power of the Node B transmit amplifier, or as an absolute power in dBm. The P-CPICH power setting has a wide impact on the RF coverage and on basic procedures such as soft/softer handover. On one hand, the P-CPICH power level should be set to a value that is high enough to ensure proper RF coverage. On the other hand, it cannot be set so high that it impacts the downlink capacity. Normally the P-CPICH power level is about 10% of the total Node B transmit power. For example, if the output power of a Node B is 43 dBm (20 watts), the P-CPICH power level is 10% of 43 dBm (or 33 dBm).

Both the P-SCH and the S-SCH are time-multiplexed with the PCCPCH and have a duty cycle of 10% of a slot, that is, 1/15 millisecond. Because the power setting specified refers to power on the chip level, it must be multiplied with the duty cycle to get the real overhead powers for these channels. The P-SCH and S-SCH power levels must be set to values that will ensure a high success rate along the different steps of the synchronization procedure and yet preserve system capacity. In general, the power levels for P-SCH and S-SCH are respectively set at −3 dB and −5 dB relative to the P-CPICH power level. For example, if the output power of the Node B is 43 dBm, the real overhead powers for P-SCH and S-SCH are 20 dBm (100 mW) and 18 dBm (63 mW), respectively.

10.2.2 Primary Common Control Physical Channel Power Level

The PCCPCH is a downlink physical channel used to carry the BCH transport channel.

It is not transmitted during the first 256 chips of each slot. As such, the duty cycle of the PCCPCH is equal to 90%. The power level of PCCPCH is also defined with respect to the P-CPICH. The PCCPCH power level should be set to a value that will ensure a high success rate in decoding SIB and MIB messages at the UE, while preserving system capacity in terms of downlink power available for dedicated channels. Normally, PCCPCH is also set at −3 dB relative to the P-CPICH. Taking 90% duty cycle into account, the real overhead power for the PCCPCH is 0.9 watt for a Node B with an output power of 20 watts.

10.2.3 Secondary Common Control Physical Channel Power Level

An SCCPCH carries FACH and PCH, which are both time-multiplexed. The SCCPCH can support three different fields in its slot format: TFCI, data, and pilot; and, there are 18 different slot formats for SCCPCH. UTRAN determines if a TFCI should be transmitted. As such, it is mandatory for the UE to support the use of TFCI. Also, the use of the pilot field is optional. When the FACH is active, DTX is enabled on the inactive PCH channel and vice versa. However, due to time-multiplexing of FACH and PCH in an SCCPCH, no power saving is achieved if at least one of the two channels are active because the power on the data part within one time slot must be the same.

If TFCI bits are transmitted, they are continuously transmitted, but the TFCI duty cycle is much lower than that of the data field. As shown in Figure 10.2, the power levels of the TFCI and pilot fields in SCCPCH can be defined as offsets relative to the power used in the data field by PO1 and PO2, respectively. Increasing the PO1 value may help improve the FACH/PCH BLER performance by reducing the TFCI decoding errors, while increasing the PO2 value may help the UE achieve synchronization and consequently perform a better reception of the SCCPCH frame.

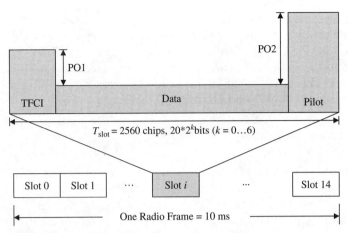

Figure 10.2 Power offsets in SCCPCH.

The power setting of FACH must be set to a value that ensures high decoding success rate of the RRC Connection Setup message sent on FACH. Experience indicates that the FACH and PCH power level should be set at 4 dB relative to the P-CPICH.

10.2.4 Paging Indicator Channel and Acquisition Indicator Channel Power Levels

The paging indicator channel (PICH) is a fixed rate physical channel carrying the paging indicators. As discussed in Chapter 7, one 10-ms radio frame consists of 300 bits. But only the first 288 bits are used to carry paging indicators. The final 12 bits are off (DTX). As such, the duty cycle for PICH is 96%. Normally, the power level of PICH is set at −6 dB relative to the P-CPICH.

The acquisition indicator channel (AICH) is also a fixed rate physical channel carrying acquisition indicators (AI). An AICH frame consists of a repeated sequence of 15 consecutive access slots (AS) with each access slot having a length of 5120 chips. Only the first 4096 chips, which is the real acquisition indicator part, are used to carry AI; the remaining 1024 chips, which are not formally part of the AICH, are off (DTX). The acquisition indicator part consists of 32 real-valued symbols. The AICH is transmitted only when RACH preambles need to be acknowledged. Therefore, the average duty cycle for the AICH is only about 1%. The power level for AICH is normally set at −5 dB relative to the P-CPICH.

10.3 Power Control for Uplink Dedicated Channels

Similar to power control for downlink dedicated channels, power control for uplink dedicated channels also involves open loop power control, which sets the initial transmit power for uplink dedicated channels, and closed loop power control, which incrementally adjusts the transmit power of uplink dedicated channels.

10.3.1 Initial Transmit Power for Uplink Dedicated Channels

An uplink dedicated channel is set up at call setup during radio link establishment. It is set up when the downlink dedicated channel synchronization is accomplished. There are two types of uplink dedicated channels: DPCCH and DPDCH. The DPCCH is always transmitted before the DPDCH. The start of DPDCH is delayed relative to the start of DPCCH by using a power control preamble procedure. The initial transmit power of the DPCCH is determined by the open loop power control mechanism.

10.3.1.1 Uplink Dedicated Channel Open Loop Power Control In open loop power control, UE sets the power level of the first uplink DPCCH as follows:

$$DPCCH\_Initial\_power = DPCCH\_Power\_offset - CPICH\_RSCP, \quad (10\text{-}5)$$

where DPCCH_Power_offset is a parameter in dB and CPICH_RSCP is measured by the UE. Once the new radio link is synchronized, the inner loop power control starts.

10.3.1.2 Uplink DPCCH Power Control Preamble An uplink DPCCH power control preamble (PCP) is a period of uplink DPCCH transmission before the start of the uplink DPDCH transmission. It is used to ensure that the inner loop power control has converged before actual data transmission starts. The downlink DPCCH must also be transmitted during an uplink DPCCH power control preamble.

The length of the uplink DPCCH power control preamble is a UE-specific higher layer parameter signaled by the RNC and can range from 0 to 7 radio frames. If the length of the power control preamble is greater than zero, the power control procedure used during the power control preamble differs from the ordinary power control that is used after the power control preamble.

During the uplink DPCCH power control preamble, the change in uplink DPCCH transmit power is given by [5]

$$\Delta_{DPCCH} = \Delta_{TPC} \times TPC\_cmd, \qquad (10\text{-}6)$$

where the parameter TPC_cmd is derived according to Algorithm 1 to be discussed later in Subsection 10.3.3.2.1. After the uplink DPCCH power control preamble, the ordinary power control with power control algorithm determined by the power control algorithm (PCA) value and step size Δ_{TPC} is used. The DPCCH power control preamble mode in the UE is configured by the RNC through RRC signaling described in [6].

10.3.2 Outer Loop Power Control for Uplink Dedicated Channels

Uplink outer loop power control is responsible for updating the SIR_{target} in the Node B in order to provide the BLER performance that meets the requested quality of service. The power control loop runs in the RNC with a speed up to 100 times per second and may use various parameters to update the SIR_{target}. These parameters include the quality estimate of the data received at the Node B, the number of soft handover legs, the interference or load level in the uplink channel, and the measurement reports from the UE. The new SIR_{target} is then sent to Node B over the Iub/Iur links. It is used by the inner loop power control as a reference against the SIR_{est}.

In the case of inter-RNC soft handover, the serving RNC performs the uplink outer loop power control and determines the SIR_{target} for all the Node Bs involved in the connection.

The 3GPP specification does not specify how to calculate the SIR_{target} value for providing the required BLER. It is up to network equipment manufacturers to define their own outer loop power control mechanism. In general, the outer loop power control involves the following tasks:

- Each Node B involved in the call performs an estimate of the radio link quality on the uplink DCH and sends the results to the serving RNC using the DCH Framing Protocol (uplink direction).

- The frame selector in the serving RNC provides unique quality estimate values to the outer loop power control algorithm, based on the radio link quality estimate sent from Node Bs.

- The outer loop power control algorithm adjusts the SIR_{target} by using the quality estimate values and other parameters as well.

- The serving RNC sends the new SIR_{target} value to the Node Bs involved using the DCH Framing Protocol (downlink direction).

Upon reception of the new SIR_{target} value, the Node B applies it to the relevant uplink inner loop power control function, completing the outer loop power control function.

10.3.2.1 Radio Link Quality Estimate The radio link quality estimate for the uplink is carried out at the Node B. A variety of metrics described in the 3GPP specification [9] can be used to indicate the quality of the radio link. The most frequently used metrics are BLER and BER.

BLER is defined as the average transport block error rate estimated on a transport channel containing a CRC check in each transport block. The Node Bs provide the results of the CRC check to the serving RNC through the Iub interface (or Iub plus Iur for the case of soft handover), using the cyclic redundancy check indicator (CRCI) field of the DCH frames. A CRCI bit is present for every transport block included in the data frame. The CRCI bit is set to 1 if CRC check failed and set to 0 if CRC check passed or there is no CRC present. The serving RNC estimates the BLER by averaging the CRCI after frame selection.

For transport channels with high quality requirements, BLER may not be the best metric for evaluation of the channel quality. In this case, the transport-channel bit error rate (BER) and physical-channel BER are better metrics to be used. These metrics are calculated at the Node B and then forwarded to the serving RNC.

The transport channel BER is defined as the bit error rate of the DPDCH before decoding. Since it is the bit error rate before decoding, it is actually the symbol error rate (SER), though the term *transport channel* BER is used in the 3GPP specification. The SER for a given dedicated channel is obtained according to the following steps:

1. Decode and re-encode the received symbols.
2. Compare the re-encoded symbols with the received symbols.
3. Count the number of symbols that do not match in step 2.
4. Derive the transport channel BER based on the number of mismatches.

Similarly, the physical channel BER is defined as the bit error rate on the DPCCH before decoding. Therefore, it is actually the symbol rate, too. The SER on the DPCCH can be measured by decoding and re-encoding of the TPC and TFCI information. For detailed information on transport channel and physical channel BERs, the readers may refer to [1] and [9].

10.3.2.2 Frame Selector Function Frame selection happens when UE is in handover status. Each Node B involved in the call performs CRC checks on each transport block. The resulting CRCI and quality estimate (QE) obtained by each Node B are sent to the serving RNC over the DCH framing protocol. The frame selector in the serving RNC uses the CRCI and QE values to estimate a CRCI and QE and forwards them to the uplink outer loop power control algorithm.

10.3.2.3 Uplink Outer Loop Power Control Algorithm The uplink outer loop power control algorithm uses the CRCI and QE values to determine whether it is necessary to update the SIR_{target}. In the serving RNC, the uplink outer loop function is made up of one or more physical layer entities with each entity being related to one transport bearer, which is a DCH framing protocol pipe. In the case where more than one DCH is related to one transport bearer, one single physical layer entity is in charge of that group of DCHs. Each physical layer entity can request an update of the uplink SIR_{target}. As a result, an uplink outer loop power control command is transmitted to each Node B involved in the call to update the uplink SIR_{target}.

10.3.3 Inner Loop Power Control for Uplink Dedicated Channels

The uplink inner loop power control changes the UE transmit power in order to achieve the target SIR of the uplink DCH channel. Since the control loop can run as high as 1500 times per second, it involves only the UE and the Node B. The inner loop power control procedure includes four steps:

1. The UE transmits user data on uplink DCH to Node B.
2. The Node B performs the SIR estimation and compares the SIR_{est} with the SIR_{target}.
3. Depending on the results of the comparison, the Node B sends the transmit power control commands to the UE using a dedicated downlink control channel (TPC in DPCCH).
4. Based on the information in the TPC command, the UE decides to either increase or decrease the transmit power. However, the UE transmit power must remain within the power range between P_{Min} and P_{Max}.

10.3.3.1 SIR Estimation and TPC Command Generation The Node B performs SIR estimation and obtains the SIR_{est} value for each radio frame slot. The SIR_{est} is the ratio of the pilot bits, power in the uplink

DPCCH to noise. 3GPP specification does not specify how the SIR_{est} is measured. It is Node B vendor-specific.

Base on the SIR_{est}, the Node B generates the TPC commands according to the following rule:

If $SIR_{est} \geq SIR_{target}$, the TPC command = 0, requesting the UE to decrease transmit power. If $SIR_{est} < SIR_{target}$, the TPC command = 1, requesting the UE to increase transmit power.

10.3.3.1.1 TPC Bit Patterns The TPC command is transmitted over the air in the TPC field of the downlink DPCCH using one of the three TPC bit patterns given in Table 10.3. Depending on the slot format, different TPC bit patterns are used. The relation between the slot format and N_{TPC} values can be found in [4].

10.3.3.1.2 Uplink Inner Loop Timing When the UE receives the TPC command from Node B, it updates its uplink DPCH output power at the beginning of the uplink pilot field of the first available slot. Depending on the propagation delay between UE and Node B, the pilot field where the update starts may belong to a slot that is a few slots later than the slot that the SIR_{est} was calculated at the Node B.

10.3.3.2 Uplink Inner Loop Power Control Algorithms The uplink inner loop power control adjusts the UE transmit power such that the received uplink SIR meets the SIR target, SIR_{target}.

After receiving one or more TPC commands from the Node Bs in a slot, the UE derives a single TPC command, TPC_cmd, for that slot using Algorithm 1 or Algorithm 2 [5]. Which algorithm to use is determined by the UE-specific higher-layer parameter, PowerControlAlgorithm, which is controlled by UTRAN. If PowerControlAlgorithm indicates algorithm1, then the physical layer parameter, power control algorithm (PCA), takes the value 1 and Algorithm 1 is used for deriving the TPC command. If PowerControlAlgorithm indicates algorithm2, then PCA takes the value 2 and Algorithm 2 is used instead.

10.3.3.2.1 Algorithm 1 (Courtesy of ETSI) Algorithm 1 is the fundamental inner loop algorithm running at a rate of 1500 times per second. It works differently depending on whether the UE is in soft handover or not.

TABLE 10.3 TPC Bit Patterns (Courtesy of ETSI)

TPC Bit Patterns			Transmitter Power Control Command
$N_{TPC} = 2$	$N_{TPC} = 4$	$N_{TPC} = 8$	
11	1111	11111111	1
00	0000	00000000	0

When a UE is not in soft or softer handover, it receives only one TPC command in each slot, either 0 or 1. In this case, the UE derives the TPC_cmd for that slot as follows:

- If the received TPC command is equal to 0, then TPC_cmd is −1.
- If the received TPC command is equal to 1, then TPC_cmd is 1.
- TPC_cmd equal to −1 means power decrease, while TPC_cmd equal to 1 means power increase.

When the UE is in softer handover, the TPC commands received from different links in a slot are the same because the TPC commands are sent from the same Node B. The UE is informed via RRC signaling and combines these TPC commands into one single TPC command, enhancing the detection of the correct TPC word.

When the UE is in soft handover, the process is a little bit complicated. First, the UE must conduct a soft symbol decision W_i on each of the power control commands TPC_i, where $i = 1, 2, ..., N$, and N is the number of TPC commands from different Node Bs. Then, the UE derives a combined TPC command, TPC_cmd, as a function γ of the soft symbol decisions W_i [5]:

$$\text{TPC\_cmd} = \gamma(W_1, W_2, ..., W_N) \qquad (10\text{-}7)$$

TPC_cmd can take the value 1 or −1. According to [5], the function γ given above must fulfill the following criteria:

- If the TPC commands are random and uncorrelated, and each of them has equal probability of being transmitted as 0 or 1, the probability of TPC_cmd being equal to 1 must be greater than or equal to $1/(2^N)$, and the probability of TPC_cmd being equal to −1 must be greater than or equal to 0.5.
- If the TPC commands from all of the Node Bs involved are reliably 1, then TPC_cmd is equal to +1.
- If a TPC command from any of the Node Bs involved is reliably 0, then TPC_cmd is equal to −1.

In soft handover situations, the UE may receive *up* commands on some legs and *down* command on others. In this case, based on the algorithm described above, the UE should decrease transmit power. The UE will increase transmit power only if it receives *up* commands for all legs.

10.3.3.2.2 Algorithm 2 (Courtesy of ETSI) In Algorithm 2, the inner loop speed is decreased to 300 Hz. This algorithm enhances the reliability of the TPC commands and allows turning off of the power control by transmitting an alternating series of TPC commands. However, due to the slower inner loop power control speed, it may affect the tracking of fast fading channels.

Algorithm 2 also works differently depending whether the UE is in soft handover or not as described in [5].

When a UE is not in soft or softer handover, it receives only one TPC command in each slot and processes the received TPC commands on a five slot cycle. The sets of five slots must be aligned to the frame boundaries and no overlap between the sets is allowed. The TPC_cmd value in the 5th slot decides whether the UE transmit power should increase, decrease, or remain the same. The UE derives the TPC_cmd value for each slot according to the following rules:

- For the first 4 slots, TPC_cmd = 0.
- For the 5th slot, the UE uses hard decisions on each of the 5 received TPC commands as follows:
  - If all 5 hard decisions within a set are 1, then TPC_cmd = 1 for the 5th slot, indicating power increase.
  - If all 5 hard decisions are 0, then TPC_cmd = −1 for the 5th slot, indicating power decrease.
  - Otherwise, TPC_cmd=0 for the 5th slot, indicating no power change.

When UE is in softer handover, the same procedure as that used for Algorithm 1 is applied. The UE combines the TPC commands into one single TPC command.

When UE is in soft handover, it follows the general scheme described below for combining the TPC commands from different Node Bs.

The UE makes a hard decision on the value of each TPC_i, where $i = 1, 2, \ldots, N$ and N is the number of TPC commands from different Node Bs. The UE makes this hard decision for 5 consecutive slots, resulting in N hard decisions for each of the 5 slots. Again, the sets of 5 slots must align with the frame boundaries and there is no overlap between the sets.

The TPC_cmd for the first 4 slots is zero. At the end of the 5th slot, the UE determines the value of TPC_cmd for the 5th slot in the following way.

For each of the N sets of 5 TPC commands, the UE first determines one temporary TPC command, TPC_temp_i as follows:

- If all 5 hard decisions in a set are 1, $TPC\_temp_i = 1$.
- If all 5 hard decisions in a set are 0, $TPC\_temp_i = -1$.
- Otherwise, $TPC\_temp_i = 0$.

Then, the UE derives a combined TPC_cmd for the 5th slot, as a function γ of all TPC_temp$_i$. That is,

$$\text{TPC\_cmd (5th slot)} = \gamma(\text{TPC\_temp}_1, \text{TPC\_temp}_2, ..., \text{TPC\_temp}_N),$$
(10-8)

where TPC_cmd (5th slot) takes the values 1, –1, or 0 according to the following rules:

- TPC_cmd = –1 if any of the TPC_temp$_i$ = –1.
- TPC_cmd = 1 if $\dfrac{1}{N}\sum_{i=1}^{N}\text{TCP\_temp}_i > 0.5$.
- TPC_cmd = 0 otherwise.

10.3.3.3 Uplink Inner Loop Power Control Execution After the TPC_cmd is derived, the UE adjusts the transmit power of the uplink DPCH(s) according to the following rules:

- If TPC_cmd is equal to 1, then the transmit power of the uplink DPCCH and DPDCHs is increased by Δ_{TPC} dB.
- If TPC_cmd is equal to –1, then the transmit power of the uplink DPCCH and DPDCHs is decreased by Δ_{TPC} dB.
- If TPC_cmd is equal to 0, then the transmit power remains unchanged.

For Algorithm 1, the step size Δ_{TPC} is a physical layer parameter derived from the UE-specific higher-layer parameter TPC_StepSize, which is controlled by UTRAN. According to 3GPP specification [6], the UE transmitter must support power adjustment steps Δ_{TPC} of 1 dB, 2 dB and 3 dB. For example, if the value of TPC_StepSize is dB1, then Δ_{TPC} takes the value 1 dB and if the value of TPC_StepSize is dB2, then Δ_{TPC} takes the value 2 dB. For Algorithm 2, Δ_{TPC} always takes the value 1 dB.

10.3.3.3.1 Maximum UE Transmit Power Since the performance of UMTS systems is limited by interference, the UE transmit power must not be higher than a maximum value. This value is set by the *maximum allowed uplink Tx power* sent from the serving RNC to the UE through RRC signaling as described in [6]. If there are several radio links involved in a connection, such as in soft or softer handover, the maximum UE transmit power is calculated at the serving RNC as:

$$\text{Max\_UL\_Tx\_Power} = \text{Min}(\text{Max\_UL\_Tx\_Power1}, .., \text{Max\_UL\_Tx\_Power}N),$$
(10-9)

where N is the number of radio links that are involved in the call. This limit is calculated at call setup, at addition of a new link, or at deletion of an existing link. When receiving the relevant RRC message, the UE keeps the uplink transmit power below the power value indicated by the message.

10.3.3.3.2 Minimum UE Transmit Power The minimum UE transmit power is not set by the radio network but specified by the 3GPP specification [3]. The minimum output power is defined as the average power in a time slot measured with a filter that has a root-raised cosine filter response with a roll off $\alpha = 0.22$, and a bandwidth equal to the chip rate. The minimum output power, as specified in [3], must be less than −50 dBm.

10.3.3.4 Uplink DPCCH/DPDCH Power Gains As already discussed in Chapter 7, spreading factor and rate matching depend on the amount of data to be transmitted over a radio frame. As such, the signal energy per symbol in the uplink DPDCH varies with the instant data rate for constant DPDCH power level.

The uplink DPCCH is used by the uplink inner loop as a reference sequence on which the SIR_{target} has to be achieved. If the SIR_{target} remains constant, and if the power ratio between the uplink DPDCH and DPCCH were to be constant, the SIR on the DPDCH would be a function of the instant data rate. This is not acceptable. Therefore, a power gain concept between the uplink DPDCH and DPCCH must be employed to provide a DPDCH QoS that is independent of the instant data rate. The relative power gain between the DPCCH and DPDCH will compensate for the fluctuations of processing gain.

The mechanism of power gain is based on the fact that the uplink DPCCH and DPDCH(s) are transmitted on different codes as illustrated in the Figure 7.28, which is repeated here as Figure 10.3 for easy reference. After spreading with OVSF codes, the signals are weighted by power gain factors G_c and G_d for DPCCH and DPDCH, respectively. At any instant, at least one of the two gain factors has a magnitude of 1.0. The gain factor values are quantized into 16 levels as shown in Table 10.4.

A set of G_c and G_d values are defined per transport format combination (TFC) in a transport format combination set (TFCS). They are transmitted by the RNC to the UE through RRC signaling and to the Node B through NBAP messages. There are two ways of controlling the gain factors of the DPCCH code and the DPDCH codes for different TFCs. The first way is that the G_c and G_d are signaled for the TFC. The signaled values are used directly for weighting of DPCCH and DPDCH(s). The second way is that the G_c and G_d are computed for the

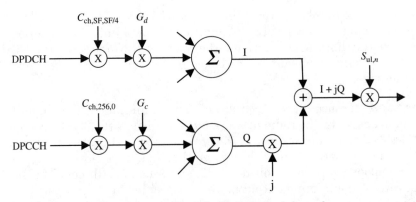

Figure 10.3 Spreading and scrambling of uplink physical channels.

TFC based on the signaled settings for a reference TFC. For a detailed description on the calculation of the gain factors, the readers may refer to [5].

Combinations of the above two methods may be used to associate G_c and G_d values to all TFCs in the TFCS. Update of the relative gain factors between DPDCH and DPCCH is on a radio frame basis depending on the current TFC used. It is independent of the uplink inner loop power control that updates the DPCCH absolute power level with steps of $\pm \Delta \text{TPC}$ dB for every time slot.

TABLE 10.4 Quantization of the Gain Factors

Signaling Values for G_c and G_d	Quantized Amplitude Ratios for G_c and G_d
15	1.0
14	14/15
13	13/15
12	12/15
11	11/15
10	10/15
9	9/15
8	8/15
7	7/15
6	6/15
5	5/15
4	4/15
3	3/15
2	2/15
1	1/15
0	Switch off

10.4 Power Control for Uplink Common Channels

The uplink common channel is the random access channel. Power control for this channel is a mix of an open loop and a coarse inner loop power control based on the preamble power ramp up technique.

10.4.1 Initial Preamble Power

Before transmission of the RACH preamble part, the open loop power control calculates the initial preamble power using (7-9), which is repeated here for easy reference.

$$\text{Preamble\_Initial\_Power} = \text{Primary CPICH TX power} - \text{CPICH\_RSCP} + \text{UL interference} + \text{Constant Value}, \tag{10-10}$$

where CPICH_RSCP is the received signal code power measured by the UE, Primary CPICH TX power is the P-CPICH transmit power, UL interference is the interference corresponding to the received signal strength indicator (RSSI) measurements at the Node B, and Constant Value is a term defined by a UTRAN parameter that sets the target SIR for the UE to achieve on the RACH. The last three parameters are signaled by UTRAN.

10.4.2 Successive Preamble Power

If the UE receives no acknowledgement message from Node B on the AICH following the transmission of the initial RACH preamble, it increases the transmit power of the next RACH preamble within the same power ramping cycle as follows:

$$\text{Next\_preamble\_power} = \text{Previous\_preamble\_power} + \text{PowerRampStep}, \tag{10-11}$$

where Previous_preamble_power is the transmit power of the previous RACH preamble within the same power ramping cycle, and PowerRampStep is the power step increase, which is defined by a UTRAN parameter.

In a power ramping cycle, UE retransmits RACH preambles with increasing powers given by (10-11) until it receives an ACK or NACK, or until the maximum number of repetitions defined by a UTRAN parameter called MaxRetranPreamble is reached. The maximum power for the RACH preamble is limited by UE capabilities.

At Node B, there is an RACH preamble detector. A UTRAN parameter sets the detection threshold of the detector. The ratio of the received preamble power to the interference level during the preamble period must be greater than this threshold in order for the RACH preamble to be detected and acknowledged at Node B.

10.4.3 RACH Message Part Power

The RACH message part power must be higher than the last preamble power in order to improve the detection of the RACH message part. The RACH message part consists of a control part and a data part. The RACH message control part power is given by

$$P_{\text{RACH\_Msg\_Part\_Control}} = \text{ACK\_Preamble\_Power} + \text{PowerOffsetPpm}, \quad (10\text{-}12)$$

where ACK_Preamble_Power is the power of the last transmitted RACH preamble that has been acknowledged by the Node B on AICH, and PowerOffsetPpm is power offset between the acknowledged RACH preamble and the RACH message control part.

The power of the RACH message data part is controlled by the gain factors that define the relative power of the control and data parts in a way similar to the uplink dedicated physical channels. More specifically, the RACH message data part power is given by

$$P_{\text{RACH\_Msg\_Part\_Data}} = \text{Round up integer of } \{10 \log[1 + (G_d/G_c)^2]\}, \quad (10\text{-}13)$$

where G_c is the control part gain factor and G_d is the data part gain factor.

The total power of the message part is the power sum of the control part and the data part. It is limited by the maximum power allowed on RACH that is determined by sIB3MaxAllowedULTxPower and sIB4MaxAllowedULTxPower broadcasted on BCCH through SIB Type 3 and SIB Type 4. If the calculated total power is greater than the maximum allowed power, the total power is set to the maximum allowed power and the ratio between control and data parts remains the same.

10.5 Power Control in Compressed Mode

In compressed mode, some frames are compressed and contain transmission gaps. There may be no TPC bits in the transmission gaps. The goal of power control in compressed mode is to restore the SIR to a value close to the target SIR as fast as possible after each transmission gap. Power control in compressed mode is specified in [5]. This section will only discuss the general approaches of power control in compressed mode. Both downlink power control and uplink power control are discussed.

10.5.1 Downlink Power Control in Compressed Mode

In compressed mode, transmission of downlink DPDCH(s) and DPCCH is stopped during the transmission gaps in the compressed frames. The power of DPCCH and DPDCH in the first slot following the transmission gap should be set to the same value as that in the last slot before the transmission gap.

According to [5], in compressed mode except during downlink transmission gaps, UTRAN estimates the kth TPC command and adjusts the current downlink power $P(k-1)$ to a new power $P(k)$ as

$$P(k) = P(k-1) + P_{TPC}(k) + P_{SIR}(k) + P_{bal}(k). \qquad (10\text{-}14)$$

The parameters in (10-14) are identical to that in (10-3), except that there is a new parameter, $P_{SIR}(k)$, which is the kth power adjustment due to downlink target SIR variation.

For a single TPC (DPC_Mode = 0), if no uplink TPC command is received, $P_{TPC}(k)$ derived by the Node B is set to zero. Otherwise, $P_{TPC}(k)$ is calculated the same way as in normal mode described in Subsection 10.1.2.2.3, with a new power control step size Δ_{STEP}. This new step size Δ_{STEP} is equal to $\Delta_{RP\text{-}TPC}$ for the slots in the recovery period length after each transmission gap, and equal to Δ_{TPC} for all other slots. The special step size, $\Delta_{RP\text{-}TPC}$, is called the recovery period power control step size, and is equal to $\text{Min}(3\text{dB}, 2\Delta_{TPC})$.

The recovery period length (RPL) is the period following the resumption of simultaneous uplink and downlink DPCCH after a compressed mode gap in either direction and is equal to Min(TGL, 7) slots, where TGL is the transmission gap length. If a transmission gap starts again before the RPL ends, then the recovery period must end at the start of the gap, and the value of RPL is reduced accordingly.

For a TPC triplet (DPC_Mode = 1), the 3-slot sets over which the TPC commands are processed must remain aligned to the frame boundaries in the compressed frame. If there is an incomplete set of TPC commands, the UE must transmit the same TPC commands in all slots of the incomplete set.

The power offset $P_{SIR}(k)$ is equal to $\delta P_{curr} - \delta P_{prev}$, where δP_{curr} and δP_{prev} are the value of δP in the current slot and the most recently transmitted slot, respectively. The value of δP is calculated as follows:

$$\begin{aligned}\delta P = {}& \text{Max}\,(\Delta P1\_\text{compression}, \ldots, \Delta Pn\_\text{compression}) \\ & + \Delta P1\_\text{coding} + \Delta P2\_\text{coding},\end{aligned}$$

$$(10\text{-}15)$$

where n is the number of different TTI lengths among the TTIs of all transport channels in the CCTrCh, and $\Delta P1\_coding$ and $\Delta P2\_coding$ are calculated from the uplink parameters DeltaSIR1, DeltaSIR2, DeltaSIRafter1, and DeltaSIRafter2, which are signaled by higher layers.

$\Delta P1\_coding$ takes the value DeltaSIR1 if the first transmission gap in the transmission gap pattern starts within the current frame, and takes the value DeltaSIRafter1 if it starts in the preceding frame. Similarly, $\Delta P2\_coding$ takes the value DeltaSIR2 if the second transmission gap in the transmission gap pattern starts within the current frame, and takes the value DeltaSIRafter2 if it starts in the preceding frame. For all other cases, both $\Delta P1\_coding$ and $\Delta P2\_coding$ are equal to 0 dB.

$\Delta Pi\_compression$ is equal to 3 dB for downlink frames compressed by reducing the spreading factor by 2, and equal to $10 \log\{15 \times F_i / (15 \times F_i - \text{TGL}_i)\}$ if there is a transmission gap created by the puncturing method within the current TTI with a length of F_i frames. TGL_i is the total number of transmission gap slots in the F_i frames. For all other cases, $\Delta Pi\_compression$ is equal to 0 dB.

If several compressed mode patterns exist simultaneously, an individual δP offset is calculated for each individual compressed mode pattern. Then, all δP offsets are summed together and applied to the frame.

10.5.2 Uplink Power Control in Compressed Mode

In uplink power control in compressed mode, the serving cells estimate the signal-to-interference ratio SIR_{est} of the received uplink DPCH, generate TPC commands, and transmit the commands once per slot, except during downlink transmission gaps, as follows:

- If $\text{SIR}_{est} \geq \text{SIR}_{cm\_target}$, then the serving cells transmit TPC command 0.
- If $\text{SIR}_{est} < \text{SIR}_{cm\_target}$, then the serving cells transmit TPC command 1.

SIR_{cm_target} is the target SIR during compressed mode and is given by

$$\text{SIR}_{cm\_target} = \text{SIR}_{target} + \Delta\text{SIR}_{PILOT} + \Delta\text{SIR1\_coding} + \Delta\text{SIR2\_coding}, \quad (10\text{-}16)$$

where $\Delta\text{SIR1\_coding}$ and $\Delta\text{SIR2\_coding}$ are obtained from uplink parameters DeltaSIR1, DeltaSIR2, DeltaSIRafter1, and DeltaSIRafter2, which are signaled by higher layers.

$\Delta\text{SIR1\_coding}$ takes the value DeltaSIR1 if the first transmission gap in the transmission gap pattern starts within the current uplink frame, and takes the value DeltaSIRafter1 if the current uplink frame is right behind a frame in which the first transmission gap in the transmission gap pattern starts. Similarly, ΔSIR2_coding takes the value DeltaSIR2 if the second transmission gap in the transmission gap pattern starts within the current uplink frame, and takes the value DeltaSIRafter2 if the current uplink frame is immediately behind a frame in which the second transmission gap in the transmission gap pattern starts. For all other cases, both ΔSIR1_coding and ΔSIR2_coding are equal to 0 dB.

ΔSIR$_{\text{PILOT}}$ is defined as

$$\Delta\text{SIR}_{\text{PILOT}} = 10 \text{ Log} (N_{\text{pilot},N}/N_{\text{pilot,curr\_frame}}), \qquad (10\text{-}17)$$

where $N_{\text{pilot,curr\_frame}}$ is the number of pilot bits per slot in the current compressed uplink frame, and $N_{\text{pilot},N}$ is the number of pilot bits per slot in a normal uplink frame without a transmission gap.

If several compressed mode patterns exist simultaneously, ΔSIR1_coding and ΔSIR2_coding are calculated individually for each compressed mode pattern. Then, the sum of ΔSIR1_coding offsets and the sum of ΔSIR2_coding offsets are used in (10.16).

In compressed mode, TPC commands on the downlink may be missing for some slots because of the transmission gaps. If no downlink TPC command is transmitted, the corresponding TPC_cmd derived by the UE is set to zero.

The number of pilot bits per slot in the uplink DPCCH may be different for the compressed and non-compressed frames. Therefore, it may be necessary to change the uplink DPCCH transmit power to compensate for the change in the total pilot energy. Thus, at the beginning of each slot, the UE must derive the value of a power-offset Δ_{PILOT}, which is given by

$$\Delta_{\text{PILOT}} = 10 \text{ Log} (N_{\text{pilot,prev}}/N_{\text{pilot,curr}}), \qquad (10\text{-}18)$$

where $N_{\text{pilot,prev}}$ and $N_{\text{pilot,curr}}$ are the pilot bit numbers in the most recently transmitted slot and the current slot, respectively. For slots in the transmission gaps, Δ_{PILOT} is equal to 0 dB.

In every slot during compressed mode except the slots in the recovery period after the transmission gap, the UE adjusts the transmit power of the uplink DPCCH with a step of Δ_{DPCCH} given by

$$\Delta_{\text{DPCCH}} = \Delta_{\text{TPC}} \times \text{TPC\_cmd} + \Delta_{\text{PILOT}}. \qquad (10\text{-}19)$$

At the start of the first slot after an uplink or downlink transmission gap, the UE must change the uplink DPCCH transmit power by an amount Δ_{DPCCH} (in dB) with respect to the uplink DPCCH power in the most recently transmitted uplink slot. In this case, Δ_{DPCCH} is given by

$$\Delta_{\text{DPCCH}} = \Delta_{\text{RESUME}} + \Delta_{\text{PILOT}}. \qquad (10\text{-}20)$$

The value of Δ_{RESUME} (in dB) is determined by the UE based on the Initial Transmit Power mode (ITP). The ITP is a UE-specific parameter signaled by the network with other compressed mode parameters. For a detailed description of ITP modes, interested readers may refer to references [5] and [9].

During the recovery period, there are two possible modes for the power control algorithm as given in Table 10.5 [5]. The recovery period power control mode (RPP) is signaled with other compressed mode parameters.

For recovery period power control mode 0, the ordinary transmit power control described in Subsection 10.3.3.3 is applied and the step size is not changed during the recovery period.

For recovery period power control mode 1, power control Algorithm 1 is applied with a step size $\Delta_{\text{RP-TPC}}$ instead of Δ_{TPC} during recovery period length slots. The change in uplink DPCCH transmit power at the start of each of the recovery period length slots immediately following the transmission gap, except for the first slot after the transmission gap, is given by

$$\Delta_{\text{DPCCH}} = \Delta_{\text{RP-TPC}} \times \text{TPC\_cmd} + \Delta_{\text{PILOT}}. \quad (10\text{-}21)$$

After the recovery period, ordinary transmit power control resumes using the algorithm specified by the value of PCA and with step size Δ_{TPC}.

10.6 Summary

Power control, including open loop power control and closed loop power control, applies to both common and dedicated channels for both uplink and downlink. Open loop power control sets the initial transmit power of the channels. Closed loop power control comprises an outer loop and an inner loop. The outer loop sets a target SIR and delivers it to the inner loop, while the inner loop adjusts the code channel power such that the measured SIR meets the target SIR. Both open loop and closed loop

TABLE 10.5 Recovery Period Power Control Modes During Compressed Mode (Courtesy of ETSI)

Recovery Period Power Control Mode	Description
0	Transmit Power Control is applied using the algorithm determined by the value of PCA with step size Δ_{TPC}.
1	Transmit Power Control is applied using Algorithm 1 with step size $\Delta_{\text{RP-TPC}}$ during RPL slots after each transmission gap.

power controls apply to dedicated channels. However, there is no closed loop power control for common channels. The function of power control for common channels is to set the power levels of these channels.

Outer loop power controls autonomously run at the RNC for the uplink, and at UE for the downlink. The 3GPP standard does not specify how outer loop power control should be implemented. It is vendor-specific.

The inner loop power control is carried out by the UE and the Node B together. For downlink, the UE measures the downlink power, calculates the SIR, compares it with the target SIR, and then sends TPC bits on the uplink DPCCH channel to the Node B. The Node B reads the received TPC bits and then adjusts the power of the downlink DCH channel accordingly. For uplink, the Node B measures the uplink power, calculates the SIR, compares it with the target SIR, and then sends TPC bits in the downlink DPCH channel to UE. The UE reads the received TPC bits and then adjusts the power of the uplink DCH channels accordingly.

For downlink dedicated channels, the open loop power control sets the initial power during call setup and when new legs are added, while the closed loop power control adjusts the power incrementally. In the open loop power control, the RNC provides the initial downlink transmit power information to the Node B through NBAP signaling during radio setup or radio link addition procedures. Closed loop power control comprises outer loop power control and inner loop power control. Outer loop power control, which runs at 100 Hz, is responsible for setting the SIR target, and inner loop power control, which runs at 1500 Hz, is responsible for adjusting the transmit power to make sure that the received SIR meets the SIR target.

Power control for downlink common channels deals with power settings of the downlink common physical channels. The transmit power of all downlink common channels are controlled by the RNC. Generally, there is no control loop for all of these channels. Normally, the P-CPICH power level is set to about 10% of the total Node B transmit power. The power levels for the P-SCH and S-SCH are respectively set at −3 dB and −5 dB relative to the P-CPICH power level. PCCPCH is also set at −3 dB relative to the P-CPICH. SCCPCH carries the time-multiplexed FACH and PCH. Normally, the FACH and PCH power level should be set at 4 dB relative to the P-CPICH. The power level of PICH is set at −6 dB relative to the P-CPICH, while the power level for AICH is set at −5 dB relative to the P-CPICH.

Power control for the uplink dedicated channels is similar to that for the downlink dedicated channels. It also involves open loop power control, which sets the initial transmit power, and closed loop power control, which incrementally adjusts the transmit power.

Power control for the uplink common channels involves calculations of the initial preamble power, successive preamble power, and RACH message power of the random access channel.

The function of power control in compressed mode is to restore the SIR to a value close to the target SIR as fast as possible after each transmission gap. In compressed mode, downlink DPDCH(s) and DPCCH are not transmitted during the transmission gaps in the compressed frames. The power of DPCCH and DPDCH in the first slot after the transmission gap should be set to the same value as that in the last slot before the transmission gap. On the uplink, the serving cells estimate the signal-to-interference ratio SIR_{est} of the received uplink DPCH, generate TPC commands, and transmit the commands once per slot, except during downlink transmission gaps. The UE derives the TPC commands and acts accordingly.

References

[1] 3GPP TS 25.133, v5.8.0, "Requirements for support of radio resource management (FDD)," (Release 5).

[2] 3GPP TS 25.433, v5.6.0, "UTRAN Iub Interface NBAP Signaling," (Release 5).

[3] 3GPP TS 25.101, v5.6.0, "UE radio transmission and reception (FDD)," (Release 5).

[4] 3GPP TS 25.211, v5.3.0, "Physical channels and mapping of transport channels onto physical channels (FDD)," (Release 5).

[5] 3GPP TS 25.214, v5.6.0, "Physical layer procedures (FDD)," (Release 5).

[6] 3GPP TS 25.331, v5.7.1, "Radio resource Control (RRC); protocol specification," (Release 5).

[7] 3GPP TS 25.104, v6.3.0, "Base Station (BS) Radio transmission and reception (FDD)," (Release 6).

[8] 3GPP TS 25.427, v5.5.0, "UTRAN Iub/Iur interface user plane protocols for DCH data streams," (Release 5).

[9] 3GPP TS 25.215, v5.5.0, "Physical layer measurements (FDD)," (Release 5).

Chapter 11

HSDPA Overview

High speed downlink packet access (HSDPA) is standardized in 3GPP Release 5. It improves system capacity, increases user data rates in the downlink direction, and can provide a packet-based data service with data transmission rate around 10 Mbps in a 5 MHz bandwidth. This improved performance is based on several key features that HSDPA employs, including adaptive modulation and coding (AMC), fast scheduling, hybrid automatic repeat request (HARQ), fast cell search, and advanced receiver design.

In 3GPP standards, Release 4 specifications provide efficient IP support, empowering provision of services through an all-IP core network. Release 5 specifications propose HSDPA to provide data rates around 10 Mbps to support packet-based multimedia services. Release 6 supports multiple-input multiple-output (MIMO) with a data rate up to 20 Mbps. HSDPA is evolved from and backward compatible with Release 99. In Release 99 and Release 4, it is possible to send data on DPCH, FACH, and DSCH. However, DPCH is inefficient in code usage and setup time; DSCH is more code-efficient, but the capacity is not good enough; and none of the Release 99 and Release 4 channels can exploit short transmission intervals and fast resource allocation. The Release 99 networks that have already been deployed can support HSDPA by adding or turning on the HSDPA features. HSDPA has higher capacity in terms of higher system throughput and higher user data rates. HSDPA-related information is given in various 3GPP technical specifications [1]-[7].

This chapter briefly addresses some key concepts of HSDPA. These include key features of HSDPA, HSDPA channels, physical layer procedure, HSDPA configurable parameters, and general considerations for HSDPA deployment.

11.1 HSDPA Key Features

With HSDPA, two of the most fundamental WCDMA features, variable spreading factor and fast power control, are disabled and replaced with adaptive modulation and coding, extensive multi-code operation and fast retransmission. In addition, HSDPA uses short transmission intervals and fast scheduling, which allow the system to benefit from the short-term variations, with the scheduling decisions being made in the Node B. Other key HSDPA features include simple code allocation, efficient power allocation, no downlink soft handover, and support of various UE categories.

11.1.1 Adaptive Modulation and Coding

In Release 99, the WCDMA networks use fast power control for radio link adaptation. This fast power control is carried out at 1500 Hz, or, in other words, on a per slot basis. Link adaptation is required because the SIR of the received signal at the UE varies over time due to fast fading and geographic location in a particular cell. To overcome this fading effect and to improve the system capacity and peak data rates, the transmission power to the UE is adjusted according to the signal variations through link adaptation.

In HSDPA the transmission power is kept constant over a TTI. It uses adaptive modulation and coding as an alternative method to power control in order to improve the spectral efficiency. In other words, link adaptation in HSDPA adapts the modulation scheme and coding according to the quality of the radio link. In addition to QPSK, HSDPA uses higher order modulation schemes like 16-quadrature amplitude modulation (16-QAM). The modulation adapted is based on the radio channel conditions. HSDPA adopts QPSK modulation for data rates up to 7.2 Mbps, and 16-QAM modulation for higher data rates. QPSK can support 2 bits/symbol, while 16-QAM can support 4 bits/symbol. As such, 16-QAM has twice the peak rate capability as compared to QPSK, using the channel bandwidth more efficiently. HSDPA also uses an adaptive code rate, based on channel quality feedback from the UE and any available Node B resources, including codes and power. The spreading factor remains fixed, but the coding rate can be 1/4, 1/2, 5/8, or 3/4. The Node B receives the channel quality indicator (CQI) report and power measurements on the associated channels. Based on this information it then determines the transmission data rate. Users close to the base station are generally assigned higher order modulations with higher code rates, such as 16-QAM and 3/4 code rate. The bottom line is that link adaptation ensures that the highest possible data rate is achieved both for users close to the Node B with good signal quality using a higher coding rate, and for users at the cell edge with poor signal quality using a lower coding rate.

Table 11.1 shows the maximum and the minimum HSDPA peak user data rates for different modulation schemes and coding rates. Take one

TABLE 11.1 HSDPA User Data Rate

No. of HS-PDSCH Codes	Modulation Scheme	Number of Bits per TTI After Rate Matching	Minimum Peak User Data Rate (kbps) ($R_{code} = 1/3$)	Maximum Peak User Data Rate (kbps) ($R_{code} = 1$)
1	QPSK	960	160	480
2	QPSK	1920	320	960
3	QPSK	2880	480	1440
4	QPSK	3840	640	1920
5	QPSK	4800	800	2400
6	QPSK	5760	960	2880
7	QPSK	6720	1120	3360
8	QPSK	7680	1280	3840
9	QPSK	8640	1440	4320
10	QPSK	9600	1600	4800
11	QPSK	10560	1760	5280
12	QPSK	11520	1920	5760
13	QPSK	12480	2080	6240
14	QPSK	13440	2240	6720
15	QPSK	14400	2400	7200
1	16-QAM	1920	320	960
2	16-QAM	3840	640	1920
3	16-QAM	5760	960	2880
4	16-QAM	7680	1280	3840
5	16-QAM	9600	1600	4800
6	16-QAM	11520	1920	5760
7	16-QAM	13440	2240	6720
8	16-QAM	15360	2560	7680
9	16-QAM	17280	2880	8640
10	16-QAM	19200	3200	9600
11	16-QAM	21120	3520	10560
12	16-QAM	23040	3840	11520
13	16-QAM	24960	4160	12480
14	16-QAM	26880	4480	13440
15	16-QAM	28800	4800	14400

HS-PDSCH code in Table 11.1 as an example. For the QPSK modulation scheme, there are 960 bits per TTI after rate matching. It means that the bit rate is 960 bits/2ms or 480 kbps. It also means that the bit rate is 480 kbps after encoding and before QPSK modulation. Therefore, the minimum peak user data rate, 160 kbps (480/3 = 160), is obtained by using an encoder with a coding rate of 1/3, while the maximum peak user data rate, 480 kbps, is obtained by using an encoder with a coding rate of 1. For QPSK modulation, two bits generate a symbol. Therefore, a bit rate of 480 kbps before QPSK modulation is equivalent to a symbol rate of 240 ksps after QPSK modulation. Since the spreading factor for HS-PDSCH is fixed at 16, a 240-ksps symbol rate is spread to a chip rate of 3.84 Mcps.

For 16-QAM modulations, four bits generate a symbol. With the chip rate being kept constant at 3.84 Mcps, the bit rate before 16-QAM modulation is 960 kbps. Therefore, the minimum peak user data rate is 320 kbps while the maximum peak user data rate is 960 kbps.

For multiple HS-PDSCH codes, the peak user data rate is equal to the peak user data rate for one HS-PDSCH times the number of HS-PDSCH codes.

11.1.2 Transmission Time Interval

In Release 99, the transmission time interval can be 10 ms, 20 ms, 40 ms, or 80 ms. The radio frame is fixed at 10 ms in length. At every radio frame boundary, the physical layer requests data from the MAC Layer. For a TTI longer than 10 ms, the data must be segmented into 10 ms a piece with each piece being multiplexed into a 10-ms radio frame in the CCTrCh. In HSDPA, the transmission time interval is fixed at 2 ms, which contains 3 time slots. In other words, HSDPA is transmitted in subframes of 2 ms. There is no multiplexing of transport channels for HSDPA. The shorter TTI can reduce the burden on hardware complexity at higher data rates. It also reduces the link adaptation delays, increases the granularity in the scheduling process, and tracks the time varying radio channel better.

11.1.3 Scheduling

In HSDPA, downlink packet transmission scheduling is purely based on channel quality feedback from the UEs. The packet scheduler resides at Node B. As such, the scheduling is done in Node B; there is no interaction with the RNC. This makes the packet scheduling decisions almost instantaneous. In addition, the scheduling is fast because it is carried out on a per TTI (2 ms) basis.

An option for fair scheduling is a round-robin method in which every user is served in a sequential manner so that all the users get the

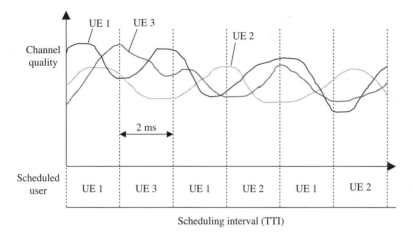

Figure 11.1 Fast scheduling based on channel quality.

same average allocation time. Another option is proportional fair packet scheduling, in which the order of service is based on the instantaneous channel quality relative to average bit rate. Since the selection is based on relative conditions, every user gets approximately the same amount of allocation time depending on its channel condition. Still another option is maximum C/I packet scheduling where user selection is purely based on the maximum C/I (best channel quality). Figure 11.1 illustrates fast scheduling based on channel quality. For example, at the beginning of the first TTI, UE1 has the best channel quality. Therefore, it is scheduled for service in that TTI. At the beginning of the second TTI, UE3 becomes the best in channel quality. Therefore, it is scheduled for service in that TTI.

11.1.4 Retransmission

As discussed in Subsection 11.1.1, the AMC uses an appropriate modulation and coding scheme according to the channel conditions. Even after AMC, errors may still exist in the received packets due to the fact that the channel may vary when the packet is in the middle of transmission. An automatic retransmission request (ARQ) scheme can be used to recover from these link adaptation errors. When the transmitted packet is received erroneously, the UE rapidly requests retransmission of that erroneous packet. In Release 99, the request is processed by the RNC. In HSDPA, the request is processed in the Node B, providing the fastest possible retransmission. Figure 11.2 shows the difference in retransmission control between HSDPA and Release 99. In Release 99, transport

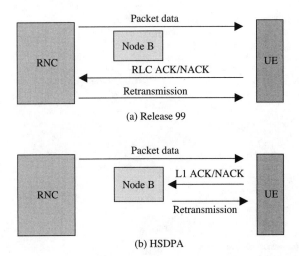

Figure 11.2 Retransmission control in Release 99 and HSDPA.

channels are terminated at the RNC. The RNC performs functions such as RLC ARQ, scheduling, and transport format selection for dedicated and shared channels, while Node B does none of these functions. In HSDPA, an HSDPA medium access control (MAC-hs) entity is installed at Node B. With MAC-hs, retransmissions can be controlled directly by the Node B, resulting in faster execution and shorter retransmission delays.

HSDPA employs hybrid automatic retransmission request (HARQ) in link level retransmission. The HARQ protocol used for HSDPA is stop and wait (SAW). In SAW, a Node B sends a TTI block and waits until it receives an acknowledgement or negative acknowledgement from the UE. In order to utilize the time during its waiting for the acknowledgements, a number of SAW-ARQ processes may be set for UE, with different processes transmitting in separate transmission time intervals. The number of SAW processes is signaled to the UE in HS-SCCH using three bits. The maximum number of SAW-ARQ processes is eight. In practice, it is normally set at six.

UE requests retransmissions when it receives erroneous data using acknowledgement and negative acknowledgement (ACK and NACK). Once the UE receives the retransmission, it combines the information from the original transmission with that of the retransmission before trying to decode the message. There are two main schemes for HARQ: chase combining and incremental redundancy.

Chase combining involves the retransmission of the same data packet that was received with errors. When the UE receives the retransmitted packet, it combines the original packet and the retransmitted packet before decoding the data packet. This method wastes more bandwidth than the incremental redundancy method, since the entire error packet must be retransmitted.

Incremental redundancy correctly selects the transmitted bits from both the original transmission and the retransmission, in order to minimize the need for further repeat requests. This method is used to get maximum performance out of the available bandwidth. In incremental redundancy, the retransmitted block consists of only the correction bits for the original data packet. The additional redundant bits are sent incrementally when the first, second, third, and so forth retransmissions are received with errors.

11.1.5 Code Allocation and Code Multiplexing of Packet Transmissions

HSDPA uses a fixed spreading factor of 16. Within each 2-ms TTI, a maximum of 15 consecutive OVSF codes can be allocated to 15 parallel HS-PDSCH channels for the HS-DSCH. These channels may all be assigned to one user during the TTI, or may be split among several HSDPA users. In other words, it is possible to schedule transmissions to a single UE or to multiple UEs at the same time. It can also transmit to a small number of UEs with good channel quality. In essence, the transmission can be time division multiplexing (single user in one TTI) or code division multiplexing (multiple users in one TTI). Figure 11.3 illustrates an example of the sharing of HS-PDSCHs among users.

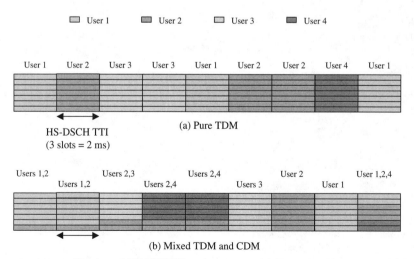

Figure 11.3 Sharing of HS-PDSCHs among users (a) Pure time division multiplexing (b) Mixed time division multiplexing and code division multiplexing.

11.1.6 Power Allocation

HSDPA allows for allocation of any power left over from dedicated and overhead channels on a 2 ms basis. There is no power control for HS-PDSCH. The allocation of downlink power can be static or dynamic. Figure 11.4 shows the power allocation strategies. In static power allocation, the sum of the power for HS-PDSCHs and HS-SCCHs within a TTI is kept constant. In this case, the total available cell power is not fully utilized. As shown in Figure 11.4 part (a), there is some power unused by the dedicated channels with power control. On the contrary, as shown in Figure 11.4 part (b), in a dynamic power allocation strategy, the HS-PDSCHs and HS-SCCHs use up all the power left over from dedicated and overhead channels within a TTI. There is no available cell power unused in this case.

11.1.7 No Downlink Soft Handover

Although one Node B can serve either one or more than one UE, each UE is served by only one Node B. Downlink soft handover would be in conflict with Node B-based scheduling and HARQ, since soft handover allows multiple Node Bs to serve a UE at the same time. As such, there is no downlink soft handover for HSDPA. This feature is similar to that of 1x evolution data optimized (1x-EVDO), where there is no downlink soft handover either. Both HSDPA and 1x-EVDO use scheduling schemes

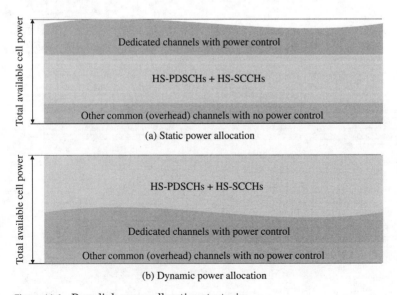

Figure 11.4 Downlink power allocation strategies.

for downlink transmissions. The difference is that for 1x-EVDO one base station can serve only one mobile station at a time, while for HSDPA one Node B can serve multiple UEs at one time.

11.1.8 HSDPA UE Capability

HSDPA supports 12 different UE categories with different available maximum data. The available maximum data rate ranges from 0.9 Mbps to 14.0 Mbps. The HSDPA UE capability is independent of the Release 99 UE capabilities. If HS-DSCH is configured for the UE, the downlink DCH capability of the UE is still limited to the Release 99 UE capability. Typically, the UE downlink DCH capability is 32, 64, 128, or 384 kbps.

The UE capability does not affect the timing of an individual TTI transmission. However, it does define how often the Node B can transmit packets to the UE. The UE capabilities include a value (1, 2, and 3) that defines the minimum inter-TTI intervals. Value 1 indicates that consecutive TTIs may be used, while values 2 and 3 respectively indicate that there is a minimum of one and two empty TTIs between packet transmissions. Table 11.2 shows the HSDPA UE categories and their associated capabilities. It should be noted that for HSDPA operation, the UE does not report individual capability values but only the category. In addition, although a maximum number of 15 codes can be allocated to 15 HS-PDSCHs, individual UE may receive a maximum of 5, 10, or 15 codes, depending on the UE capability.

TABLE 11.2 HSDPA UE Capability Categories

Category	Codes	Minimum Inter-TTI Interval	Maximum Transport Block Size	Total No. of Soft Channel Bits	Achievable Peak Data Rate (Mbps)
1	5	3	7298	19200	1.2
2	5	3	7298	28800	1.2
3	5	2	7298	28800	1.8
4	5	2	7298	38400	1.8
5	5	1	7298	57600	3.6
6	5	1	7298	67200	3.6
7	10	1	14411	115200	7.2
8	10	1	14411	134400	7.2
9	15	1	20251	172800	10.2
10	15	1	27952	172800	14.0
11	5	2	3630	14400	0.9
12	5	1	3630	28800	1.8

11.2 HSDPA Channels

HSDPA adds one additional transport channel and three additional physical channels to the existing Release 99 and Release 4 channels. The additional transport channel is the high-speed downlink shared channel (HS-DSCH). The additional physical channels are the high-speed physical downlink shared channel (HS-PDSCH) and high-speed shared control channel (HS-SCCH) for downlink, and the high-speed dedicated physical control channel (HS-DPCCH) for uplink.

11.2.1 HS-DSCH

The HS-DSCH carries the HSDPA payload. The new aspects of HS-DSCH include: fast adaptation of transport format to channel conditions, fast scheduling, taking priorities and channel quality into account, a short transmission time interval with dynamic transport block size, hybrid ARQ with chase combining and incremental redundancy of retransmissions, and mapping of up to 15 physical channels.

In HS-DSCH, there is no transport channel multiplexing, only one transport block per TTI. This is different from the concept of CCTrCh. The transport block size is dynamically determined and there are a total of 254 possible transport block sizes to choose from. The choice is adaptive to channel quality. The transport channel downlink TTI is always 2 ms. A scheduler resides in the Node B and conducts the scheduled transmission based on the channel quality feedback provided by the UE. The transport format and resource combination (TFRC) define the modulation type, transport block size, and channelization code set that will be used to transmit the HS-DSCH. It is equivalent to TF/TFCI in Release 99.

11.2.2 HS-PDSCH

The HS-PDSCH carries the actual information payload. The channel mapping of the HS-DSCH to the HS-PDSCH was shown in Figure 2.14, which is repeated here as Figure 11.5 for easy reference. The HS-PDSCH has a number of specific characteristics compared with Release 99 channels. The transmission time interval is defined to be 2 ms or 3 slots that are short compared to the 10-, 20-, 40-, or 80-ms TTIs supported in Release 99.

The spreading factor for HS-PDSCH is always 16. It carries 960 bits per TTI for QPSK and 1920 bits per TTI for 16-QAM. There is no power control, with the amount of allocated power determined by Node B, based on available Node B power and channel quality feedback. Up to 15 HS-PDSCHs can be used to carry HS-DSCH. As such, 15 consecutive OVSF codes can be used for the set of HS-PDSCHs.

HS-PDSCHs can be code multiplexed. This means that a number of UEs can be served at the same time, using different OVSF codes for different UE.

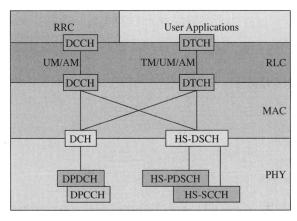

Figure 11.5 Channel mapping of HS-DSCH.

11.2.3 HS-SCCH

The HS-SCCH carries physical layer control information. UTRAN allocates HS-SCCHs corresponding to the maximum number of code-multiplexed users. However, if the HS-DSCH carries no data, then no HS-SCCH is transmitted. There may be many HS-SCCHs per cell, but the UE only needs to monitor a maximum of four HS-SCCHs at any given time. In other words, the UE only has to support a maximum of four HS-SCCHs.

The HS-SCCH spans 3 slots for each TTI and uses OVSF codes with a spreading factor of 128. However, the code is not fixed. It is transmitted in two parts, with slot 1 as the first part and slots 2 and 3 as the second part. The first part signals channelization codes and modulation such as the starting point in the code tree, number of codes, and modulation scheme (QPSK or 16-QAM). It is coded with $R = 1/3$ convolutional code and scrambled with the UE ID. The first part is needed prior to HS-PDSCH demodulation.

The second part signals transport block size and HARQ parameters. Although there are a total of 254 possible transport block sizes that are available, only 63 are used for a given number of HS-PDSCH codes. The HARQ parameters include ARQ process number, redundancy version, and new data indicator. The ARQ process number shows which ARQ process the data belongs to. Redundancy version information allows proper decoding and combining with possible earlier transmissions. The new data indicator indicates whether the transmission is to be combined with the existing data in the buffer, or the buffer should be flushed and filled with new data. The second part is also coded with $R = 1/3$ convolutional code. In addition, there is a 16-bit CRC, which spans the first and the second

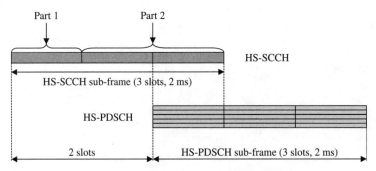

Figure 11.6 Timing relation between HS-SCCH and HS-PDSCH.

parts and is masked with the UE ID. The UE ID is needed to determine if an HS-SCCH or which HS-SCCH contains control information for the UE. The second part is required prior to HS-PDSCH decoding.

The HS-SCCH and its corresponding HS-PDSCH are not transmitted exactly simultaneously. They are offset from each other by two slots. Figure 11.6 shows the timing relation between HS-SCCH and HS-PDSCH. Figure 11.7 shows the timing relation between HS-SCCH and Release 99 channels. Note that PCCPCH frames are time aligned with blocks of 5 HS-SCCH subframes.

11.2.4 HS-DPCCH

The HS-DPCCH is the new uplink physical channel for HSDPA. It spans 3 time slots with a fixed spreading factor of 256. It is divided into two parts as shown in Figure 11.8.

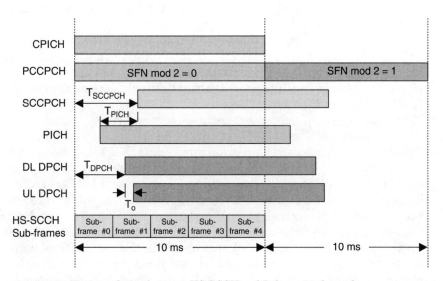

Figure 11.7 Timing relation between HS-SCCH and Release 99 channels.

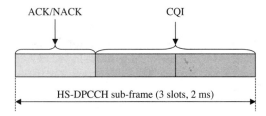

Figure 11.8 HS-DPCCH structure.

The first part, consisting of one slot called the ACK/NACK channel, carries ACK/NACK information, which presents the results of the CRC check after the packet decoding and combining. The ACK/NACK information is channel coded using a (10,1) repetition code. Power for ACK/NACK is offset relative to the DPCCH with different ACK/DPCCH and NACK/DPCCH ratios. The duty cycle of ACK/NACK depends on downlink data activity.

The second part of the HS-DPCCH, consisting of two slots called the CQI channel, carries the CQI information to indicate the estimated transport block size, modulation type and number of simultaneous codes that can be received correctly in the downlink direction. In other words, it indicates which TFRC is currently possible. The definition of CQI value is described in [4]. The channel quality information is coded using a (20,5) code. Its power is also offset relative to the DPCCH. The duty cycle of the CQI channel is variable up to 100% and is configurable through RRC/NBAP.

ACK/NACK and CQI can be repeated in multiple subframes. They are controlled by RRC/NBAP signaling and are useful in soft handover scenarios.

The timing of the HS-DPCCH is tied to the downlink HS-PDSCH. As shown in Figure 11.9, an HS-DPCCH subframe starts 7.5 slots (19200 chips) + 0...255 chips after the corresponding HS-PDSCH subframe is

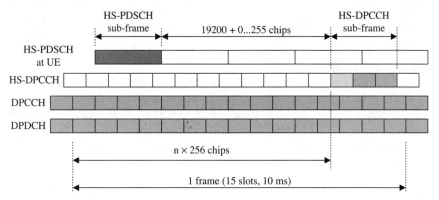

Figure 11.9 HS-DPCCH timing.

received on downlink. In other words, the UE has an available processing time of approximately 5 ms. Also shown in Figure 11.9 is the relative timing of the HS-DPCCH subframe and the uplink DPCCH and DPDCH frame boundary. The time offset between the beginning of the HS-DPCCH subframe and the DPCCH and DPDCH slot boundary is a multiple of 256 chips.

11.3 HSDPA Physical Layer Procedure

The HSDPA physical layer operation involves both the UE and the Node B. The scheduling and transport format selection is handled by the MAC-hs entity in the Node B. In order to obtain HSDPA service, UE must be configured for HSDPA operation. The HSDPA physical layer procedure begins when the UE starts to measure the channel quality and ends after the UE sends ACK/NACK to the Node B. More specifically, the procedure includes the following sequential events:

1. The UE measures the channel quality, reports CQI, and repeats these actions periodically. The repetition factor of CQI is signaled to the UE and the Node B from higher layers.

2. The Node B evaluates the overall condition for each user and selects the UE to serve in a particular TTI. The factors that affect the UE selection include channel quality, amount of data in the buffer, time lapse since last service, retransmission status, and so forth.

3. The Node B transmits the HS-SCCHs after it has selected the UE to serve in a particular TTI. The HS-SCCH is transmitted two slots before the associated HS-PDSCHs are transmitted. It carries physical layer control information, including the number of codes, modulation scheme (QPSK or 16-QAM), transport block size, and HAQR parameters such as ARQ process number, redundancy version, and new data indicator.

4. The Node B transmits a set of HS-PDSCHs, which partially overlaps with HS-SCCHs.

5. The UE monitors the HS-SCCHs and decodes the HS-SCCH intended for it. If the UE has detected control information intended for it on one of the HS-SCCHs in the preceding subframe (TTI), the UE only needs to monitor and decode the same HS-SCCH in the current subframe. Otherwise, the UE has to monitor all HS-SCCHs. If the UE detects that one of the monitored HS-SCCHs carries control information intended for it, the UE must decode the HS-SCCH parameters to determine whether it needs to decode HS-PDSCHs and also to obtain the ARQ information.

6. The UE decodes HS-PDSCHs and sends an ACK or NACK to Node B based on the results of the CRC check. The UE must repeatedly send the ACK/NACK information over N_acknack_transmit consecutive HS-DPCCH subframes. N_acknack_transmit is the repetition factor of ACK/NACK and is a physical layer parameter signaled to the UE and the Node B from higher layers.

11.3.1 CQI Reporting

The exact definition of CQI is stated in 3GPP TS25.214. It basically says that in an observation period, the UE reports the highest tabulated CQI value for an appropriately formatted single HS-DSCH subframe that has a BLER of no greater than 10%. The subframe is received in a 3-slot period ending one slot before the beginning of the first slot that transmits the reported CQI. Here "appropriately formatted" means that the HS-DSCH subframe in question was formatted with transport block size, number of HS-PDSCH codes and modulation corresponding to the reported CQI value or a lower value.

The UE does not necessarily need to average over exactly three slots. It just needs to schedule measurements such that the CQI is determined according to the above definition. Therefore, it allows for different implementations of the CPICH measurement, but should result in compatible CQI reports for all vendors. However, when the UE measures the CPICH strength, the measurement should take place in a 3-slot reference period ending one slot before CQI value is sent.

The reported CQI value ranging from 0 to 30 is an index to a reference TFRC in a CQI table. The CQI tables for different UE categories are given in [4]. The indexed TFRC provides the highest data rate for which the UE can guarantee a BLER of less then 10%, if the channel stays as it was in the reference period and if the default HS-PDSCH Tx power stays the same.

The CQI value may vary even when the channel is static because of additive white Gaussian noise (AWGN). For static channel conditions, the UE must report CQIs such that if the median of the reported CQI values is actually used for selecting the TFRC at the transmitter, the resulting packet error rate would be 10% or less.

11.4 HSDPA Related Parameters

There are three parameter classes: cell parameters, UE parameters, and fixed parameters. The cell parameters are configured in the Node Bs for each individual cell. These parameters are related to radio resource management and are configured either by NBAP common procedures, such as Cell Setup and Cell Reconfiguration procedures, or by operation and maintenance procedures. The UE parameters are assigned either by a Node B or by an RNC for each individual UE. These UE-specific parameters are used in NBAP, RNSAP, and RRC signaling messages. The fixed parameters are defined by the specifications and are not configurable.

11.4.1 Cell Parameters

The cell parameters include the HS-PDSCH FDD code parameter, HS-SCCH FDD code parameter, HS-PDSCH and HS-SCCH scrambling code parameter, HS-PDSCH and HS-SCCH total power parameter, and common measurement parameters for HSDPA.

The HS-PDSCH FDD code parameter specifies HS-PDSCH channelization code allocations. A cell can be allocated up to 15 codes with a spreading factor of 16. The HS-SCCH FDD code parameter specifies HS-SCCH channelization code allocations of a cell.

HS-PDSCH and HS-SCCH scrambling code information indicates the scrambling code on which HS-DSCH and HS-SCCH are supported. HS-PDSCH and HS-SCCH must always use the same scrambling code. However, this scrambling code may be different from the one used by the associated DPCH.

The HS-PDSCH and HS-SCCH total power parameter indicates the maximum power available for HS-PDSCH and HS-SCCH transmissions for a Node B. If the parameter does not set the maximum power limit for HS-PDSCH and HS-SCCH transmissions, all power unused by the common channels and the power-controlled dedicated channels within a TTI may be used for HS-PDSCH and HS-SCCH transmissions.

Common measurement parameters for HSDPA include transmitted carrier power not used for HS-PDSCH or HS-SCCH transmission, HS-DSCH required power, and HS-DSCH provided bit rate.

11.4.2 UE Parameters

The UE parameters are included in the UE information elements, radio bearer information elements, transport channel information elements, physical channel information elements, and measurements information elements.

The UE information elements contain information about UE capabilities related to HS-DSCH, UE category, and HS-DSCH radio network temporary identity, which is a 16-bit UE identity used for CRC calculation and evaluation.

The radio bearer IEs include radio bearer mapping information, which provides mappings between downlink logical channels and HS-DSCH

MAC-d flows. It is also possible to map downlink logical channels onto the HS-DSCH and DCH at the same time.

11.4.2.1 Transport Channel Information Transport channel information [7] includes downlink transport channel type, transport format set, MAC-d flow identity, MAC-d PDU size index, MAC-hs reset indicator, HARQ information, priority queue information, MAC-hs reordering buffer size, and initial capacity allocation.

Downlink transport channel type indicates whether the HS-DSCH is included in the downlink transport channels. If the downlink transport channel type IE is equal to DCH, DCH+DSCH, or DCH + HS-DSCH this IE is mandatory present. Otherwise it is not needed.

Transport format set information indicates whether the 2-ms TTI for an HS-DSCH has been added to the system. In general, it contains many mandatory present information elements for both common and dedicated transport channels. Typical examples include RLC size, number of transport blocks and TTI list, and the choice logical channel list.

MAC-d flow identity is an index ranging from 0 to 7 used to identify a MAC-d flow. The MAC-d PDU size index is an integer ranging from 0 to 7. It defines the mapping of the actual MAC-d PDU sizes configured for the HS-DSCH to the size index included in the MAC-hs header.

MAC-hs reset indicator is a Boolean indicating whether the MAC-hs in the UE should be reset or not. It is set by the drift RNC, sent to the serving RNC by RNSAP signaling and then sent to the UE. If it is set to TRUE, the MAC-hs entity needs to be reset.

HARQ information provides the number of parallel HARQ processes to be established in the UE. The largest number of parallel HARQ processes is eight. It also specifies the memory size for each HARQ process. Depending on choice memory partitioning, the memory size for each HARQ process can be equal or different. If choice memory partitioning is implicit, the UE shall apply memory partitioning of equal size across all HARQ processes. If choice memory partitioning is explicit, memory size is different for each HARQ process and indicated in terms of the number of soft channel bits.

Priority queue information includes priority queue ID, MAC-hs window size, scheduling priority indicator, discard timer, and MAC-hs guaranteed bit rate.

MAC-hs reordering buffer size indicates the total buffer size defined in UE capability minus the RLC AM buffer in kilobytes.

HS-DSCH initial capacity allocation parameters provide flow control for each scheduling priority class for the HS-DSCH frame protocol over Iub. These parameters include scheduling priority indicator, maximum MAC-d PDU size, and HS-DSCH initial window size.

11.4.2.2 Physical Channel Information   Physical channel information [7] includes HS-SCCH information, HS-SCCH code change indicator, HS-SCCH code change grant, HS-SCCH power offset, uplink DPCH power control information, and serving HS-DSCH radio link indicator.

HS-SCCH information specifies the HS-SCCH channelization codes for the HS-SCCHs assigned to the UEs. Node B can allocate up to 4 HS-SCCHs to UE. The HS-SCCH code change indicator indicates whether the HS-SCCH code change is needed or not. It is used by NBAP to request the RNC for a change of HS-SCCH codes. HS-SCCH code change grant indicates that modification of HS-SCCH codes is granted by the RNC. HS-SCCH power offset specifies the power offset of the HS-SCCH relative to the pilot bits of the downlink DPCCH for a user. It ranges from −32 to 31.75 dB with a step of 0.25 dB, and is indicated optionally to the Node B from the serving RNC.

Uplink DPCH power control information includes ACK power offset, NACK power offset, and ACK/NACK repetition factor. The ACK power offset parameter is used to calculate the power offset to be used for transmitting ACK on the HS-DPCCH. Similarly, the NACK power offset is used to calculate the power offset for NACK. The ACK/NACK repetition factor is used to indicate the number of consecutive HS-DPCCH subframes on which an ACK or a NACK can be transmitted. It is signaled to the UE and the Node B from higher layers.

Serving HS-DSCH radio link indicator is a Boolean parameter provided for each radio link in the active set. It is used to identify the cell that serves the HS-DSCH.

11.4.2.3 Measurement Information   Measurement information includes measurement feedback information and HS-DSCH-required power per UE weight. Measurement feedback information includes default power offset, channel quality indicator feedback cycle, CQI repetition factor, and CQI power offset. HS-DSCH-required power per UE weight indicates the HS-DSCH percentage of the total HS-DSCH plus HS-SCCH power consumed for a specific UE.

Default power offset specifies the power offset of the HS-PDSCH code channel relative to the P-CPICH used by the UE for calculating the channel quality. Channel quality indicator feedback cycle indicates the intervals at which CQI measurement reports are sent on the uplink HS-DPCCH. It can be 2, 4, 8, 10, 20, 40, 80, or 160 milliseconds. The measurement feedback can also be shut off completely if the channel quality indicator feedback cycle is set to 0.

CQI repetition factor indicates the number of times the same CQI can be sent to the Node B in consecutive subframes on the HS-DPCCH. CQI power offset is a parameter used for calculating the power offset to be used for transmission of the CQI on the HS-DPCCH.

11.4.3 Fixed Parameters

Fixed parameters include transport block size tables, mapping between channelization code set indexes and the actual HS-PDSCH assignments, slot formats, spreading factors, and coding parameters for HS-PDSCH, HS-SCCH, and HS-DPCCH. The transport block size is derived from the transport format resource indicator (TFRI) value signaled on the HS-SCCH. The mapping between the TFRI value and the transport block size is defined in [6].

11.5 General Considerations for HSDPA Deployment

There are many issues that need to be considered when deploying HSDPA. The key issues include deployment strategy, deployment option, user mobility, and impact of HSDPA on Release 99.

11.5.1 Deployment Strategies

Depending on the traffic volume, traffic distribution, and available spectrum, different deployment strategies may be adopted. Some service providers prefer to deploy HSDPA on the same carrier as an existing Release 99, while others prefer to use a separate carrier for HSDPA. When the traffic volume is low, sharing the same carrier with Release 99 is relatively cost effective. However, when the demand of high data throughput increases, more power is required for the HSDPA users, resulting in degradation of existing Release 99 services if they share the same carrier. Use of a separate carrier for HSDPA wastes spectrum initially, but will not have any impact on existing Release 99 services when the data traffic increases later on. Therefore, it is suggested that at the initial deployment stage, HSDPA may share the same carrier with Release 99. When the data traffic increases and requires high data throughput, a separate carrier, if available, may be allocated for HSDPA services alone.

11.5.2 Deployment Options

There are two deployment options for HSDPA. One option is a 1:1 overlay with an existing Release 99 network, while the other is an independent non-coincident HSDPA layer. The former option is normally preferred since its deployment cost is substantially lower than the latter option.

11.5.3 User Mobility

HSDPA may use 16-QAM for modulation. 16-QAM provides two times the throughput of QPSK modulation; but, it requires much higher accuracy in both magnitude and phase than QPSK modulation. This indicates that in order to be effective, 16-QAM is adopted only for radio links with good RF environment. As such, achievement of higher data throughput is generally limited to stationary or slow moving users with line-of-sight to the Node B. In addition, extremely high data throughput of HSDPA is possible only at locations relatively close to the Node B.

11.5.4 Impact of HSDPA on Release 99 Networks

The impact of HSDPA on Release 99 can be addressed from both service and technical viewpoints. From a service point of view, when HSDPA is deployed on the same carrier as an existing Release 99, it might affect the service depending on the mix and density of the services. Many existing Release 99 networks support substantial circuit-switched services, such as 12.2-kbps AMR voice and 64-kbps video telephony. These services will not likely be affected by the implementation of HSDPA. However, for packet-switched services, both existing and new services need to be considered when HSDPA is introduced. This includes data traffic control and distribution between Release 99 and HSDPA. The associated change in billing that will affect the service demand must also be considered.

From a technical point of view, introduction of HSDPA on the same carrier as an existing Release 99 will increase the transmitted power on the downlink and hence degrade the Ec/Io. Field test results indicate that a degradation of about 2 dB in Ec/Io when HSDPA traffic is introduced. Deterioration of Ec/Io will affect the Release 99 cell edge coverage and aggravate the pilot pollution areas, resulting in degradation of Release 99 services. In extreme cases, it is required to redo RF optimization and/or modify the original RF network design in order to minimize the impact of HSDPA on Release 99 networks. To avoid this problem, it is suggested that HSDPA traffic be taken into account when planning a Release 99 network.

In addition, there is no downlink soft handover for HSDPA. The frequency of change of best serving cell should be kept to a minimum to maintain a high throughput. This indicates that the number of active set pilots should be kept to a minimum to reduce the probability of best serving cell change. This, in turn, may involve redesign and/or re-optimization of the RF network to reduce the soft handover areas.

Some service providers have deployed Release 99 networks on top of GSM/GPRS networks with 1:1 overlay. These networks especially need to be re-optimized to achieve the expected high throughput and QoS in targeted service areas when HSDPA is introduced.

11.6 Summary

HSDPA can provide a packet-based data service with data transmission rate around 10 Mbps in a 5 MHz bandwidth. In HSDPA, variable spreading factor and fast power control are disabled and replaced with adaptive modulation and coding, extensive multi-code operation, and fast retransmission. Furthermore, HSDPA employs short transmission intervals (2 ms) and fast scheduling. These features allow the system to benefit from the short-term variations with the scheduling decisions being made at the Node B. Other major HSDPA features include simple code allocation, efficient power allocation, no downlink soft handover, and support of various UE categories.

All Release 99 transport channels terminate at the RNC. The RNC performs RLC ARQ, scheduling, and transport format selection for dedicated and shared channels, while the Node B does not perform any of these functions. In HSDPA, an HSDPA medium access control (MAC-hs) is installed at the Node B. As such, the Node B can perform more functions, including link adaptation, scheduling, HARQ, and resource allocation for HS-DSCH. Packet data retransmissions are now normally carried out by the Node B instead of the RNC. With MAC-hs, retransmissions are controlled by the Node B directly. Therefore, the retransmission execution is faster and the retransmission delays are shorter.

HSDPA adds one additional transport channel and three additional physical channels to the existing Release 99 and Release 4 channels. The additional transport channel is the high-speed downlink shared channel (HS-DSCH). The additional physical channels are the high-speed physical downlink shared channel (HS-PDSCH) and high-speed shared control channel (HS-SCCH) for the downlink, and the high-speed dedicated physical control channel (HS-DPCCH) for the uplink.

The HS-DSCH carries HSDPA payload and offers many new features. These new features include: fast adaptation of transport format to channel conditions, fast scheduling, short transmission time interval with dynamic transport block size, hybrid ARQ with chase combining and incremental redundancy of retransmissions, and mapping of up to 15 physical channels.

The HS-PDSCH, which carries the information payload has a number of specific characteristics compared with Release 99 channels. The transmission time interval is 2 ms or 3 slots. The spreading factor for the HS-PDSCH is always 16. It carries 960 bits per TTI for QPSK and 1920 bits per TTI for 16-QAM. Up to 15 HS-PDSCHs can be used to carry the HS-DSCH. Therefore, 15 consecutive OVSF codes can be used for a set of HS-PDSCHs. HS-PDSCHs can be code multiplexed. It means that a number of UEs can be served at the same time by using different OVSF codes.

The HS-SCCH carries physical layer control information. UTRAN allocates a number of HS-SCCHs corresponding to the maximum number of code-multiplexed users. If the HS-DSCH carries no data, then no HS-SCCH is transmitted. The UE only needs to monitor a maximum of four HS-SCCHs at any given time. HS-SCCH spans 3 slots for each TTI and uses OVSF codes with a spreading factor of 128. It is transmitted in two parts, with slot 1 as the first part, and slots 2 and 3 as the second part. The first part, required prior to HS-PDSCH demodulation, signals channelization codes and modulation. The second part, required for HS-PDSCH decoding, signals transport block size and HARQ parameters.

The HS-DPCCH spanning 3 time slots with a fixed spreading factor of 256 is divided into two parts. The first part, consisting of one slot called the ACK/NACK channel, carries ACK/NACK information. The second part, consisting of two slots called the CQI channel, carries CQI information to indicate the estimated transport block size, modulation type, and number of simultaneous codes that can be received correctly in the downlink direction.

The physical layer operation for HSDPA involves the following steps:

- The UE measures channel quality and reports CQI.
- The Node B evaluates the overall condition for each user and selects the UE to serve in a particular TTI.
- The Node B transmits the HS-SCCHs after it has selected the UE to serve in a particular TTI.
- The Node B transmits a set of HS-PDSCHs, which partially overlaps with HS-SCCHs.
- The UE monitors the HS-SCCHs and decodes the HS-SCCH parameters to determine whether it needs to decode HS-PDSCHs and also to obtain the ARQ information.
- The UE decodes HS-PDSCHs and sends an ACK or NACK to Node B based on the results of the CRC check.

The HSDPA-related parameters can be categorized into three classes: cell parameters, UE parameters, and fixed parameters. The cell parameters are configured in the Node Bs for each individual cell. These parameters are related to radio resource management and are configured either by NBAP common procedures or by operation and maintenance procedures. The UE parameters are assigned either by a Node B or by an RNC. These UE-specific parameters are used in NBAP, RNSAP, and RRC signaling messages. The fixed parameters are specified by the specifications and cannot be configured.

The key issues that need to be considered when deploying an HSDPA system include deployment strategy, deployment option, user mobility, and impact of HSDPA on Release 99.

References

[1] 3GPP TS 25.308, v5.7.0, "High speed downlink packet access (HSDPA; overall description; stage 2," (Release 5).

[2] 3GPP TR 25.877, v5.1.0, "High Speed Downlink Packet Access: Iub/Iur protocol aspects," (Release 5).

[3] 3GPP TS 25.212, v5.4.0, "Multiplexing and channel coding (FDD)," (Release 5).

[4] 3GPP TS 25.214, v5.6.0, "Physical layer procedures (FDD)," (Release 5).

[5] 3GPP TS 25.306, v5.11.0, "UE Radio Access Capabilities," (Release 5).

[6] 3GPP TS 25.321, v5.7.0, "Medium Access Control (MAC) protocol specification," (Release 5).

[7] 3GPP TS25.433, v5.6.0, "UTRAN Iub interface NBAP signaling," (Release 5).

Chapter 12

WCDMA RF Network Planning

The goal of WCDMA RF planning is to generate an RF plan such that the base stations (Node Bs) deployed according to the plan can achieve target coverage and capacity. RF planning involves a variety of technical and non-technical issues, such as site survey, site selection, site acquisition, link budget calculations, antenna selection, coverage prediction, and so forth. Coverage prediction is so complicated that an RF planning tool is required to calculate the coverage. Most WCDMA network equipment vendors and many engineering service/consulting firms have developed their own RF planning tools to carry out the RF planning.

There is no intention to discuss any particular RF planning tool here. Rather, this chapter covers only the basic principles of WCDMA RF network planning. Major subjects discussed include capacity and coverage, link budgets, load control, overhead channel power allocation, scrambling code allocation criteria, and antenna selection. The RF planning process is also summarized at the end of this chapter. It should be noted that many concepts discussed in this chapter also apply to CDMA systems in general. Therefore, throughout this chapter, the generic terms base station and mobile station are used in many occasions to reflect that the concepts under discussion apply to both WCDMA and CDMA systems. In addition, reverse link and uplink are used interchangeably, as are forward link and downlink.

12.1 Capacity and Coverage

Capacity and coverage are fundamental to WCDMA RF network planning and operation. Just like CDMA2000, WCDMA is an interference-limited system. The capacity of a cell is limited by the amount of total interference in the air interface, while cell coverage is dependent on the link budget. Higher capacity means the cell is serving more UEs, while low capacity means the cell is serving fewer UEs. More UE in a cell means the cell load and the interference level are both higher. Higher interference levels translate to smaller link budgets that decrease the coverage. Therefore, there is a correlation between capacity and coverage. Higher capacity means smaller coverage and lower capacity means larger coverage. There is always a trade-off between capacity and coverage.

In addition, for a given coverage, it is possible to increase the capacity by lowering the service quality requirements for various services. This indicates that the capacity limits are soft rather than hard. Hard capacity means the capacity is limited by the hardware or by available channels such as GSM systems. Soft capacity provides the flexibility that a cell can support more UEs at the expense of slightly degrading the call quality of all UEs in the cell.

12.2 Uplink Analysis

Uplink is also known as reverse link. In RF planning, link budget is a key element for the dimensioning process, which provides the cell capacity and the cell radius of a WCDMA cell. As such, within a predefined coverage boundary and with a certain traffic density distribution, the required number of Node Bs can be calculated with reasonable accuracy. Before establishing a link budget, a few technical parameters need to be defined and explained. These include bit rate, traffic load, processing gain, required E_b/I_o, propagation environment, receiver sensitivity, pole capacity, noise rise due to interference, antenna gain and feeder cable loss, fade margin, fast fade margin, body loss, UE transmit power, and Node B receiver noise figure. These parameters are described in the following subsections.

12.2.1 Bit Rate

Bit rate is the base-band user information rate. For AMR voice, the bit rate can be 12.2 kbps or 7.95 kbps. For data, there are a variety of bit rates including 32 kbps, 64 kbps, 128 kbps, 144 kbps, 384 kbps, and 2 Mbps for Release 99, and up to a theoretical value of 14.4 Mbps for HSDPA. Bit rate is needed for calculating the information bit energy from the wideband power. It is also needed for calculating the approximate processing gain.

12.2.2 Traffic Load

For voice and circuit-switched data, the traffic load is expressed in Erlangs. An Erlang is equivalent to 3600 seconds of call. For example, if a user talks on the phone continuously for 30 minutes, the traffic load of that user is 0.5 Erlang.

For packet-switched data, the traffic load is provided in bits per second. The major difference between circuit-switched traffic and packet-switched traffic is that packet-switched traffic, such as high-speed packet data, can tolerate a higher BLER than circuit-switched data because packet-switched data packets can be retransmitted. The capability of tolerating a higher BLER translates to a lower minimum Eb/Io requirement for packet-switched services. Quality of service (QoS) of packet data, such as mean delay and average throughput, should be considered when the traffic load is given as a combination of user numbers and traffic per user.

12.2.3 Processing Gain

For all spread spectrum systems, such as WCDMA and CDMA systems, processing gain alleviates self-interference. Strictly speaking, processing gain should be defined as the ratio of chip rate to air interface symbol rate, plus coding gain at a specific error rate. However, in RF engineering planning, processing gain is generally defined as the ratio of chip rate to data rate. For example, for 12.2 kbps voice, the processing gain is given as $3,840,000/12,200 = 314.75$ or 24.98 dB.

12.2.4 Required Eb/Io

The required Eb/Io is the minimum value at which a Node B receiver can successfully demodulate the received signal for a given BLER—under a specified RF propagation environment. Eb is the received bit energy and Io is the received noise density, including thermal noise and interference.

Under a given propagation environment, the value of the required Eb/Io depends on the types of traffic, the error protection capability of the coding scheme, and the performance of the Node B receiver. For example, for a given Node B receiver, the required Eb/Io value could be higher than 6.5 dB for 12.2-kbps voice and lower than 3.5 dB for 64-kbps packet-switched data. The required Eb/Io also plays an important role in link budget and pole capacity.

12.2.5 Propagation Environment

The propagation environment affects the Eb/Io requirement and hence the coverage and capacity. Propagation environment and UE speed characterize the channel type. There are pedestrian, vehicular, and composite channels. A composite channel may contain different components of pedestrian and vehicular channels. For example, a composite channel may be a combination of 25% AWGN, 37% pedestrian at 3 km per hour, 13% pedestrian at 30 km per hour, 13% vehicular at 30 km per hour, and 12% vehicular at 100 km per hour. Different types of channels have different requirements of Eb/Io at various data rates.

In general, for a defined area to be covered, several types of propagation channels may be used. Therefore, the radio link budget should be calculated for all types of propagation channels. The final channel should be a combination of related channels. The combination may come from field measurement or from the service providers.

12.2.6 Node B Receiver Noise Figure

Noise figure, by definition, is given as $(S_i/N_i)/(S_o/N_o)$, where S_i and N_i are the input signal and noise, respectively; whereas S_o and N_o are the output signal and noise, respectively. Receiver noise figure is determined by the hardware of the receiver and it also reflects the performance of the receiver. A smaller noise figure results in better receiver sensitivity and hence better link budget. Different types of receivers may have different noise figures. Typically, WCDMA Node B receivers have a noise figure of around 3 dB.

12.2.7 Receiver Sensitivity, Pole Capacity, and Loading

Receiver sensitivity is a measure of the receiver's ability to detect a signal. It is defined as the minimum power required per receive diversity branch at the Node B receiver input, and normally expressed in units of dBm. The receiver input is the point where the feeder cable is connected to the receiver.

Receiver sensitivity can be derived from the definition of the received Eb/Io. By definition, the received Eb/Io at the Node B receiver is given as

$$\text{Eb/Io} = (S/R)/\{FN_{th} + \alpha(1 + \beta)(N - 1)S/W\}, \qquad (12\text{-}1)$$

where

- Eb = bit energy
- Io = power spectral density of thermal noise plus interference
- S = received signal strength
- R = information rate
- F = noise figure of the Node B receiver
- N_{th} = thermal noise spectral density (dBm/Hz)
- α = channel activity factor (voice activity)

- β = other cells to serving cell interference ratio
- N = number of active users within the cell
- W = channel bandwidth

To achieve a certain BLER, the received Eb/Io must reach a certain threshold, say Eb/Io = d, which is called the required value of Eb/Io. Defining the processing gain as $G = W/R$ and solving (12-1) for N, the result is

$$N = G/\{\alpha(1+\beta)d\} + 1 - FN_{th}W/\{\alpha(1+\beta)S\}. \quad (12\text{-}2)$$

Pole capacity is defined as the maximum allowable value of N. In (12-2), N will reach its maximum if the third term at the right hand side approaches zero (S approaches infinity). In other words, the number of users N will approach its pole capacity when the receiver sensitivity approaches infinity. By setting S to infinity, the pole capacity, N_{max}, is obtained as

$$N_{max} = G/\{\alpha(1+\beta)d\} + 1. \quad (12\text{-}3)$$

From (12-3), it's obvious that pole capacity is determined by processing gain, channel activity, other cells to serving cell interference ratio, and the required Eb/Io. For example, assume G, d, α, and β are equal to 314.75 (for 12.2 kbps traffic), 3.5, 0.55, and 0.85 (for three-sector Node Bs), respectively, then the pole capacity is equal to 89.

Pole capacity is not the capacity that a cell can reach practically. It is the maximum theoretical air interface capacity of a WCDMA cell and corresponds to infinite noise rise, which is equivalent to 100% cell loading. Cell loading is defined as a fraction of the pole capacity and is given by

$$\mu = N/N_{max}. \quad (12\text{-}4)$$

In a practical WCDMA system, cell loading normally runs from 20% to 70%, depending on the traffic density. For WCDMA RF planning, it is recommended that the cell loading be set at around 50% so that there is room for cell breathing to cope with loading changes. Cell breathing is a phenomenon that the cell radius shrinks as loading increases and expands as loading decreases.

Using (12-3) and (12-4) and solving (12-1) for S, the receiver sensitivity can be obtained as

$$S = (FN_{th})Rd\,[1 + \alpha(1+\beta)d/G]/(1-\mu). \quad (12\text{-}5)$$

As can be seen from (12.5), the receiver sensitivity is a function of the receiver noise figure, required Eb/Io, loading, information rate, channel activity factor, and other cell to serving cell interference ratio.

Example: assume $F = 3$ dB, $d = 5.44$ dB, $R = 12.2$ kbps, $\alpha = 0.55$, $\beta = 0.85$, and $\mu = 50\%$, then the receiver sensitivity S is equal to -121.7 dBm.

12.2.8 Noise Rise Due to Interference

The receiver sensitivity given by (12-5) can also be expressed in terms of pole capacity and cell loading. That is, (12-5) can be rewritten as

$$S = (FN_{th})Rd\,[N_{max}/(N_{max} - 1)]/(1 - \mu). \qquad (12\text{-}6)$$

Normally N_{max} is much larger than 1. Therefore, (12-6) can be approximated as

$$S = (FN_{th})Rd/(1 - \mu). \qquad (12\text{-}7)$$

From (12-7), you can see that with a constant information rate, the noise rise, or the desensitization of the receiver is a function of cell loading. The noise rise is normally expressed in dB and is given by

$$\text{Noise rise due to loading} = -10 \log(1 - \mu). \qquad (12\text{-}8)$$

In the reverse link budget calculation, this noise rise is regarded as the interference margin due to loading. The higher the loading, the larger the interference margin needed in the reverse link, resulting in a smaller link budget and hence a smaller coverage area. Figure 12.1 shows the noise rise as a function of cell loading.

As can be seen in Figure 12.1, the noise rise is small when the cell loading is low, and the noise rise is large when the cell loading is high. Since noise rise reduces the link budget, a lower cell loading allows

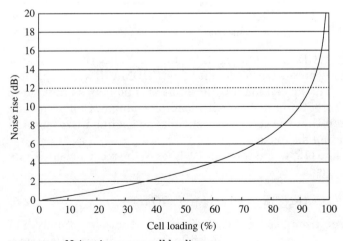

Figure 12.1 Noise rise versus cell loading.

for a larger link budget and hence results in a larger cell radius and a larger coverage area. On the contrary, a higher cell loading allows for a smaller link budget and thus results in a smaller cell radius and smaller coverage area. The cell radius and cell coverage area change as the cell loading changes. This phenomenon is called cell breathing. Figure 12.2 shows the relative coverage versus cell loading. The horizontal axis represents cell loading and the vertical axis represents the coverage area relative to no load.

It is also observed from Figure 12.1 that noise rise increases more rapidly at heavy loading than at light loading. If the cell loading is over 75%, the noise rise increases so rapidly that a small amount of loading increase causes a substantial increase in the noise rise. This will render the system unstable, and therefore cell loading of more than 75% is not recommended for WCDMA systems. On the other hand, for cell loading below 20%, there is no significant noise rise. However, in this case, the cell capacity is too small for practical systems and it is not recommended either. A reasonable cell loading for RF planning purpose is around 50%, at which the cell coverage area is about 70% of that at no load for a typical suburban morphology.

12.2.9 Antenna Gain, Feeder Cable Loss, and Body Loss

Antenna gain, feeder cable loss, and body loss are three components that also need to be included when calculating the link budget. For directional antennas, the antenna gain taken into account is the gain of the main lobe of the antenna pattern. For omni antennas, the antenna gain

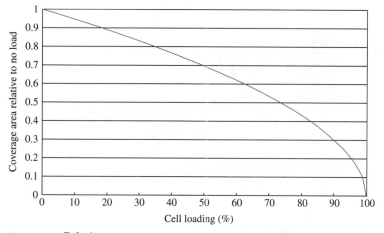

Figure 12.2 Relative coverage area versus percentage loading.

is the same in every horizontal direction. Typical antenna gain ranges from 14 dBi to 20 dBi for directional antennas, and from 11 dBi to 13 dBi for omni antennas.

Feeder cable loss per unit length depends on the cable size, cable type, and operation frequency. For a typical cell site, the total feeder cable loss is generally assumed to be 3 dB for RF planning purpose.

Body loss applies to a handheld mobile station and needs to be considered only for voice. This is because a person, when making or receiving a voice call, often places the mobile station close to one of the ears. The signal will experience a body loss if the mobile station and the base station are at the opposite sides of the person. For data services, a high-speed data mobile station is likely to be PC mounted and hence there is no body loss. The body loss is assumed to be 3 dB for RF planning purpose.

12.2.10 Shadow Fading, Coverage Probability, and Shadow Fade Margin

In RF network planning, shadow fading and coverage probability must be taken into account by adding a shadow fade margin to the link budget. Shadow fading is part of over-the-air path loss. Shadow fading manifests itself in the variations of measured signal strength. Figure 12.3 shows an example of path loss in dB as a function of distance. The values shown in Figure 12.3 are negative. This is because the path loss in this particular example was calculated by subtracting the transmitted

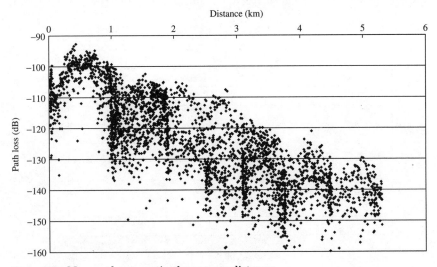

Figure 12.3 Measured propagation loss versus distance.

signal from the received signal. From the path loss data in the figure, it is evident that, at any given distance, there is a significant variation in the path loss around its mean value. This is because some man-made obstacles such as buildings, or nature formations such as vegetation have obstructed the RF path. This phenomenon is called shadow fading or slow fading. Shadow fading scatters the received signal strength. For example, the path loss at a distance of 1.89 km varies from as high as 143 dB to as low as 116 dB in the data shown in Figure 12.3.

Field measurements indicate that at any given distance, the variations of the path loss in dB are Gaussian distributed with a standard deviation of about 8 dB, and the shadow fading is often referred to as lognormal fading. This standard deviation generally remains the same for any radio-path distances. The measured standard deviation of 8 dB near the cell site is due to the nearby buildings surrounding the cell site, while the same standard deviation obtained at a distant location is due to the large variation along various radio paths. The 8 dB standard deviation is used for calculating the fade margin for an outdoor environment in RF planning.

Coverage probability can be defined in terms of cell area coverage probability or cell edge coverage probability. There is a one-to-one correspondence between the two. Cell area coverage probability is defined as the percentage of time that the path loss at any given point within the cell boundary is less than the mean path loss at the cell edge. Or, it is defined as the percentage of points within the cell boundary that have a path loss smaller than the mean cell-edge path loss at any given time.

Similarly, cell edge coverage probability is defined as the percentage of time that the path loss at any given point along the cell edge (boundary) is less than the mean cell-edge path loss. This definition can also be interpreted as the percentage of points along the cell edge that have a path loss smaller than the mean cell-edge path loss at any given time. Here is a specific example to explain the cell-edge coverage probability. Suppose the mean path loss at the original cell edge, shown as circle C1 in Figure 12.4, is 150 dB, which serves as the criterion for coverage. That is, a point is considered covered if its path loss is less than 150 dB; otherwise, it is not covered. Due to the shadow fading discussed above, the path loss at any point along C1 will be greater than 150 dB fifty percent of time and less than 150 dB fifty percent of time. In this case, the cell edge coverage probability along C1 is 50%. Now, if the cell edge is moved inward to circle C2, the chance for the path loss at any point along C2 to be smaller than 150 dB will be more than 50%. In other words, the cell edge coverage probability along C2 is more than 50%. The path loss difference between C1 and C2 is the path loss margin that must be included in the link budget in order to achieve a

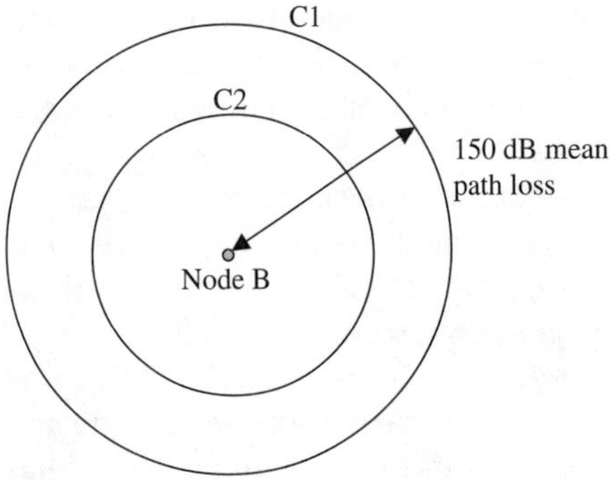

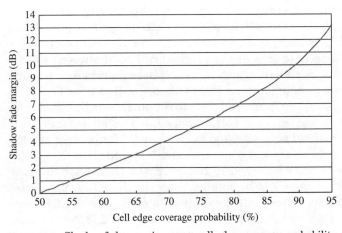

Figure 12.4 Mean path loss from Node B to the cell boundary.

cell edge coverage probability of higher than 50%. This additional path loss margin is called the shadow fade margin.

The amount of fade margin required depends on the cell edge coverage probability and the shadow-fading standard deviation. Assuming the cell edge coverage probability is P, and the shadow-fading standard deviation is σ, the shadow fade margin ρ can be calculated by $\rho = z\sigma$ such that $F(z) = P$, where $F(z)$ is the cumulative distribution function (CDF) of a Gaussian distribution function with zero mean. Figure 12.5 illustrates the shadow fade margin versus cell edge coverage probability, assuming $\sigma = 8$. For example, if cell edge coverage probability is 90%, the shadow fade margin is 10.32.

12.2.11 Fast Fade Margin

Fast fade margin is needed due to fast power control. This is especially true for slow moving mobile stations. When a mobile station is

Figure 12.5 Shadow fade margin versus cell edge coverage probability.

moving slowly like a pedestrian, the fast power control is able to perfectly overcome the fading by increasing the mobile station transmit power. As such, the average Eb/Io requirement for such a channel is small. However, due to perfect power control for fade compensation, the required Eb/Io varies widely as the fading condition changes in the channel. So does the required mobile station transmitted power. This indicates that a fast fade margin on top of the average Eb/Io must be added to the link budget to guarantee the service reliability at the cell edge. Typically, this fast fade margin is 2 to 3 dB.

When a mobile station is moving fast, like a fast-moving car, the average Eb/Io requirement is high in order to compensate for the all the fades in the channel that power control is not fast enough to keep up with. As such, the variations of the required Eb/Io are relatively small. So are the changes of the required mobile station transmitted power. Therefore, the margin required to account for the mobile station transmitted power is relatively small. Typically, it is about 0.5 to 1 dB. Considering both slow-speed mobile stations and high-speed mobile stations, the overall fast fade margin used for RF network planning purposes is about 1.5 dB.

12.2.12 Soft Handover Gain

Soft handover gain is another factor that must be considered in the reverse link budget. The soft handover gain calculation has been developed in detail in [1]. By definition, a UE is in soft handover when it is simultaneously communicating with more than one Node B. In other words, there are at least two different communication links simultaneously associated with the UE, each with a Node B. When one link fades at a certain moment, the other link(s) may not fade at the same moment. RNC always selects the best link for communicating with the UE. As a result, the UE does not need to increase the transmit power to combat the fade as it needs to in a non-handover (single link) situation. The difference in the UE transmitted power between the non-handover and handover situations is the soft handover gain.

The magnitude of soft handover gain depends on the following factors: probability of cell edge coverage, number of simultaneous links between the UE and the Node Bs, standard deviation of the shadow fading, and the correlation between the links. In general, a high probability of cell edge coverage, a large standard deviation of the shadow fading, and a large number of simultaneous links between the UE and the Node Bs will all result in a large soft handover gain. On the other hand, high correlations between the links will result in a small soft handover gain. Typical values for soft handover gain may range from 3 dB for rural areas to as high as 6 dB for dense urban areas.

12.2.13 UE Transmit Power

UE transmit power is defined in the 3GPP specification [2] and is shown in Table 12.1.

Normally, Power Class 4 is used for voice and Power Class 3 is used for data. The power classes define the nominal maximum UE output power. That is, the power in a bandwidth of at least $(1+\gamma)$ times the chip rate of the radio access mode, where $\gamma = 0.22$ is the roll-off factor defined in Section 6.8.1 of [2]. The period of measurement for the transmit power is at least one timeslot.

12.2.14 Penetration Loss

Although penetration loss is not considered in the link budget for outdoor coverage, it is an additional margin to be included in the link budget for indoor coverage. Penetration loss varies with building material and building structure. It is also a function of radio frequency. Normally, higher frequencies have lower penetration losses. For RF network planning, penetration loss should be measured on a city-by-city basis, because the building structures and material may vary widely from city to city. Typically, penetration loss ranges from 20 to 30 dB for dense urban areas, 15 to 25 dB for urban areas, 10 to 20 dB for suburban areas, and below 10 dB for rural areas.

12.2.15 Uplink Budget

Link budget, by definition, is the maximum allowable path loss between the mobile station and the base station. Uplink budget is the basis for calculating the cell coverage radius from the uplink aspect. It is determined by mobile station transmit power, body loss, mobile station antenna gain, mobile station cable loss, base station antenna gain, base station cable loss, base station receiver sensitivity, shadow fade margin, fast fade margin, and soft handover gain. All of these factors have been discussed earlier. Taking all of the above factors into account, the uplink budget is given by

TABLE 12.1 UE Power Classes (Courtesy of ETSI)

Power Class	Nominal Maximum Output Power	Tolerance
1	+33 dBm	+1/–3 dB
2	+27 dBm	+1/–3 dB
3	+24 dBm	+1/–3 dB
4	+21 dBm	± 2 dB

$$L_{path} = P_{tx,mobile} + G_{mobile} - L_{cable\_mobile} - L_{body} + G_{BTS} - L_{cable\_BTS}$$
$$- L_{fade\_margin} - L_{fast\_fade\_margin} + G_{handover} - S_{BTS}, \qquad (12\text{-}9)$$

where $P_{tx,mobile}$ is the mobile station transmit power, G_{mobile} is the mobile station antenna gain, L_{cable_mobile} is the mobile station cable loss, L_{body} is the body loss, G_{BTS} is the base station antenna gain, L_{cable_BTS} is base station cable loss, L_{fade_margin} is the shadow fade margin, $L_{fast\_fade\_margin}$ is the fast fade margin, $G_{handover}$ is the soft handover gain, and S_{BTS} is the base station receiver sensitivity.

Table 12.2 illustrates a reverse link budget for a typical composite channel for various data rates. The values of the required Eb/Io in this example are assumed. They are very close to the industry average. Note that in the real RF network design, the base station equipment vendors should provide the Ec/Io values. This example only calculates the reverse link budget for outdoor (on street) coverage. For indoor coverage, the penetration loss should be taken into account by subtracting it from the outdoor link budget. In some cases, tower mounted amplifiers are used for enhancing the reverse link. The effect of tower mounted amplifiers on the reverse link is elimination of the feeder cable loss, and replacement of the base station noise figure with an effective noise figure for the subsystem consisting of the tower mounted amplifier, cable, and the base station receiver.

12.3 Propagation Models

Once the link budget (maximum allowable path loss) is determined, the cell coverage radius can be calculated by using a propagation model. There are many propagation models that have been proposed and used for RF network planning. These include the Okumura model [3], Hata model [4], COST 231 model [5], and Lee model [6]. These models are briefly described in the next four subsections.

12.3.1 Okumura Model

The Okumura model [3] is one of the most widely used models for signal prediction in urban areas. It is an empirical model applicable for the frequency ranges from 150 MHz to 1920 MHz, and for the distances from 1 km to 100 km with base station antenna heights ranging from 30 meters to 1000 meters. The Okumura model is entirely based on measured data. It is given in dB as

$$L = \text{FSPL} + A(f,r) - G(h_b) - G(h_m) - G_{area}, \qquad (12\text{-}10)$$

where FSPL is the free space path loss, $A(f,r)$ is the median attenuation relative to free space, $G(h_b)$ is the base station antenna height gain

TABLE 12.2 WCDMA Reverse Link Budget Example

Item	Units	12.2 kbps Voice	64 kbps PS	128 kbps PS	Note
(a) Bit rate, R	Bits/s	12200	64000	128000	
(b) UE Tx power available	dBm	21	21	21	UE with power class 4
(c) UE antenna gain	dBi	0	0	0	Included in (b)
(d) Body loss	dB	3	0	0	
(e) UE Tx EIRP	dBm	18	21	21	(b) – (d)
(f) Node B Rx antenna gain	dBi	17	17	17	Assumption
(g) Rx cable and connector loss	dB	4	4	4	Assumption
(h) Receiver noise figure	dB	3	3	3	Assumption
(i) Thermal noise density	dBm/Hz	–174	–174	–174	kT
(j) Required Eb/Io	dB	5.45	2.25	1.80	Typical values
(k) Pole capacity, N_{max}		89	36	20	Equation (12.3) with $\alpha = 0.55$ and $\beta = 0.85$.
(l) Cell loading, N/N_{max}	%	50	50	50	Assumption
(m) Noise rise	dB	3.01	3.01	3.01	$-10*\log(1 - N/N_{max})$
(n) Bit rate	dB	40.86	48.06	51.07	$10*\log(R)$
(o) Receiver sensitivity	dBm	–123.6	–118.1	–115.2	(h)+(i)+(j)+(m)+(n)+$10*\log[N_{max}/(N_{max}-1)]$
(p) Shadow fade margin	dB	10.3	10.3	10.3	Assume 90% edge coverage
(q) Fast fade margin	dB	1.5	1.5	1.5	Assumption
(r) Soft handover gain	dB	4	4	4	
(s) Maximum allowable path loss, outdoor	dB	146.8	144.3	141.4	(e)+(f)–(g)–(o)–(p)–(q)+(r)

factor, $G(h_m)$ is the mobile station antenna height gain factor, G_{area} is gain due to type of RF environment, f is the frequency, r is the distance, h_b is the base station antenna height, and h_m is the mobile station antenna height. Each term at the right-hand side of (12-10) is described below:

$$\text{FSPL} = 20 \log(4\pi r f/c), \qquad (12\text{-}11)$$

where c is the speed of light.

$$G(h_b) = 20 \log(h_b/200) \text{ for } 30 \text{ m} < h_b < 1000 \text{ m} \qquad (12\text{-}12)$$
$$G(h_m) = 10 \log(h_m/3) \text{ for } h_m \leq 3 \text{ m},$$
$$\quad 20 \log(h_m/3) \text{ for } 3 \text{ m} < h_m < 10 \text{ m} \qquad (12\text{-}13)$$

G_{area} depends on frequency as well as type of RF environment. At 2 GHz, it is about 33 dB for open area, 27 dB for quasi-open area, and 13 for suburban area. The median attenuation $A(f,r)$ is a function of frequency and distance. At 2 GHz, it is about 22 dB for a distance of 1 km and 33 dB for a distance of 10 km.

12.3.2 Hata Model

Masaharu Hata established the Hata model in 1980 by using the information in the field strength curves produced by Okumura [3]. The Hata model is a set of empirical formulas for path loss. For urban areas, the Hata model path loss is given by

$$L_{\text{urban}} = 69.55 + 26.16 \log(f) - 13.82 \log(h_b) - a(h_m)$$
$$+ [44.9 - 6.55 \log(h_b)] \times \log(r), \qquad (12\text{-}14)$$

where r is the distance between the base station and the mobile station in kilometers, f is in MHz, h_b and h_m are in meters, and $a(h_m)$ is the correction factor for effective antenna height. The correction factor $a(h_m)$ is given by

$a(h_m) = [1.1 \log(f) - 0.7] h_m - [1.56 \log(f) - 0.8]$ dB for medium sized cities,

$8.29 [\log(1.5 h_m)]^2 - 1.1$ dB for large cities at $f \leq 300$ MHz,
$3.2 [\log(11.75 h_m)]^2 - 4.97$ dB for large cities at $f > 300$ MHz.

For suburban areas, the Hata model path loss is given by

$$L_{\text{suburban}} = L_{\text{urban}} - 2[\log(f/28)]^2 - 5.4. \qquad (12\text{-}15)$$

For open rural areas, the Hata model path loss is given by

$$L_{\text{rural}} = L_{\text{urban}} - 4.78 [\log(f)]^2 + 18.33 [\log(f)] - 40.94. \qquad (12\text{-}16)$$

The Hata model is valid for the following conditions: 150 MHz $< f <$ 1500 MHz, $r <$ 20 km, 30 m $< h_b <$ 200 m, and 1 m $< h_m <$ 10 m.

12.3.3 COST 231-Hata Model

The European Cooperative for Scientific and Technical Research project has studied the propagation in the PCS band. The study has resulted in two models: COST 231-Hata model and COST 231-Walfish-Ikegami model. The COST 231-Hata model extends the Hata model to 1500–2000 MHz. As such, it is also called the extended Hata model. The model is given as

$$L = 46.3 + 33.9 \log(f) - 13.82 \log(h_b) - a(h_m)$$
$$+ [44.9 - 6.55 \log(h_b)] \log(r) + C, \quad (12\text{-}17)$$

where the constant C is a function of morphology. For example, for medium city and suburban areas, the value of C could range from –6 to 0, and for metropolitan centers, it could range from 0 to 3. The model applies to the following range of parameters: 1500 MHz $< f <$ 2000 MHz, 1 km $< r <$ 20 km, 30 m $< h_b <$ 200 m, and 1 m $< h_m <$ 10 m.

Although both Hata and COST 231-Hata models are limited to base station antenna heights greater than 30 meters, they can still be used for lower base station antenna heights provided that the surrounding buildings are well below the base station antenna.

The COST-Walfish-Ikegami model applies to the cases in which the base station antennas are either above or below the rooftops. It takes roof heights, street widths, and street orientation (relative to radio path) into account. This model works best when the base station antennas are much higher than the rooftop. It does not work well when the base station antennas are much lower than the rooftop because it does not consider street canyon effect and diffraction at street corners.

12.3.4 Lee Model

The Lee model is a mobile point-to-point model. In a general form, it is given in dB as [6]

$$L = K + g \log(r) - 20 \log(h_b) - 10 \log(h_m) - G_{\text{BTS}} - G_m, \quad (12\text{-}18)$$

where G_{BTS} and G_m are the base station antenna gain and mobile station antenna gain, respectively. The values of K and g vary and need to be measured for different RF environments.

12.4 Downlink Analysis

Downlink analysis differs from uplink analysis in a number of ways. First, the purpose of downlink analysis is not to determine the coverage area, but to ensure that the base station has sufficient power to support all the users within the coverage area dictated by the uplink. Second, in downlink, the base station power is shared among multiple users, as opposed to the mobile station transmit power that is dedicated to that mobile station in uplink. Third, in downlink, the interference of other cells at the mobile station is greater than the other cell interference to the serving cell in uplink, because a mobile station at the cell edge may be subjected to a significant amount of interference from the neighboring base stations. Fourth, the downlink analysis does not use the Eb/Io requirement; it uses the fractional downlink power, Ec/Ior, which is a function of mobile station geometry. The geometry is defined as the ratio of the total received power from the cells inside the active set to the sum of receiver noise and the total received power from all cells not inside the active set.

For downlink, it is very complicated and unpractical to give a specific link budget, since the Eb/Io requirement is very sensitive to the RF condition as well as mobile station speed. Normally, the downlink analysis is done using simulations, which assume a predefined number of mobile stations randomly distributed within the coverage area determined by the uplink budget and calculate the percentage of the mobile stations that can be served by the system. The simulation is an iterative process in which the base station power is modified once for each iteration cycle until it reaches a steady state. A predefined number of mobile stations are equivalent to a predefined capacity. Therefore, the downlink analysis, in essence, considers both the capacity and coverage. If the base stations do not have enough power to support all the mobile stations within the coverage area, either the capacity of the base stations or the coverage radius from the uplink budget needs to be scaled back until the base stations can support all the mobile stations in question. The output of the simulations should provide, for a given UMTS service, the maximum cell radius for a given traffic value per cell.

Due to the complexities in downlink analysis, many wireless equipment vendors and service providers often use dimensioning tools to estimate the rough cell count with a predefined coverage boundary and a predefined traffic capacity. Discussion of dimensioning tools is beyond the scope of this book.

12.5 Overhead Channel Power Allocation

Power allocation for downlink overhead channels is also an important issue in RF network planning. The downlink overhead channels include CPICH, SCCPCH, PCCPCH, P-SCH, S-SCH, and PICH. The pilot power allocation directly affects the coverage probability of the pilot channel. Normally the pilot power is about 10% of the base station transmit power. This value is based on the assumptions that the Ec/Io requirement is −15 dB, and the other-cells-to-same-cell interference ratio is 4 dB. By definition, Ec/Io is defined as

$$\text{Ec/Io} = [\eta P_{\text{BTS}} L(r)/W]/(N_{\text{th}} + I_{\text{oc}} + I_{\text{sc}}), \qquad (12\text{-}19)$$

where Ec is energy per chip, Io is the total interference density, P_{BTS} is the base station transmit power, W is the carrier bandwidth, $L(r)$ is the transmission loss, N_{th} is the thermal density, I_{oc} is the other cell interference density, and I_{sc} is the same cell interference density. Solving (12-19) for η results in

$$\eta = (Ec/Io)[N_{th}W + I_{sc}W(1+ I_{oc}/I_{sc})]/[P_{BTS}L(r)]. \quad (12\text{-}20)$$

The same cell interference is given by

$$I_{sc}W = (1-\eta)P_{BTS}L(r). \quad (12\text{-}21)$$

Substituting (12-21) into (12-20) and solving for η again, obtains the pilot power fraction as

$$\eta = (Ec/Io)[N_{th}W/P_{BTS}L(r) + (1 + I_{oc}/I_{sc})]/[1 + (Ec/Io)(1 + I_{oc}/I_{sc})]. \quad (12\text{-}22)$$

From Table 12.2, you can see that the link budget for 12.2 kbps voice is greater than that of any packet-switched data rate. The transmission loss $L(r)$ for 12.2 kbps voice is about −132.8 dB (base station antenna gain-body loss−reverse link budget). Assuming a base station transmit power of 43 dBm, the value of $P_{BTS}L(r)$ is 43−132.8 = −89.8 dBm. The thermal noise in the carrier frequency band is $N_{th}W$ = −108.2 dBm. It is obvious that $N_{th}W \ll P_{BTS}L(r)$. Therefore, (12-22) can be simplified to

$$\eta = (Ec/Io)(1 + I_{oc}/I_{sc})/[1 + (Ec/Io)(1 + I_{oc}/I_{sc})]. \quad (12\text{-}23)$$

Assume Ec/Io = −15 dB as the criterion for pilot coverage and I_{oc}/I_{sc} = 4 dB, the pilot power fraction is, from (12-23), about 10%. This value holds for all data rates and services.

The power for other overhead channels is set relative to the pilot power. In common practice, the recommended power settings for other overhead channels are as follows:

- P-SCH: −3 dB relative to pilot with 10% duty cycle
- S-SCH: −5 dB relative to pilot with 10% duty cycle
- PCCPCH: −3 dB relative to pilot with 90% duty cycle and 10% BLER
- SCCPCH: 4 dB relative to pilot with a duty cycle of 20% (10% FACH and 10% PCH) and a BLER of 10%
- PICH: −6 dB relative to pilot with a 96% duty cycle and a BLER of 5%
- AICH: −5 dB relative to pilot with 1% duty cycle

TABLE 12.3 Power Settings for the Overhead Channels

Channel	Power Relative to Pilot (dB)	Absolute Power in dBm	Duty Cycle	Absolute Power in Watts
CPICH	–	33	100%	2.000
P-SCH	–3	30	10%	0.100
S-SCH	–5	28	10%	0.063
PCCPCH	–3	30	90%	0.900
PICH	–6	27	96%	0.480
AICH	–5	28	1%	0.006
SCCPCH (FACH)	4	37	10%	0.500
SCCPCH (PCH)	4	37	10%	0.500
Total power for overhead channels				4.549

Assume the base station transmit power is 43 dBm, the power settings for the overhead channels are summarized in Table 12.3.

From Table 12.3, you can observe that the total power for the overhead channels is 4.55W or 22.75% of the base station transmit power.

12.6 Scrambling Code Planning

The goal of scrambling code planning is to assign scrambling codes to the cells such that the UE will correctly identify the cell scrambling code during the initial acquisition procedure. To achieve this goal, the scrambling codes for the neighboring cells must have as small cross-correlation values with each other as possible. As pointed out in Chapter 7, the cross-correlation between any two scrambling codes is very low regardless of the timing offset between them. It is very reasonable to assume that the scrambling codes are nearly orthogonal and therefore impose no significant requirement on scrambling code planning.

As already discussed in Chapter 7, there are 512 primary scrambling codes, which are divided into 64 code groups. Each of these groups consist of eight primary scrambling codes. In WCDMA networks, the neighboring cells can be assigned scrambling codes in a variety of ways. The method of assignment will affect the processing requirements and synchronization performance at the UE [7]. For example, for a given cell, if the number of its neighboring cells is less than 8, the cell, together with its neighboring cells, could use different primary scrambling codes belonging to the same code group. They could also use different scrambling codes with each code coming from a different code group. The first approach shifts the processing burden in the initial acquisition procedure from frame synchronization and code-group identification to scrambling-code identification, while the second approach does just the reverse. In general, a good scrambling code planning approach would be a trade-off between the processing burden on the UE and the synchronization time.

12.6.1 Scrambling Code Planning Example

Assume there is a cluster consisting of 19 Node Bs with each Node B having 3 cells (sectors). A total of 57 scrambling codes are required. These 57 scrambling codes may come from 57 code groups with one scrambling code from one code group. Since the scrambling codes are orthogonal to one another, the assignment of the 57 code groups to the cells may be arbitrary. The only requirement is that the once the assignment pattern is fixed for one cluster, the same pattern should be repeated for other clusters. This will guarantee that the same code group in different clusters will not be placed close to each other, hence reducing the probability that the same scrambling code is used for two cells that are not physically far away from each other. Figure 12.6 shows an example of code group assignment to the 57 cells in a cluster, in which the cells are arranged in hexagonal rings around a center Node B.

From the geometry of Figure 12.6, the code group reuse distance is equivalent to seven tiers of cells. This distance is quite sufficient for providing path loss to alleviate interference and avoid ambiguity on cell identification in most practical networks, assuming flat terrain. For hilly terrains, the reuse distance requirement may be more stringent or relaxed, depending on individual cases.

As a matter of fact, the downlink scrambling code group allocation scheme exemplified by the above example is straightforward and easy to implement. The arrangement of 57 different code groups in the 57 cells

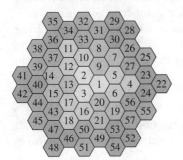

Figure 12.6 Example of scrambling code group assignment.

shown above may be random as long as the assignment pattern is kept the same for the clusters that are adjacent to each other. It should be noted that in practice the coverage area is not truly hexagonal. Also, in some cases, a Node B may have more than three cells, reducing the reuse distance. Furthermore, network operators may like to reserve more scrambling code groups for hierarchical cell structure and/or variable cell numbers per Node B. All these factors need to be taken into account during the scrambling code planning.

12.7 Base Station Antennas

Base station antennas are very important to the performance of a wireless network. Based on RF environment and coverage objectives, RF engineers should define the requirements for antennas to be used. The requirements must spell out the antennas electrical and mechanical specifications. These specifications include antenna gain, horizontal beam width, vertical beam width, frequency range, input impedance, power rating, voltage standing wave ration (VSWR), polarization, front-to-back ratio, down tilt, physical size, weight, wind load, connector types, installation accessories, and so forth.

12.7.1 Antenna Gain and Beam Width

UMTS systems normally operate at 2 GHz frequency range. The antennas at this frequency range have a higher gain than 850-MHz antennas of comparable physical size. Since the propagation loss at 2 GHz is higher than 850 MHz, it is recommended to use high gain antennas for WCDMA networks. This will not only compensate for part of the propagation loss, but also provide some indoor coverage for small and medium size buildings. Selection of antennas also depends on morphology. For urban areas, antennas with a gain of 15–19.5 dBi, a horizontal beam width of 65° to 90°, and a vertical beam width of 6° to 12° are recommended. For suburban and rural areas, antennas with a gain of 12–17 dBi, a horizontal beam width of 90° to 110°, and a vertical beam width of 6° to 12° are recommended. In some rural areas where capacity requirements are not high, omni antennas with a gain of 9 to 12 dBi can also be used. For freeways, the antennas used normally have a horizontal beam width of 30° to 60° and a gain of 15 to 18 dBi.

In the previous discussion, dBi was used as the antenna gain unit. As a matter of fact, antenna gain can also be expressed in terms of dBd. The unit dBi specifies the antenna gain with regards to an isotropic antenna, which does not actually exist in reality but serves as a theoretical reference. The unit dBd specifies the antenna gain with regards to a dipole antenna, which is the most basic antenna. For example, a short segment of metal wire can serve as a dipole antenna. There is a 2.15 dB difference between using dBi and using dBd as the reference point. For example, a 15-dBi antenna means a 12.85-dBd antenna. RF engineers normally use dBi in the link budget calculation, while most antenna manufacturers use dBd in their antenna specifications.

12.7.2 Antenna Down Tilt

In wireless networks, there are two common problems, namely, coverage holes and co-channel interference from other cells. These problems are often solved by down tilting the antennas. However, when down tilting the antenna, the coverage area also shrinks.

There are two types of antenna down tilts: electric down tilt and mechanical down tilt. In some cases electric down tilt is used, while in other cases electric down tilt is used. There are also many cases that a combination of electric and mechanical down tilts is used. The difference between electric down tilt and mechanical down tilt is that electric down tilt does not change the antenna radiation pattern, while mechanical down tilt does. Mechanical down tilting shrinks the main lobe coverage area but not the side lobe coverage areas. As such, the antenna coverage area will change from circular shape to elliptic shape. This sometimes will increase the overlap areas of the cells, degrading the ability to reduce interference. Electric down tilting shrinks the coverage area uniformly in all directions. Therefore, in order to reduce the inter-cell interference, electric down tilt antennas are preferred for urban area deployment.

12.7.3 Side Lobe Suppression and Null Fill

In addition to down tilt, there are two other techniques that can improve the antenna radiation pattern, namely, upper side lobe suppression and null fill. When a typical antenna is deployed, its upper side lobe may point to the neighboring cell, especially when mechanical down tilt is used. This may cause co-channel and adjacent channel interference. Suppression of the upper side lobes can alleviate this interference. Currently there is no industry standard specified for upper side lobe suppression. However, it is a common practice that if the first upper side lobe of the antenna is 15 dB below the main lobe, the antenna is considered as an upper-side-lobe-suppressed antenna.

Null fill is another technique that is employed to improve antenna performance. In a typical antenna, there are nulls between the main lobe and first side lobe, as well as between the side lobes. These nulls cause the non-uniform radiation pattern within a cell. This results in coverage discontinuity and hence affects the performance of the network. Null fill can eliminate the nulls of the antennas. Again, there is no industry standard for null fill. However, an antenna is considered as a null-filled antenna if the difference between the null and the main lobe is less than 18 dB.

12.7.4 Dual Polarization Antenna

A dual polarization antenna is a combination of two antennas in one structure with their polarizations orthogonal to each other. The receive diversity gain is achieved by polarization diversity instead of spatial diversity. Currently there are two types of dual polarization antennas on the market: 45° slanted polarized antennas and vertical/horizontal-polarized antennas. In 45° slanted polarized antennas, the polarization of one antenna is +45° relative to vertical and the other is −45° relative to vertical. Field tests indicate that the forward link may decrease by 2 dB due to 45° slant in polarization. In a vertical/horizontal-polarized antenna, the polarization of one antenna is vertical, while the other is horizontal. The vertical polarization is used for transmission, while both polarizations are used for receiving.

Dual polarization antennas are often used in dense urban areas where there is limited space for antenna installation, and in the locations where multiple path reflections and/or scattering exist. While slanted polarized antennas are frequently used, the vertical/horizontal-polarized antennas are not due to their poor receive diversity.

12.7.5 Voltage Standing Wave Ratio and Front-to-Back Ratio

Voltage standing wave ratio (VSWR) is a parameter frequently used for determining the compatibility of the components in an antenna. By definition, VSWR is given by

$$\text{VSWR} = (V_+ + V_-)/(V_+ - V_-), \qquad (12\text{-}24)$$

where V_+ is the incident voltage signal to the antenna and V_- is the reflected voltage signal from the antenna. In many countries, the required specification for VSWR of base station antennas is VSWR ≤ 1.5. Field experience indicates that a VSWR of less than 1.35 makes the system perform much better. It is recommended that a VSWR of less than 1.4 be required for antennas used in a WCDMA system.

The front-to-back ratio is an indication of how much energy an antenna is radiating backward. It is defined as the ratio of main lobe gain and back lobe gain. A large front-to-back ratio will prevent adjacent cell interference due to backward radiation. The minimum requirement for front-to-back ratio is about 22 dB. Nowadays, most antennas have a front-to-back ratio of 25 dB or more.

12.7.6 Mechanical Specifications of Antennas

When selecting antennas, you also need to consider the mechanical parameters of the antennas, such as physical size, weight, wind load, temperature range, and installation accessories. These parameters are very important because they affect the engineering of antenna installation and the performance of the antenna and cable systems.

12.8 WCDMA RF Planning Process

An RF planning process encompasses a number of steps. Before starting RF planning, there must be a clear understanding of the goals for the network. These include the frequency band to be used, capacity of the network, and coverage boundaries. Normally the frequency band is auctioned or assigned to the service providers for specific technologies (for example, UMTS, CDMA, TDS-CDMA). The information on network capacity and coverage boundaries should be obtained from the service providers or deduced from reasonable assumptions. The capacity-related issues include the target total number of subscribers, Erlangs per user, traffic distribution, blocking rate, data traffic model, and user distribution at various data rates (i.e. 64 kbps, 128 kbps, and 384 kbps, etc.). The coverage-related issues include coverage area, morphology definition, coverage probability, and in-building penetration. Most of these issues have already been discussed in the previous sections.

An RF planning process is summarized in Figure 12.7. Firstly, the RF dimensioning tool is used to generate the cell site quantity and cell

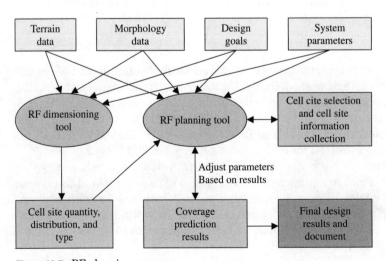

Figure 12.7 RF planning process.

site distribution. Secondly, the RF planning tool is used to predict the coverage area. Both tools use the terrain data, morphology data, design goals, and system parameters as inputs. In addition, the RF planning tool also uses the output of the RF dimensioning tool as an input.

If the predicted coverage does not meet the RF design goals, the system parameters and/or cell site parameters are adjusted, based on the predicted results. The adjustments may include changes of antenna azimuth, down tilt, antenna height, and cell site location. The adjusted parameters are fed into the RF planning tool for another run of coverage prediction. Several iterations may be required before the RF design goals are fulfilled. Once the predicted coverage meets the design goals, the final design results are documented and used as the baseline for network deployment.

12.9 Summary

RF network planning is essential prior to the deployment of any wireless network. RF network planning predicts the coverage and capacity based on a set of predefined RF parameters. It also estimates the number of base stations that are required to meet the capacity and coverage targets. For any given cell, there is always a trade-off between capacity and coverage. A higher capacity means the cell is serving more UEs, resulting in a higher cell load and thus a higher interference level. This in turn translates to a smaller link budget and hence a smaller coverage area. Link budget, which provides the cell capacity and the cell radius, is an indispensable element in RF planning. The parameters that can affect a link budget include bit rate, traffic load, processing gain, required Eb/Io, propagation environment, receiver sensitivity, pole capacity, noise rise due to interference, antenna gain and feeder cable loss, fade margin, fast fade margin, body loss, UE transmit power, and Node B receiver noise figure.

Once the link budget is determined, the cell radius can be calculated by using a propagation model. Many propagation models have been proposed during the past few decades. The most frequently used include the Okumura model, Hata model, COST 231 model, and Lee model.

Downlink analysis is also indispensable in RF network planning. It differs from uplink analysis in many ways. First, the purpose of downlink analysis is to ensure that the base station has sufficient power to support all the users within the coverage area dictated by the uplink. Second, in downlink, the base station power is shared among multiple users. Third, in downlink, the interference of other cells at the mobile station is greater than the other cell interference to the serving cell in uplink. Fourth, the downlink analysis does not use an Eb/Io requirement; it uses the fractional downlink power, Ec/Ior, which is a function of mobile station geometry.

Normally, the downlink analysis is carried out using simulations. The simulation assumes a predefined number of fixed-position base stations with coverage area dictated by the uplink budget. It also assumes a predefined number of mobile stations randomly distributed within the coverage area. It then calculates the percentage of mobile stations that can be served by the base stations. The simulation is an iterative process in which the base station power is modified once for each iteration cycle until it reaches a steady state. The downlink analysis considers both the capacity and coverage. If the base stations do not have enough power to support all the mobile stations within the coverage area, either the capacity of the base stations or the coverage radius needs to be reduced until the base stations can support all the mobile stations. Power allocation for downlink overhead channels is also very important in RF network planning. The downlink overhead channels include CPICH, SCCPCH, PCCPCH, P-SCH, S-SCH, and PICH. The pilot power allocation directly affects the coverage probability of the pilot channel. In general, the pilot power is about 10% of the base station transmit power. The power for other overhead channels is set relative to the pilot power.

In addition to power allocation for downlink overhead channels, scrambling code planning is also a part of RF network planning. Scrambling code planning is assigning the scrambling codes to the cells. The UE identifies the cell scrambling code during the initial acquisition procedure. Scrambling code planning ensures that the scrambling codes for the neighboring cells have as small cross-correlation values with each other as possible. Base station antennas always play an important role in the performance of any wireless network. Selection of base station antennas should be based on RF environment and coverage objectives. The selection requirements must spell out the antennas' electrical and mechanical specifications. These specifications include antenna gain, horizontal beam width, vertical beam width, frequency range, input impedance, power rating, voltage standing wave ratio (VSWR), polarization, front-to-back ratio, down tilt, physical size, weight, wind load, connector types, and installation accessories.

To make RF network planning succeed, an RF planning process is also very essential. A well-designed RF planning process makes RF network planning easier to carry out. Also, before starting RF planning, it's important to have a clear understanding of the network to be planned. This includes the frequency band to be used, capacity of the network, and coverage boundaries. The major tools used for RF planning are an RF dimensioning tool and an RF planning tool. The RF dimensioning tool generates the cell site quantity and distribution. The RF planning tool predicts the coverage area.

References

[1] A. J. Viterbi, A. M. Viterbi, K. S. Gilhousen, and E. Zehavi, "Soft handoff extends CDMA cell coverage and increases reverse-link capacity," IEEE J. Selected Areas of Communication, vol. 12, pp. 1281–1288, October 1994.

[2] 3GPP TS25.101, User Equipment (UE) radio transmission and reception (FDD).

[3] Yoshihisa Okumura, et al., "Field strength and its variability in VHF and UHF land-mobile radio service," Review of the Electrical Communications Laboratory, vol. 16, No. 9–10, September–October 1968.

[4] Masaharu Hata, "Empirical formula for propagation loss in land mobile radio services," IEEE Transactions on Vehicular Technology, vol. 29, No. 3, August 1980.

[5] COST 231, Digital mobile radio towards future generation systems, Final Report, COST Telecom Secretariat, European Commission, Brussels, Belgium, 1999.

[6] William C. Y. Lee, "Mobile Cellular Telecommunications," 2nd Edition, McGraw-Hill, 1995.

[7] Stamatis Kourtis, "Code Planning Strategy for UMTS FDD Networks," IEEE Vehicular Technology Conference 'VTC 2000' vol. 2. pp. 815–819.

Chapter 13

WCDMA RF Network Optimization

RF optimization is an indispensable step in the deployment of wireless networks. It is often also a continuous process well into the operation stage of the network. It involves the assessment and improvement of network performance. WCDMA RF optimization is similar to that of any other technology's wireless networks. In general, the overall WCDMA RF optimization process consists of three phases:

1. Pre-optimization phase
2. Drive-test-based optimization phase
3. Traffic-statistics-based optimization phase

In the pre-optimization phase, examinations are carried out to check whether the base station is correctly installed, hardware is working properly, and the power settings are properly set. An antenna audit is also conducted to make sure the antenna height, tilts, and azimuths are correctly set according to the design document. A well-done pre-optimization can substantially reduce the amount of work during the drive-test-based optimization phase.

The drive-test-based optimization is the primary and time-consuming optimization phase. It is performed prior to a commercial launch of the network. The goal of the drive-test-based optimization is to make sure that the network provides adequate coverage and capacity and to ensure the good performance and quality of the network.

The traffic-statistics-based optimization is performed in a commercial network with live traffic. In this phase, network equipment vendors or network operators often use software tools and specific network performance counters to identify and optimize the problem areas.

This chapter will address the RF optimization issues, discuss the drive-test-based optimization in detail, and briefly explain the traffic-statistics-based optimization. Special RF optimization techniques, such as hierarchical cell structure or in-building optimization, are not covered. The drive-test-based optimization discussed in this chapter applies to a variety of wireless networks, including brand new wireless mobility networks, expansion of existing networks, and overlay of a wireless mobility system over an existing system of a different air interface technology.

13.1 RF Optimization Overview

The two most important aspects of optimization are coverage and capacity. Optimizing a network basically is optimizing its coverage and capacity. Depending on the specific network requirements, a network may need to maximize its coverage while sacrificing its capacity. For example, an early-stage green field network, where cell count is limited, may place the emphasis on providing large coverage with low capacity. On the other hand, a network may need to maximize its capacity while keeping its coverage unchanged. For example, in dense urban areas, cells are added to provide capacity, while coverage is not an issue.

There are many factors that can affect the coverage and/or capacity of the network. The major ones include pilot pollution, missing neighbors, and around-the-corner problems. Any of these factors may result in dropped calls, origination failures, and termination failures. In addition, coverage is also deeply affected by cell breathing: cell coverage shrinks as cell loading increases.

Generally speaking, the coverage area may be defined as the area where the pilot Ec/Io is larger than a minimum threshold in the forward link (for example, −11 dB for 384 kbps circuit-switched data). It may also be defined as the area where the forward link traffic Eb/Io is greater than a required threshold, or the area where the uplink traffic Eb/Io is greater than the required value (for example, 3.4 dB for 64 kbps data). Strictly speaking, the coverage area shall be defined as the overlap area of these three areas.

The required Eb/Io (both downlink and uplink) and Ec/Io depend on services and quality defined for the specific network. The achievable Eb/Io is directly related to the path loss. Higher path losses result in lower Eb/Io values, while lower path losses result in higher Eb/Io values. The maximum allowable path loss is governed by the link budget. During RF optimization, coverage often can be improved by adjusting the antenna orientation, antenna down tilts, and Node B output power.

13.2 Issues in RF Optimization

There are many inherent issues that need to be resolved in WCDMA systems during the optimization stage. These issues include cell breathing, pilot pollution, near-far problem, around-the-corner problem, handover, and incomplete neighbor lists.

13.2.1 Cell Breathing

Cell breathing exists in all systems using CDMA technology. There is no exception for WCDMA system. Cell breathing is a phenomenon that occurs when the cell coverage changes with cell loading. An increase of cell load increases the network interference, resulting in service quality degradation at the original cell coverage border, and thus shrinks the coverage area. On the contrary, low load leads to low network interference, and thus, increases the cell coverage.

To deal with cell breathing, RF optimization is often performed using a simulated network traffic load for the forward link during the pre-commercial launch stage. The traffic load simulation generates downlink interference to simulate live traffic. On the uplink, an attenuator is often attached to the UE to simulate the traffic load.

13.2.2 Pilot Pollution

Pilot pollution occurs when there is an absence of a dominant pilot, or there are strong pilots that are not added to the active set. Such situations are often found at street intersections, elevated highways, bridges, lakes, and the upper floors of buildings.

In pilot pollution areas, the Ec/Io is low though the pilot may have sufficient Ec signal strength. However, low Ec/Io may also be due to insufficient Ec signal strength caused by excessive path loss. In this case, the problem area is regarded as a coverage hole not a pilot pollution area. As such, RF engineers should determine whether the low Ec/Io is due to pilot pollution or to excessive path loss before conducting iterative drive tests and attempting to establish a dominant pilot by adjusting the antenna and system parameters. Once pilot pollution is identified, the next step is to increase the dominant pilot coverage and reduce the coverage of the weaker pilots. This can be achieved by adjusting the antenna down tilts and/or P-CPICH channel power. During the drive-test optimization process, attention should also be paid to maintain continuous coverage through soft handover.

Multi-paths may also generate interference if some strong multi-path signals are not received and demodulated by the UE rake receiver. For example, a six-finger rake receiver may be fully occupied when it receives two pilots each with three multi-paths. Other pilots and multi-path signals become interference.

13.2.3 Near-Far Problem

When UE is near a cell site and transmits at a power level of more than required, it will create excessive interference for UE that is located far away from the cell site. Therefore, closed loop power control is required to instruct the UE to quickly throttle up or throttle down the transmit power so that the cell site will receive equal signal strengths from all UEs in the cell regardless of their distances from the cell site.

RF optimization for this problem is to make sure that all power control mechanisms are working properly. There can be a power control failure if Node B or the UE is always transmitting at full power even when the block error rate is small. If a power control failure occurs, power control parameter settings should be examined and adjusted.

13.2.4 Around-the-Corner Problem

An around-the-corner problem occurs when a UE travels into an area where there is a strong pilot from a new cell site that was not previously serving the UE. The downlink will degrade temporarily until the handover is performed. When the UE goes into handover with the new cell site, fast power control is required to quickly reduce new cell site transmit power.

RF optimization for this problem is similar to that for the near-far problem. The power control mechanisms should be examined to make sure that they are working properly. The around-the-corner problem often occurs at a street intersection or an elevated highway. To solve the around-the-corner problem, in addition to examining the power control mechanisms, the handover parameters and/or the antenna orientation and antenna down tilts of the new cell site may need to be adjusted.

13.2.5 Handover Problem

Handovers are unavoidable in any wireless network. When working properly, handovers can not only prevent call drops but also enhance the quality of the call. On the other hand, unnecessary handovers due to non-contiguous coverage or pilot pollution should be avoided because they will waste the resources, reducing the system capacity.

The handover mechanism must act fast, as unnecessary delays in handovers may cause uplink/downlink interference. Especially, when there are rapid changes in the path loss between the UE and the Node B due to fading, quick handovers are extremely important. Time delays due to resource allocation in a handover process will not only degrade call quality, but also reduce the throughput of data calls.

RF optimization for handovers is to optimize the handover performance by properly setting and adjusting the handover parameters such as thresholds and timers.

13.2.6 Incomplete Neighbor List

During the RF design stage, one of the design tasks is to create a neighbor list for each sector. Oftentimes, some neighbor lists may not be 100% complete or accurate. Pilots received by the UE cannot be added to the active set if they are not in the neighbor list. These missing neighbors will cause interference. Therefore, it is very important that all received pilots are included in the neighbor list.

Checking the neighbor lists is part of RF optimization. Normally, RF engineers use a software tool to check the reciprocity of the neighbor list. Missing neighbors can also be found during the drive test.

13.3 Pre-Optimization Preparation

Pre-optimization preparation includes a hardware check and antenna audit. The purpose is to make sure that the base station equipment is functioning properly and the antennas are installed according to the design document.

13.3.1 Hardware Check

A hardware check is usually performed after the base stations are installed and integrated. The purpose of the hardware check is to make sure that the base stations will operate properly. Although the hardware check is not a prerequisite of RF optimization, it will avoid potential hardware-caused problem during the optimization process.

13.3.2 Antenna Audit

An antenna audit involves a sequence of quality checks to ensure proper installation of the antenna and cable system. The number of audited cell sites depends on the quality and confidence level of the antenna installations. As a rule of thumb, it is recommended that 20% to 25% of cell sites in a cluster should be audited. Cell sites should be selected randomly for audit. If more than 50% of the audited antennas do not meet the installation requirements, the remaining antennas in the cluster must also be audited. Audit items include antenna height, antenna azimuth, antenna type, antenna down-tilts, jump cable arrangement, feeder cable distribution, lightning protection, grounding, weather protection, and so forth.

13.4 RF Drive-Test-Based Optimization

The RF drive-test-based optimization comprises four phases: RF optimization planning, sector verification, cluster optimization, and system verification.

13.4.1 RF Optimization Planning

Tasks in this phase include pre-drive test planning, tool preparation, equipment setup, and spectrum clearance verification.

13.4.1.1 Pre-drive Test Planning Pre-drive test planning includes cluster planning, review of RF optimization entrance criteria, database parameter verification, and determination of drive test routes.

For cluster planning, in general, approximately 20 geographically-contiguous cell sites are grouped into one cluster. The actual number in a group is based on the cell site distribution, as well as on the topographical environment. Cluster planning may also change with data rates. It often uses natural barriers such as hills or rivers for cluster separation to minimize overlap and influence between the clusters. A small cell site overlap should be maintained between clusters to ensure continuity across the boundaries. The network can first be divided into clusters with each cell site appearing in only one cluster. However, cell sites can also be in more than one cluster with a goal that handovers can be tested. After creating the clusters, maps should be drawn showing the boundaries of each cluster and lists of the cell sites in each cluster should be recorded.

A review of RF optimization entrance criteria is indispensable. This will help determine when the optimization should begin. Normally RF optimization will begin after a certain number of sites/clusters are installed and integrated.

In database parameter verifications, RF parameters such as scrambling codes, power attenuation, and all handover threshold parameters are inspected for consistency with what is in the system document. Another important step within the database verification stage is neighbor list verification. All neighbor lists are required for comparing the neighbor relations with that of the network design plots. They also need to be verified for recent updates, validity, and appropriateness. Although standards allow up to 32 members in a neighbor list, it is not recommended to fill up the list at the beginning when the neighbor list is created.

Drive test routes need to be defined for sector verification, cluster optimization, and system verification. Normally, coverage prediction plots and morphology can be used to determine the drive test routes.

13.4.1.1.1 Drive Routes for Sector Verification The drive test route for sector verification is relatively simple. It is a route that encircles the cell site with a radius equal to half of the cell radius.

13.4.1.1.2 Drive Routes for Cluster Optimization The drive routes for cluster optimization play a key role in cluster optimization. This should comprise major roads, highways, and hot spots such that important traffic areas are not overlooked. Theoretically, total time to drive all routes in a typical cluster is about 6 to 8 hours. However, it is not a hard limit. In practice, depending on the traffic condition of the drive routes and the size of the cluster, it normally takes about 3 to 4 hours to complete a cluster drive test. In addition to the cluster drive test routes, one control route per cluster is also selected for verifying system performance. A control route is a subset of the cluster drive test routes and should be limited to less than one hour. Furthermore, additional border routes are also selected to verify system performance in the regions where clusters overlap. A border route should be selected in such a way that it crosses the cluster borders without entering the inner areas of the clusters.

13.4.1.1.3 Drive Routes for System Verification The system verification drive routes are used for collection of the metrics for the exit criteria. They are a combination of the cluster control routes and the border routes.

13.4.1.2 Equipment Setup Before each drive test, the drive test equipment must be set up and calibrated. The equipment setup includes the following tasks:

- The RF drive test kit must be installed and checked.
- A test van containing drive test equipment must be fully configured.
- The UE must be programmed.
- Penetration loss and uplink loading calculations should be computed.
- The laptops required for drive test storage and data analysis must be ready.

13.4.1.3 Spectrum Clearance Verification Spectrum clearance verification ensures that there is no external interference and that there are sufficient guard bands between the WCDMA system and other systems.

Detection of interferences is time consuming and difficult once the WCDMA system is up and running. It is very important that the spectrum is cleared prior to any drive tests.

13.4.2 Sector Verification

Sector verifications basically examine the basic call-processing functions. These include call origination, call termination, and handover. It also verifies that each sector is transmitting with the appropriate power levels and the correct scrambling code. Sector verification also detects the undiscovered hardware, software, configuration, and parameter problems for each cell site in the cluster before the cluster drive test.

The sector verification tests can be conducted using a data collection tool such as CAIT3G and test UE. All data from the sector verification tests are collected and post-processed using a post-processing tool such as LDAT3G. Any errors should be corrected and the sector verification test should be repeated until all tests pass.

Nowadays, in RF network optimizations, the sector drive test is often replaced by making calls with a dummy load. A drive test for sector verification is carried out only for special requirements, such as verification of softer handover areas.

13.4.3 Cluster Optimization

As mentioned earlier, a cluster consists of approximately 20 geographically contiguous cell sites. Cluster optimization is performed on the cell sites within the cluster. It is a multi-cell site optimization. The advantages of cluster-wise optimization are as follows:

- There is a better focus on the optimization area because smaller sector numbers make it easier to track both the parameter changes and the impact of these parameters on cluster performance.
- Multiple drive test teams can optimize different clusters simultaneously. Each team can focus on its cluster with minimal impact from other teams.
- It speeds up the system tests for commercial operation. Drive test optimization on installed and integrated clusters can go on simultaneously with the installation of other clusters.

Each cluster should be tested under unloaded and loaded conditions. If there is live traffic, cells in the tested clusters must be barred for all UE except for the test UE. The unloaded cluster tests for all clusters within the network should be finished before conducting loaded cluster tests. When a set of adjacent clusters passes the optimization criteria, a drive test for the borders of the clusters should be performed to verify that the borders also pass the optimization criteria.

For cluster optimization, two types of tools are used for data collection: processing, and analysis. The first type is phone-based, such a CAIT3G. The second type is scanner-based, such as Viper. The scanner-based tool is an important tool because of its multiple pilot measurement capability. This capability is especially useful for more in-depth coverage analyses in challenging RF environments including bridges, large water bodies, and uneven terrains.

In addition to unloaded cluster optimization and loaded cluster optimization, UE origination test, UE termination test, and cluster performance verification are also carried out during cluster optimization.

13.4.3.1 Unloaded Cluster Optimization During the first cluster optimization phase, a measurement drive is performed along the drive test route under unloaded network conditions. The purpose of the drive test in this phase is to identify and optimize the problem spots based on the data collected. During this phase, the coverage holes, handover regions, pilot pollution areas, and the overshooting sites belonging to other clusters are identified. Optimization in this phase may result in corrections of neighbor lists and adjustments of the antenna azimuths, antenna down tilts, and base station transmit power. It reveals the flaws in the RF design and allows corrections to be made.

The unloaded drive test is carried out in the following steps:

- Use long call (or Markov call) for drive test measurements.
- Record the BLER using a software tool.
- Log the data using CAIT or other tools.
- Process the log files using a post-processing tool.

After the log files are processed, a variety of performance metric values along the drive routes are generated. Normally, maps of the following parameters are also generated: CPICH max finger, Ec/Io of dominant pilot, UE transmit power, UE receive power, and BLER.

13.4.3.2 Loaded Cluster Optimization The second phase of cluster optimization is performed under loaded conditions. The drive test under loaded conditions is carried out along the same drive routes as that for the unloaded drive test. Under loaded conditions, the noise floor rises, resulting in coverage shrinkage compared with unloaded conditions. Potential coverage holes appear and BLER rises, resulting in lower throughput and more dropped calls. It is loaded cluster optimization that will fix these problems observed in the field. This involves the fine tuning of RF parameters, such as the transmit power or handover parameters. Occasionally, the antenna azimuth and down tilts may need to be ready. Sometimes, the antenna height or the antenna type may also need to be adjusted.

After the changes are implemented, the team may need to re-drive the problem areas. However, it is common engineering practice not to drive a problem area more than three times. If the problem cannot be solved after three drives, a root cause analysis needs to be performed and the team proceed to drive test the next cluster.

Loaded drive tests use a software tool to simulate the load in the downlink direction because there is still no live traffic at this stage. On the uplink, an attenuator is inserted at the UE end to simulate the load. The loaded cluster test can explore the performance of the network under loaded conditions.

The loaded drive test is carried out in the following steps:

- Turn on the downlink load simulation feature at every base station in the cluster.
- Use UE to make data calls at a fixed data rate.

- Record the BLER using a software tool.
- Log the data using CAIT or other tools.
- Process the log files using a post-processing tool.

After the log files are processed, a variety of performance metric values along the drive routes are produced. Similar to the unloaded tests, maps of the following parameters are generated: CPICH max finger, Ec/Io of dominant pilot, UE transmit power, UE receive power, and BLER.

13.4.3.3 UE Call Origination Test After the loaded test, the next step is the UE call origination test. The purpose of the UE call origination test is to examine the call origination capability of the UE. During a UE call origination test, the UE continues to make 20-second calls with 10 seconds between the consecutive calls. The UE can automatically originate the calls using the CAIT software.

The procedure for conducting the UE call origination test consists of the following steps:

- Use 64 kbps as the data rate for the call origination test.
- Use CAIT or other tool to log the data.
- Use a post-processing tool to process the CAIT log files.
- Run origination programs.
- Plot origination failure maps.
- Calculate the call origination failure rate.

13.4.3.4 UE Termination Test The purpose of the UE call termination test is to examine the UE call termination capability. In this test, a landline phone continuously makes 20-second calls to the UE with 10 seconds between the consecutive calls. The procedure for the UE call termination test is as follows:

- Use 64 kbps and 384 kbps data rate for the call termination test.
- Use CAIT or other tool to log the data.
- Use a post-processing tool to process the CAIT log files.
- Run termination programs.
- Plot termination failure maps.
- Calculate the call termination failure rate.

13.4.3.5 Cluster Performance Verification Cluster performance verification is the last phase of cluster optimization. An exit drive test is conducted to verify the cluster performance, which is measured against the cluster exit criteria. The final statistics from the cluster exit drive test are then obtained and documented. The statistics may contain plots as well as data in tabular form.

13.4.4 System Verification

System verification is the last phase of drive-test based optimization activities. It mainly focuses on collecting the overall system performance statistics. System verification begins when all clusters are drive tested and major problems corrected. It is carried out under loaded conditions with all cells activated. The system test requirements are normally well-defined. During system verification, the exit criteria must be met system-wide.

Normally, system verification is a drive test covering the major business district, major highways, primary roads, and specifically defined coverage areas. The drive routes for system verification should be chosen to adequately represent the entire network but not as densely as that for cluster optimization. Problem areas that are identified during the cluster optimization, must be revisited during the system verification stage. However, the system drive routes are normally not used for optimization, because optimizing a system route can result in very good performance on the system verification drive routes but poor performance elsewhere.

The procedure and analysis for system verification are identical to those used in cluster optimization. Performance data are collected and statistics are obtained to characterize coverage and performance over the entire network. System verification is a continuation of cluster performance verification, except that it covers a larger contiguous coverage area.

Problem areas identified by the system verification are addressed on a case-by-case basis after the entire drive test is completed. An individual cluster-optimization drive test is used to fix the coverage problems by adjusting transmit powers and neighbor lists. In some cases, handover thresholds, channel power parameters or other not-frequently-tuned parameters may also require modification. A new drive test is required to ensure that the surrounding regions are still working properly after a change of parameters is made in one region.

System verification basically uses the same tool as that used in cluster optimization. The RF optimization procedure is considered completed at the end of system verification. At that stage, the network is ready for live traffic. Once significant live traffic is present on the network, additional fine tuning of system parameters may be required to accommodate uneven traffic conditions and other effects that cannot be modeled with a simulated load.

13.4.5 RF Optimization Tools

Measuring performance is very important for RF optimization. A variety of tools are required to perform the measurement. As mentioned earlier, the tools, based on their functions, can be classified into data collection and data analysis tools. There are many commercial tools available on the market. Network equipment vendors have also developed their own tools for RF optimization. Individual tools available on the market or proprietarily used by network equipment vendors will not be discussed here. Rather, some of the frequently used tools for RF optimization regardless of their origin or developer are identified here. Commercially available tools that can be used for data collection and/or data processing include CAIT (Qualcomm), VIPER (Agilent), Aircom 3G (Aircom), Actix (Actix), TEMS (Ericsson), WINDS (Qualcomm), Friendly Viewer (Qualcomm), and others. Proprietary tools and features developed by wireless network equipment vendors for drive-test-based optimization include Alcatel-Lucent's LDAT, OCNS, RF Call Trace, and so forth.

13.4.6 WCDMA Performance Metrics

Performance metrics are used for assessing the specific quality and performance of a WCDMA network. The assessments are needed for a general health check or for warranty purposes. For the sake of the warranty, it is important to make sure that the deployed network is performing consistently with the design requirements. The performance metrics consist of a series of quality indicators. There are many quality indicators that are used for function and performance tests. However, for RF optimization, only a subset of key indicators that best represents the quality and performance of a WCDMA network are used.

For voice, the key performance indicators include channel block error rate, dropped call rate, origination successful rate, and termination successful rate. For data, the key performance indicators include unloaded throughput, loaded throughput, and round trip delay. Each of these key performance indicators are briefly described in the following subsections.

13.4.6.1 Block Error Rate The BLER is measured for downlink and uplink under simulated load. A continuous call is maintained along the drive test routes. A new call is set up immediately if the call drops. Data are collected using a data collection tool such as CAIT3G during the call. These data are post-processed using a post-processing tool such as LDAT3G.

13.4.6.2 Dropped Call Rate Under simulated downlink and uplink load, a series of test calls are placed along the drive test routes. The calls that get connected and the calls that drop prior to a hold time of 90 seconds are recorded. The dropped call rate is the ratio of dropped calls to the total number of calls that entered a connected state. That is,

$$\text{Dropped call rate} = \text{Drop calls} / \text{Total calls} \times 100\%. \quad (13\text{-}1)$$

The acceptable dropped call rate is normally between 2% and 5%.

13.4.6.3 Call Origination Success Rate A series of valid test calls are placed along the drive test routes under simulated downlink and uplink load. A valid test call means a correctly dialed call to a non-busy number at the MSC. The origination success rate is defined as the ratio of the total number of successful originations to total number of valid test calls. A successful origination means a call that reaches a connected state. That is

$$\text{Origination success rate} = \text{Successful originations} /$$
$$\text{Total originations} \times 100\%. \quad (13\text{-}2)$$

The origination success rate normally must exceed 95% to be acceptable.

13.4.6.4 Call Termination Success Rate Similar to the call origination success rate measurement, a series of valid test calls are placed to the test UE along the drive test routes during the call termination success rate measurements. A valid test call means correctly dialed calls at the MSC to the test UE. The termination success rate is defined as the ratio of total number of successful terminations to the total number of valid test calls. A successful termination means a call that reaches a connected state. That is

$$\text{Terminations success sate} = \text{Successful terminations} /$$
$$\text{Total terminations} \times 100\%. \quad (13\text{-}3)$$

For acceptance, the termination success rate normally must exceed 95%.

13.4.6.5 Unloaded Throughput Unloaded throughput is measured by continuously transferring data under unloaded conditions along the drive test routes using a certain bearer. The average UE throughput is calculated. In case the bearer drops, it must immediately be re-established.

For acceptance tests, the acceptable average value of throughput for the active bearer is normally defined by the following criteria:

- Mean of data rate distribution is greater than or equal to a certain agreed data rate.
- Standard deviation of distribution is less than or equal to a certain agreed value.
- The measured average value of throughput must fall within the standard deviation of the mean.

13.4.6.6 Loaded Throughput Loaded throughput measurements are conducted in exactly the same way as unloaded throughput measurements. The above acceptance criteria for unloaded throughput can also be used for loaded throughput, except that the agreed mean data rate and standard deviation may be different.

13.4.6.7 Round Trip Delay The round trip delay can be measured through repeated pinging along the drive test routes using a small payload such as 64 bytes. This is carried out by using a computer connected through UE to a server on the GGSN. The delay measurements are taken under unloaded conditions.

The delays for the different bearers must be characterized separately. For example, the delays for 64 kbps uplink and 64 kbps downlink (64/64) and 64 kbps uplink and 384 kbps downlink (64/384) are calculated separately. The magnitude of round trip delay is normally in the millisecond range. For acceptance, it usually requires that a certain percent of the delays measured, within the design coverage area and with a certain bearer, must not exceed an agreed value. For example, 95% of the delays measured within the design coverage area with 64 kbps uplink and 384 kbps downlink bearer must not exceed 2 milliseconds.

13.5 Traffic-Statistics-Based Optimization

When the network is in commercial operation, the traffic volume will increase over time. With a certain amount of traffic on the network, it is best to perform the optimization by combining the analysis of traffic statistics and the drive test together.

13.5.1 Traffic Statistics Data Collection and Processing

Traffic statistics reflect the real operation status of the network. They can be used as a reference for network optimization and maintenance, and as a base for assessing the performance of the network. For WCDMA networks, the Operation and Maintenance Center-UTRAN (OMC-U) collects the traffic statistic data. The data are analyzed and the performance metrics are calculated based on the traffic statistics. From the performance metrics, you can discover the problem areas where optimization needs to be redone by adjusting the RF parameters.

13.5.2 Key Traffic Statistics Metrics for WCDMA Networks

Different network equipment vendors may define the traffic statistics metrics differently. In general, the key traffic statistic data for WCDMA networks are as follows:

- System access performance metrics, such as access failure rate, access blocking rate, and UE origination success rate.
- System service performance metrics, such as dropped call rate, handover success rate, and average call duration.
- System resource allocation and loading metrics, such as OVSF code usage and resource allocation blocking rate.
- Data throughput, including single user throughput and cell throughput.

Traffic-statistics-based optimization normally involves many network equipment vendors' proprietary tools and software at the OMC. It is beyond the scope of this book to discuss the details of how the traffic-statistics-based optimizations are performed for a given live network.

13.6 Summary

The overall WCDMA RF optimization process includes three phases:

1. Pre-optimization phase
2. Drive-test-based optimization phase
3. Traffic-statistics-based optimization phase

Basic tasks in the pre-optimization phase include a Node B installation check, hardware check, power setting check, and antenna audit.

The drive-test-based optimization is performed prior to a commercial launch of the network. The purpose of the drive-test-based optimization is to ensure that the network provides adequate coverage and capacity, and that the network performance meets the criteria. The drive-test-based optimization comprises four steps: pre-drive test planning, sector verification, cluster optimization, and system verification. Pre-drive test planning includes cluster planning, review of RF optimization entrance criteria, database parameter verification, and determination of drive test routes. Sector verification examines the basic call-processing functions, which include call origination, call termination, and handover. It also verifies the transmit power and scrambling code. Cluster optimization is performed on all cell sites in the cluster. Each cluster should be drive tested under unloaded and loaded conditions. If there is live traffic, cells in the tested clusters must be barred for all UE except for the test UE. System verification collects the overall system performance statistics. System verification starts when all clusters are drive tested and major problems corrected. It is conducted under loaded conditions with all cells activated. The exit criteria must be met system-wide during system verification.

Performance metrics are used for assessing the quality and performance of a WCDMA network. There are many quality indicators that are used for function and performance tests. For RF optimization, only a subset of key indicators that best represents the quality and performance of a WCDMA network is used. The key performance indicators for voice include channel block error rate, dropped call rate, origination successful rate, and termination successful rate. The key performance indicators for data include unloaded throughput, loaded throughput, and round trip delay. These indicators are applicable for both circuit-switched and packet-switched services.

The traffic-statistics-based optimization is performed when there is sufficient live traffic in the network. In this phase, software tools and specific network performance counters are used to identify and optimize the problem areas. The key traffic statistics metrics to be monitored in this phase include system access performance metrics, system service performance metrics, system resource allocation and loading metrics, and data throughput.

Chapter 14

Applications of Repeaters and Tower Mounted Amplifiers in WCDMA Networks

Repeaters take the signal they receive, amplify it and then retransmit it. Repeaters perform this function in both the downlink and uplink directions. On the downlink, repeaters receive signals from the host cell site (donor cell) either via a radio link or a fiber link. Then they amplify and rebroadcast the signals to the mobile stations via subscriber antennas. On the uplink, repeaters receive signals from the mobile stations, and then amplify and retransmit them to the host cell site. Since there is no signal processing in the repeater, it will indiscriminately amplify signals regardless of where they come from. For radio-link repeaters, signals from the host cell site, as well as those from other cell sites, are received and amplified. Therefore, a repeater should be carefully positioned to ensure that the downlink signals received by the repeater are strongly dominated by the host cell site.

An inherent problem associated with repeater usage is noise rise at the base station receiver due to injected noise from the repeater via donor link. This noise rise will cause shrinkage of the donor cell coverage area. In other words, repeater coverage must be exchanged for donor coverage. This negates the advantage of using repeaters for continuous coverage area extension. Also, the addition of a repeater is likely to reduce the capacity of the host cell site because the mobile stations connected through the repeater typically require a higher Eb/Io.

Similar to repeaters, tower mounted amplifiers (TMA) can be used for coverage improvement, especially for the case where the coverage is limited by the uplink. This very often occurs because of the limitation of the mobile station maximum transmit power and the noise rise of the base station receiver. Since the mobile station transmit power will affect the operation time of the mobile station battery, it is normally preferred to restrict the maximum mobile station transmit power to a certain level. As such, it is very important to reduce the effective noise figure of the system to improve the reverse coverage.

Use of tower-mounted amplifiers will reduce the overall effective noise figure of the system. Even more significant is that use of tower-mounted amplifiers will effectively eliminate the cable loss between the antenna and the base station, and hence increase the link budget. This will result in the improvement of coverage area, and thus reduces the number of required base stations. Another improvement is in the uplink performance. Use of tower-mounted amplifiers may improve the call quality and reduce dropped calls. The reliability of commercially available tower-mounted amplifiers has been improved substantially during the past few years. Their physical sizes have also become smaller and they have been widely used in many networks.

This chapter addresses the use of repeaters and tower-mounted amplifiers in WCDMA networks. The repeaters will be discussed first, followed by tower-mounted amplifiers. Engineering considerations on the usage of repeaters in WCDMA networks from a practical point of view will be discussed and guidelines for repeater deployment will be provided. Potential problems that may occur in the field will also be examined, along with practical solutions.

The improvement of effective overall system noise figure and link budget, and the applications of tower-mounted amplifiers will also be discussed.

The theory and concepts discussed in this chapter apply to wireless systems in general regardless of access network technology. As such, the generic terms base station and mobile station are used throughout this chapter.

14.1 Repeater Engineering Considerations

There are several important engineering aspects that need to be considered when deploying repeaters. These include coverage objective, base station desensitization, donor link, pilot discrimination, antenna isolation and repeater gain settings, timing issues, handover issues, and donor cell overload.

14.1.1 Repeater Coverage Objectives

The coverage area of a repeater must be defined. The coverage area should include the donor-to-repeater transition zone, as well as soft handover zones surrounding the repeater. Repeater coverage objective is one of the factors to determine the position of a repeater.

14.1.2 Base Station Desensitization

Any repeater added to the network will inject noise to the donor base station, desensitizing the base station receiver. The direct consequence of this noise injection is the shrinkage of the donor cell coverage. As such, calls can no longer be made in some areas that were originally covered by the donor cell. A widely accepted concept of quantifying the injection noise is the noise injection margin (NIM), which is defined in dB as

$$\text{NIM} = F_B - F_R - G_T, \qquad (14\text{-}1)$$

where F_B is the base station noise figure in dB, F_R is the repeater noise figure in dB, and G_T is the net gain of the repeater donor link in dB. G_T includes repeater gain, donor antenna gain, donor antenna cable loss, base station antenna gain, base station cable loss, and the path loss between the repeater and the base station.

In many real applications, there can be more than one repeater being served by a donor cell. In these cases, all repeaters contribute to the base station desensitization. Therefore, a limit exists for the number of repeaters that can be served by a donor cell. This limit, of course, depends on the repeater noise figure and G_T of each repeater. For simplicity, assuming that there are N parallel repeaters served by a donor cell with each one having the same NIM, the effective base station noise figure in dB seen by the mobile stations off the base station is

$$F_{B,\text{eff}} = F_B + 10 \log (1 + N \times 10^{-\text{NIM}/10}). \qquad (14\text{-}2)$$

The base station desensitization (noise rise over donor noise floor) in dB is

$$\text{BTSD} = F_{B,\text{eff}} - F_B = 10 \log (1 + N \times 10^{-\text{NIM}/10}). \qquad (14\text{-}3)$$

Figure 14.1 shows the base station desensitization versus noise injection margin for different numbers of repeaters. The injected noise is directly dictated by the repeater gain. The higher the repeater gain is set the greater the injected noise will be. This, in turn, will lead to a lower NIM, and, hence, desensitize the donor receiver more. The value used for NIM depends on the required sensitivity of the subscriber link. For rural highway range extension where good subscriber link sensitivity is required, a NIM value of 0 is typically used. At this value, the noise rise at the donor receiver is 3 dB. If the repeater gain is set too high, the NIM will go negative, resulting in more than 3 dB in base station desensitization. If the repeater gain is set too low, NIM will be large and base station desensitization will be reduced. But, the repeater coverage will shrink, defeating the purpose of using repeaters for highway range extension. For in-building coverage and hole-filling applications where high repeater gain is normally not required because of a small

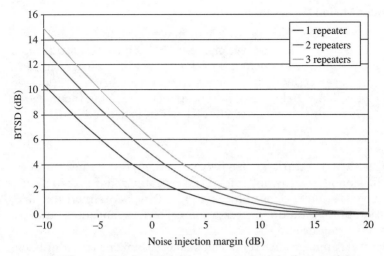

Figure 14.1 Base station desensitization versus noise injection margin.

required coverage area, a typical value of 10 dB for NIM is used. Formula (14-3) also shows that the noise rise at the base station receiver due to 10 repeaters with a NIM of 10 dB is equivalent to that due to a single repeater with a NIM of 0 dB. Therefore, it's not surprising to learn that in some markets, repeaters outnumber the cell sites by a ten to one margin, because more than 90% of the repeaters in these markets are for in-building coverage with large NIMs.

14.1.3 Repeater Noise Figure Rise

When N repeaters are connected to a donor cell with the same NIM, the mobile stations off a repeater will see an effective repeater noise figure as follows:

$$F_{R,\text{eff}} = F_R + 10 \log (N + 10^{\text{NIM}/10}). \quad (14\text{-}4)$$

It can be seen from (14-4) that if the NIM is large, the effective repeater noise figure will be large, resulting in the shrinkage of repeater coverage area. In other words, a large NIM will favor the mobile stations off the donor cell, while a small NIM will favor the mobile stations off the repeaters.

14.1.4 Donor Link Characteristics

Donor link involves the path loss between the repeater and the donor cell. For radio-link repeaters, line-of-sight (LOS) has the best link quality and is obviously the preferred link type. For a fiber-link repeater, normally, the optical transmit power is about 3 dBm, while the optical receiver sensitivity is about −8 dBm. Therefore, the maximum allowable path loss for the fiber is about 11 dB, which corresponds to a fiber length of about 25 km (assuming that the fiber loss is about 0.4 dB/km for single mode fiber). Noise generated in the optical transceivers will also cause noise rise in the base station receiver.

14.1.5 Pilot Discrimination

A radio-link repeater will amplify all pilots it receives at the donor antenna. It's very difficult for the donor antenna to spatially filter the desired pilot from many pilots. Therefore, it's very important to carefully select the repeater position such that the repeater will receive the desired pilot with significant dominance over other pilots that are also seen by the donor antenna. One way to achieve this is to pick the donor sector that is easiest to isolate from other sectors. Use of a narrow beam-width donor antenna is also a viable choice. That is why parabolic antennas are usually used as donor antennas.

For fiber-link repeaters, no donor antenna is required. The dominant pilot is transmitted from the donor cell to the repeater through a fiber link. There is no pilot discrimination problem.

14.1.6 Antenna Isolation and Gain Setting

Repeater gain is limited by the isolation between the donor antenna and the subscriber antenna to avoid positive feedback. Common engineering practice calls for the isolation to be at least 15 dB higher than the repeater gain. For example, if the repeater gain is set at 90 dB, the antenna isolation required is 105 dB. If the isolation is not sufficient, the repeater may oscillate or have to be operated with less gain. Oscillation will introduce spurious emission to the network, while operation with less gain will reduce the repeater coverage. Sufficient isolation may be achieved by a careful choice of antenna type, as well as vertical and horizontal separation as well.

Repeater gain is also limited by the need to tightly control donor coverage shrinkage. As discussed in Subsection 14.1.2, the base station noise floor rise is affected by repeater gain. The repeater gain setting must be precisely set to avoid excessive noise floor rise at the donor base station receiver.

14.1.7 Handover Issue

When a repeater is added for continuous coverage extension, the total coverage area is extended. At the same time the handover region between the donor base station and the neighbor cells is also extended. A mobile station will stay in the handover state longer than before. Therefore, a repeater should be placed at a location where there is minimum overlapping with neighbor cells.

14.1.8 Donor Cell Traffic Overload

When a repeater is added to a base station, the base station will serve more mobile stations than before. Some of the traffic that was served by the neighbor cells may also be shifted to the base station that has repeaters. The donor base station may be overloaded quicker than the cells surrounding it as the traffic increases in that area.

14.1.9 Narrow-Band Interference Amplified by Repeaters

If a narrow-band interference exists, the repeater may amplify it on the uplink, affecting the base station reception. It can cause the call origination success rate to drop and may sometimes block the call origination attempts. This has happened in some markets where repeaters were used for area coverage extension. It was found that when the repeater was turned on, the call origination success rate decreased and at times origination attempts were blocked entirely.

Field tests have also shown that when a repeater is added to the system, the mobile station transmit power must be increased in order to overcome the noise injected by repeaters. In addition, the reverse link frame error rate increases as the repeater gain increases. These field experiences have proven that repeaters can degrade the performance of the networks. Furthermore, it also shows that low-quality mobile stations are more severely impaired in their ability to establish and maintain calls in an area served by an improperly adjusted repeater.

14.2 Major Repeater-Related Problems in WCDMA Networks

There are many repeater-related problems that may occur in WCDMA networks. In the repeater coverage area, the potential problems include: inability to make a call, high dropped call rate, high mobile station transmit power, small coverage area, frequent handovers, inability to make handovers with other cells, long access time, and so forth. In the donor cell coverage area, the problems include high dropped call rate, high mobile station transmit power, and shrinkage of the coverage area because the repeaters raise the base station noise level. For other cells adjacent to repeaters, the dropped call rate may rise due to interference caused by the repeaters. Other problems that may occur include origination failures and pilot pollution.

In the following subsections, the causes of the repeater-related problems and their solutions will be discussed.

14.2.1 Inability to Make a Call in the Repeater Coverage Area

The causes of not being able to make a call in the repeater coverage area can be that the Ec/Io received by the donor antenna is too low and/or the repeater forward link gain is too high. If the Ec/Io received by the repeater donor antenna is less than −15 dB, you need to adjust the orientation of the donor antenna (for radio link repeater), change donor cell (for fiber link repeater), and/or move the repeater to a better location. You can also reduce the repeater forward link gain such that the pilot signal level in the repeater coverage area is comparable to that in the donor cell coverage area. The repeater reverse link gain should also be reduced accordingly.

14.2.2 High Dropped Call Rate in the Repeater Coverage Area

High dropped call rates in the repeater coverage area can be due to an incorrect donor cell, interference existing in the surrounding areas, or oscillation of the repeater (positive feedback of the repeater). To correct the incorrect donor cell problem, you can adjust the repeater donor antenna orientation (for radio link repeater), change donor cell (for fiber link repeater), and/or move the repeater to a better location. To avoid the interference existing in the surrounding areas, you should avoid collocating repeaters with other systems' equipment. You also need to locate and isolate the interference sources such as TV channel expanders or, cable TV in-line amplifiers. To prevent the unwanted positive feedback, you need to maintain a sufficient isolation between donor and subscriber antennas. An isolation of 15 dB plus the repeater forward link gain is normally required.

14.2.3 High Mobile Station Transmit Power in the Repeater Coverage Area

High mobile transmit power in the repeater coverage area is normally caused by the interference coming from the surrounding areas, or by in-band spurious emissions and inter-modulations that are generated by the forward link repeater amplifier and then fed back on the uplink through the repeater diplexer. The solutions to these problems are to locate and eliminate the interference sources, move the repeater to a non-interference location if possible, and eliminate in-band spurious emissions and inter-modulations by using high quality repeaters.

14.2.4 Repeater Coverage Area Smaller Than Expected

There are several things that may cause a repeater coverage area to be smaller than expected. These include low repeater antenna height, large repeater donor link path loss, and in-band spurious emissions and inter-modulations generated by the forward link repeater amplifier. If fiber-link repeaters are used, the fiber connector mismatch can introduce unexpected optical loss, resulting in smaller repeater input power in the downlink direction.

The solutions for the problems mentioned previously include raising the repeater antenna height, moving the repeater closer to the donor cell to reduce NIM (but be aware of the noise rise of the base station receiver), eliminating in-band spurious emissions and inter-modulations by using high quality repeaters, and replacing the mismatched fiber connectors with good ones if fiber-link repeaters are used.

14.2.5 Frequent Handovers Within the Repeater Coverage Area

If the repeater receives signals from multiple base station sectors, it may make frequent handovers within the repeater coverage area. To avoid this problem, you should adjust the repeater donor antenna orientation and/or move the donor antenna location, limiting the donor cell to one.

14.2.6 Inability to Make Handovers with Other Cells

If the repeater broadcasts the donor cell signal to an area that is not normally served by the donor cell, the mobile stations in that area may not be able to make handovers with other cells. A typical example is that the donor cell is more than 1 tier away. If this happens, it will render the relevant neighbor lists incomplete. The solution to this problem is modifying the neighbor list of the donor cell and repeater-neighboring cells.

14.2.7 Long Access Time in the Repeater Coverage Area

Long access time in the repeater coverage area may be due to

- Repeater noise level being too high (effective repeater noise figure is too high)
- Uplink interferences
- Repeater reverse link gain is set too low relative to the repeater forward link gain.

The solutions to solve this problem are to

- Decrease NIM by moving the repeater closer to the donor cell.
- Locate and isolate the interference sources such as TV channel expanders.
- Increase the repeater reverse link gain to within 5 dB of or equal to the repeater forward link gain.

14.2.8 Donor Cell Coverage Area Shrinkage

Donor cell coverage area shrinkage may occur if the effective donor cell receiver noise figure rises due to the addition of the repeaters, or if uplink in-band spurious emissions are occurring. To solve the problem, you can move the repeaters further away from the donor cell or tune down the repeater reverse link gain.

14.2.9 High Dropped Call Rate and High Mobile Station Transmit Power in the Donor Cell Coverage Area

High dropped call rate and high mobile station transmit power in the donor cell coverage area may occur if the donor cell receiver noise level is raised too high or uplink narrow band interference is amplified by the repeater. To lower the donor cell receiver noise figure, you may increase the NIM by moving the repeater further away from the donor cell or reducing the repeater reverse link gain. The only way to get rid of the uplink narrow band interference is to locate and eliminate the narrow-band interference sources.

14.2.10 Pilot Pollution

Pilot pollution occurs when there is no dominant pilot. This may occur if a repeater is placed in the cell overlapping areas where there is no dominant pilot. This may also occur if the repeater adds a new but not dominant pilot to the area.

If a repeater is placed in the cell overlapping areas, you should use a narrow-beam donor antenna and adjust the donor antenna orientation to achieve line-of-sight connection. You can also reduce the repeater forward link and reverse link gain to shrink its coverage area, or modify the neighbor lists of the donor cell and repeater-neighboring cells.

14.3 Repeater Deployment Guidelines

Repeater deployment includes three stages: selection of repeater locations, repeater installation, and general adjustments of repeaters after installation. Some general guidelines for each of these three stages are described in the following subsections.

14.3.1 Selection of Repeater Locations

Selection of a repeater location involves the following steps:

- Define the repeater coverage area.
- Pick the donor cell and select a repeater location where the donor antenna can achieve line-of-sight communication with the donor cell (sector) antenna.
- Limit the distance between the repeater and the donor cell to 25 kilometers for radio-link repeaters and 15 kilometers for fiber-link repeaters, though the theoretical maximum allowable distance is higher than that. Do not deploy repeaters in areas where the frequency band (up and downlinks) is not clean.
- Adjust the orientation of the donor antenna to make sure that there is only one strong Ec/Io (dominant pilot) received by the repeater donor antenna. The Ec/Io should be monitored at the output port of the donor antenna.

14.3.2 Repeater Installation

The following precautions should be followed during installation:

- Make sure the donor and subscriber antennas are installed properly at the designated positions.
- Apply the same precautions and necessary steps applied to the installation of base station antenna/cable systems.
- Make sure there is enough isolation between the donor and subscriber antennas. The isolation between the donor and subscriber antennas must be greater than the repeater forward link gain by at least 15 dB to ensure that there is no positive feedback between the two antennas. The isolation can be calculated by measuring the received signal level at the donor antenna output port when a continuous wave signal is fed into the subscriber antenna. Typically, it should be larger than 65 dB.

14.3.3 General Adjustments of Repeaters After Installation

After repeater installation, the general adjustments required include:

- In the forward link direction, measure the repeater-received power at the output of the donor antenna.

- Calculate the path loss between the repeater and the base station, and calculate the repeater forward link gain based on the repeater output power.
- Check whether the calculated repeater forward link gain is less than the calculated path loss by 5 to 10 dB. If not, reduce the repeater forward link gain to meet this criterion.
- Calculate the required isolation between the donor and subscriber antennas.
- Measure the isolation between donor and subscriber antennas to see if it meets the isolation requirement. If not, adjust the antenna separation.
- Tune the repeater reverse link gain to a value equal to or smaller than the forward link gain. Normally, this will depend on applications. For outdoor applications, the reverse link gain can be smaller than the forward link gain by at most 5 dB. For indoor applications, this difference can be as much as 10 dB.

14.4 Tower Mounted Amplifiers

Tower mounted amplifiers are also known as tower top amplifiers. A TMA unit consists of high Q band-pass filters, a low noise amplifier (LNA), a current extractor, and an LNA bypass circuit.

In many wireless networks, including WCDMA networks, the coverage radius is often limited by the maximum mobile station transmit power and the base station receiver sensitivity, which is tied to the noise figure of the base station receiver. Due to the fact that the mobile station transmit power will impact the lifetime of the mobile station battery, you will normally limit the maximum mobile station transmit power to a certain level. As such, it is very important to improve the coverage by reducing the effective noise figure of the base station receiver. Use of a TMA will make the effective base station noise figure smaller than the original one. Even more important is that use of a TMA will essentially eliminate the cable loss between the base station antenna and the base station receiver. Therefore, the link budget is increased, and hence the cell coverage radius is also increased. As a result, the required number of base stations decreases, expediting the commercialization of the network. In addition, a TMA can also improve the reverse link performance, enhancing the call quality and reducing dropped calls. Reliability has been improved substantially in the current commercial TMAs. The physical size has also become smaller. As such, TMAs are widely used in wireless networks nowadays.

14.4.1 Analysis on the Improvement of Base Station Noise Figure and Link Budget

When a TMA is inserted into the reverse link, the TMA, the feeder cable, and the base station receiver are connected in series and the whole combination is regarded as an equivalent base station receiver. The effective noise figure of this equivalent base station receiver is given by

$$F_{\text{eff}} = F_{\text{TMA}} + (F_{\text{Cable}} - 1)/G_{\text{TMA}} + (F_{\text{BTS}} - 1)/G_{\text{TMA}}G_{\text{Cable}}, \quad (14\text{-}5)$$

where F_{TMA}, F_{Cable}, and F_{BTS} are the noise figure of the TMA, feeder cable, and base station receiver, respectively. G_{TMA} and G_{Cable} are the gain of the TMA and feeder cable, respectively. It should be noted that the gain of feeder cable is negative in dB. Assuming the noise figure and gain of the TMA is 1.7 dB and 12 dB, respectively, and the noise figure of the base station receiver is 3 dB, the effective noise figure calculated by using (14.5) is 2.22 dB, which is 0.78 dB smaller than the original base station receiver noise figure. In other words, use of a TMA will improve the receiver sensitivity by 0.78 dB.

Link budget is the basis for calculating the coverage radius. Before a TMA is used, the reverse link budget is given by

RLB = Mobile's EIRP − body loss − base station receiver sensitivity
+ base station antenna gain − feeder cable loss − fade margin
− fast fade margin + soft handover gain. (14-6)

After a TMA is inserted into the reverse link, the link budget becomes

RLB = Mobile's EIRP − body loss
− effective base station receiver sensitivity
+ base station antenna gain − fade margin
− fast fade margin + soft handover gain. (14-7)

A comparison of (14-6) and (14-7) reveals that after a TMA is inserted into the reverse link, the base station receiver sensitivity is replaced with effective base station receiver sensitivity in the reverse link budget equation. At the same time, the feeder cable loss is no longer included in the equation. Therefore, the reverse link budget with TMA is larger than that without TMA. Using the numbers given earlier as an example, the reverse link budget with TMA is 3.78 dB larger than that without TMA. This translates into an increase of about 25% in cell radius or 56% in coverage area.

14.4.2 Applications of TMAs

TMAs are normally used for those cell sites that are uplink limited. For downlink limited cell sites, TMAs cannot provide any advantages.

In practice, TMAs are used in cell sites that have small capacity but require large coverage, such as cell sites in rural areas. In some special applications such as sea coverage, TMAs are often used in conjunction with power boosters on the downlink to extend the coverage beyond the normal range (for example, 150 km). From an engineering standpoint, before a TMA is added, you must consider the necessity of using it, examine its electrical and mechanical specifications to verify whether they meet the requirements, and check whether the site is ready for adding the TMA.

14.5 Summary

Repeaters amplify the signals on both uplink and downlink. A repeater will indiscriminately amplify signals regardless of where they come from because there is no signal processing in the repeater. An intrinsic issue associated with repeater usage is the noise rise at the donor cell base station receiver due to injected noise from the repeater via the donor link. This noise rise will shrink the donor cell coverage area. As such, repeater coverage is exchanged for donor cell coverage. When deploying repeaters, the important engineering issues that need to be considered include coverage area, base station desensitization, donor link, pilot discrimination, antenna isolation, repeater gain settings, and handovers.

Various repeater-related problems may occur in WCDMA networks that employ repeaters. In the repeater coverage area, the potential problems include inability to make a call, high dropped call rate, high mobile station transmit power, small coverage area, frequent handovers, mobile station cannot make handovers with other cells, and long access time. In the donor cell coverage area, the potential problems include high dropped call rate, high mobile station transmit power, and shrinkage of the coverage area. For other cells adjacent to repeaters, the dropped call rate may rise due to interference caused by the repeaters. Other problems that may occur include origination failures and pilot pollution.

Various solutions may be employed to solve or alleviate the aforementioned problems. Typical solutions include: adjusting the repeater's reverse and forward link gains, relocating the repeaters, moving the repeaters closer to or further away from the donor cell, adjusting the donor antenna orientation and/or height, increasing the antenna isolation between the donor and subscriber antennas, searching for and eliminating interference sources, and modifying the neighbor list of the donor cell and repeater-neighboring cells.

Tower mounted amplifiers are used for coverage improvement when the coverage is limited by the uplink. This occurs very often in WCDMA networks. Use of tower mounted amplifiers can reduce the overall effective noise figure of the system. It effectively eliminates the cable loss between the antenna and the base station, and hence increases the link budget, extending the coverage.

For downlink limited cell sites, tower mounted amplifiers do not provide any benefits. Practically, TMAs are best used in cell sites that have small capacity but require large coverage, such as cell sites in rural areas. However, in some special applications such as sea coverage, TMAs can be used in conjunction with downlink power boosters to extend the coverage beyond the normal range.

Chapter 15

Intersystem Interferences

Networks of different technologies may coexist in the same geographical areas. For example, UMTS and CDMA2000 may be deployed in the same region where GSM and PHS are already in place. Depending on frequency band allocations and antenna positions and orientations, intersystem interference between two coexisting systems may occur. The intersystem interference basically refers to the transmitted power of one system affecting the receiver of the other system operating at adjacent bands. As such, the interference can be evaluated in terms of adjacent channel performance.

15.1 Adjacent Channel Performance

To evaluate adjacent channel performance, it is necessary to define the adjacent channel performance parameters. The three key adjacent channel performance parameters are adjacent channel interference power ratio (ACIR), adjacent channel leakage power ratio (ACLR), and adjacent channel selectivity (ACS). ACIR is defined for simulation purpose, ACLR is defined for transmitter performance requirements, and ACS is defined for receiver performance requirements. Each of these parameters can be defined for any frequency offset. Therefore, the term "adjacent channel" may refer to the channel closest to the assigned channel, the second adjacent channel, or the third adjacent channel, and so forth.

15.1.1 Adjacent Channel Interference Power Ratio

ACIR is defined as the ratio of the total power transmitted from a source (base station or UE) to the total interference power affecting a victim receiver, resulting from both transmitter and receiver imperfections. It is used for system performance simulation.

In a system performance simulation, in which the transmitter and receiver use adjacent RF carriers, an assumption has to be made on the amount of power that is leaking from the transmitter to the receiver on the adjacent channel, due to imperfect transmitter mask and imperfect receiver filter.

15.1.2 Adjacent Channel Leakage Power Ratio

Due to transmitter nonlinearity, the spectrum mask from a transmitter may leak into adjacent channels. This spectrum mask leakage is a very important system parameter and is used for evaluating coexisting performance of systems on adjacent channels. It is also important for transmitter design. An overly rigorous requirement on adjacent channel leakage can restrict the implementation of efficient and low-complexity transmitters for the UEs. To quantify the spectrum mask leakage, in generic terms, the ACLR is defined as the ratio of the transmitted power to the power measured after a receiver filter is performed in the adjacent RF channel. In 3GPP specification [1], more specifically, ACLR is defined as the ratio of the RRC filtered mean power, centered on the assigned channel frequency to the RRC filtered mean power centered on an adjacent channel frequency. The minimum requirement for base station ACLR is specified in Table 15.1. In other words, the ACLR must be higher than the value specified in Table 15.1.

Similarly, the minimum requirement for UE ACLR specified in the 3GPP specification [2] is given in Table 15.2, which lists the minimum required ACLR value for the adjacent channels with power greater than −50 dBm.

3GPP specification [2] also requires that the minimum ACLR requirement must still be met in the presence of switching transients. Also, the minimum ACLR requirement reflects what can be achieved with present state-of-the-art technology, and they must be reconsidered as the state-of-the-art technology progresses.

TABLE 15.1 Base Station ACLR (Courtesy of ETSI)

BS Adjacent Channel Offset Below the First or Above the Last Carrier Frequency Used	ACLR limit
5 MHz	45 dB
10 MHz	50 dB

TABLE 15.2 UE ACLR (Courtesy of ETSI)

Power Class	Adjacent Channel Frequency Relative to Assigned Channel Frequency	ACLR Limit
3	+ 5 MHz or − 5 MHz	33 dB
3	+ 10 MHz or − 10 MHz	43 dB
4	+ 5 MHz or − 5 MHz	33 dB
4	+ 10 MHz or −10 MHz	43 dB

15.1.3 Adjacent Channel Selectivity

Due to imperfections of receiver filters, a receiver may receive interference from the adjacent channel. It is the filter side lobes in the adjacent channel that cause the power from the main lobe of the transmitted interference source to affect receiver performance.

To quantify the receiver filter's suppression of the main lobe of the transmitted signal in adjacent channels, adjacent channel selectivity (ACS) is defined. ACS is a measure of a receiver's ability to receive a signal at its assigned channel frequency in the presence of a modulated signal in the adjacent channel. ACS is defined as the ratio of the receiver filter attenuation on the assigned channel frequency to the receiver filter attenuation on the adjacent channel(s). The minimum ACS requirement as specified in [1] is given in Table 15.3. The interference signal is offset from the wanted signal by the frequency offset Fuw; the interference signal must be a WCDMA signal, and the BER must not exceed 0.1%.

Similarly, for UE, the minimum requirement for ACS as specified by 3GPP specification [2] is given in Table 15.4. The ACS must be better than the value indicated in Table 15.4, as the test parameters specified in Table 15.5 maintain that the BER must not exceed 0.1%.

Note: In Table 15.5, $\hat{I}_{or}$ is the received power spectral density of the downlink signal as measured at the UE antenna connector. The I_{oac} (modulated) signal consists of the common channels needed for tests and 16 dedicated data channels.

TABLE 15.3 Adjacent-Channel Selectivity for Base Station (Courtesy of ETSI)

Parameter	Level Wide Area BS	Level Medium Range BS	Level Local Area BS	Unit
Data rate	12.2	12.2	12.2	kbps
Wanted signal mean power	−115	−105	−101	dBm
Interfering signal mean power	−52	−42	−38	dBm
Fuw offset (Modulated)	5	5	5	MHz

TABLE 15.4 Adjacent-Channel Selectivity for UE (Courtesy of ETSI)

Power Class	Unit	ACS
3	dB	33
4	dB	33

15.1.4 Relation Between ACIR, ACLR, and ACS

With ACIR, ACLR, and ACS (defined previously), it is obvious that ACIR depends on ACLR and ACS. More specifically, ACIR is a function of ACLR and ACS given as follows:

$$1/\text{ACIR} = 1/\text{ACLR} + 1/\text{ACS}. \quad (15\text{-}1)$$

On the uplink, the UE transmitter normally dominates the uplink interference because ACLR_{UE} is much smaller than $\text{ACS}_{Node\ B}$. This implies that the ACIR is approximately equal to ACLR_{UE} for the uplink. As such, an uplink interference simulation is basically a simulation of the UE ACLR performance.

On the downlink, the limiting design factor is normally the UE receiver, which dominates the downlink interference because ACS_{UE} is much smaller than $\text{ACLR}_{Node\ B}$. This implies that the downlink ACIR is approximately equal to ACS_{UE}. Therefore, a downlink interference simulation is mainly a simulation of UE ACS performance.

When the base stations of multiple systems are collocated, the base station transmitting power of one system may affect the base station receiver of another system. In this case, both the ACLR of the base station transmitter and the ACS of the base station receiver may contribute to the intersystem interference equally.

15.2 Interferences Between UMTS and CDMA2000 Systems

Assume UMTS and CDMA2000 are deployed in the adjacent frequency band. For example, UMTS uses 1920 MHz to 1935 MHz for uplink and 2110 MHz to 2125 MHz for downlink, whereas CDMA2000 uses

TABLE 15.5 Test Parameters for Adjacent-Channel Selectivity (Courtesy of ETSI)

Parameter	Unit	Level
DPCH_Ec	dBm/3.84 MHz	−103
$\hat{I}_{or}$	dBm/3.84 MHz	−92.7
I_{oac} mean power (modulated)	dBm	−52
F_{uw} (offset)	MHz	+5 or −5
UE transmitted mean power	dBm	20 (for Power class 3) 18 (for Power class 4)

1935 MHz to 1950 MHz for uplink and 2125 MHz to 2140 MHz for downlink. As shown in Figure 15.1, the potential interference scenarios between UMTS 2100 FDD and CDMA2000 include:

1. **Scenario 1:** The UMTS UE transmitted power affects the adjacent-band CDMA2000 base station receiver.
2. **Scenario 2:** The CDMA2000 MS transmitted power affects the adjacent-band UMTS base station receiver.
3. **Scenario 3:** The UMTS base station transmitted power affects the adjacent-band CDMA2000 mobile station receiver
4. **Scenario 4:** The CDMA2000 base station transmitted power affects the adjacent-band UMTS UE receiver.

These four scenarios have a common characteristic, namely, the transmitted power of one system affects the receiver of the other system operating at the adjacent band. Therefore, the interference can be evaluated in terms of adjacent channel performance discussed in the previous section.

Figure 15.2 illustrates the effects of interference from one transmitter to the adjacent-band receiver. For example, the effects of CDMA2000 base station to UMTS UE interference include the following two types (if there is only one carrier): interference shown by the horizontal rectangle and interference shown by the vertical rectangle. The horizontal rectangle represents the CDMA2000 base station out-of-band emissions falling into the UMTS UE receiving band. These emissions cannot be attenuated by the UMTS UE receiver filters and could cause UE receiver desensitization. The vertical rectangle represents the CDMA2000 base station carrier powers attenuated by the UMTS UE receiver filters. The carrier powers could cause overload or blocking at the UMTS UE receiver.

If there are multiple CDMA2000 carriers, third-order inter-modulation products may also be generated within the UMTS UE receiving frequency band.

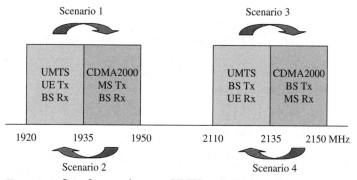

Figure 15.1 Interferences between UMTS and CDMA2000 systems.

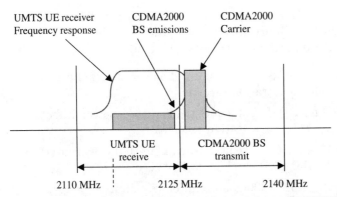

Figure 15.2 Interference from CDMA2000 base station to UMTS FDD UE.

As discussed in the previous section, the ACIR is a function of ACLR and ACS. In this example, the CDMA2000 base station interferes with the UMTS UE. Therefore, the ACIR is a function of the CDMA2000 base station ACLR and UMTS UE ACS. The effects of the four previous interference scenarios can be analyzed by a simulation approach.

Normally, the Monte Carlo simulation is used to investigate the effects of mutual interference between the UMTS 2100 FDD and CDMA2000 systems operating in adjacent frequency bands. With the Monte Carlo simulation, the impact of mutual interference on the UMTS downlink and uplink capacity, as well as the CDMA2000 downlink and uplink capacity, can be revealed.

In any market or city where UMTS and CDMA2000 systems coexist, the UMTS and CDMA2000 systems may have different cell site layouts. Their cell sites may collocate in some areas and may not in others. From the interference standpoint, the best case is that UMTS and CDMA2000 cell sites collocate and the worst case is that UMTS and CDMA2000 cell sites are located at each other's cell edge. Therefore, for simulation purpose, only these two cases need to be considered.

In the collocation case, UMTS base stations are collocated with adjacent-band CDMA2000 base stations. There is no near-far problem in this case and the mutual interference impact is minimal. It is recommended that that adjacent-frequency-band systems share the same antenna tower to prevent mutual interference and to reduce cost. In practice, collocating antennas is a commonly-used approach for alleviating adjacent-frequency-band interference.

In the worst non-collocation case, in which UMTS and CDMA2000 cell sites are located at each other's cell edge, the near-far problem occurs when an affected UMTS UE (CDMA2000 MS) is far away from its serving UMTS base station (CDMA2000 base station) and close to an interfering CDMA2000 base station (UMTS base station). To assess the capacity impact accurately, the interference from the UMTS system to the CDMA2000 system and the interference from the CDMA2000 system to the UMTS system must be considered simultaneously.

Simulation results of intersystem interference between UMTS and CDMA2000 indicate that, for the collocation case, the mutual interference impact on UMTS and CDMA2000 is negligible for both downlink and uplink. For the worst non-collocation case, mutual interference between UMTS and CDMA2000 on the downlink is not an issue either; however, the impact of mutual interference on the UMTS uplink and CDMA2000 uplink is something that requires attention. To minimize the mutual interference between UMTS and CDMA2000 systems, UMTS and CDMA2000 should collocate wherever possible.

15.3 Interferences Between UMTS and PHS Systems

Assume the PHS system is operating in the frequency band 1900-1920 MHz, which is actually the case in some countries. Also assume that the UMTS system is operating in the frequency band 1920-1935 MHz for the uplink. If PHS and UMTS coexist in the same market (city), potential interference scenarios, as shown in Figure 15.3, can include the following:

1. **Scenario 1:** PHS base station transmitter interferes with UMTS base station receiver.
2. **Scenario 2:** PHS MS transmitter interferes with UMTS base station receiver.
3. **Scenario 3:** UMTS UE transmitter interferes with PHS base station receiver.
4. **Scenario 4:** UMTS UE transmitter interferes with PHS MS receiver.

Among these four scenarios, interferences in the third and fourth scenarios are insignificant and will not be discussed. Interference in the first scenario is the most critical and is investigated by an analytical approach. Also, for both the first and second scenarios, simulation results based on the transmission and reception characteristics of UMTS FDD and PHS at appropriate frequency offsets [2, 3] are presented.

15.3.1 Analysis of the Interference Between PHS Base Station Transmitter and UMTS Base Station Receiver

When UMTS FDD coexists with a PHS system in the same geographical area, the PHS base station transmitter may interfere with the UMTS base station receiver because their frequency bands are so close.

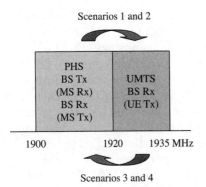

Figure 15.3 Interferences between UMTS and PHS systems.

However, the UMTS base station transmitter will not interfere with the PHS base station receiver because their frequency bands are so far apart. The interference might cause the UMTS system performance degradation including voice quality, coverage, and/or capacity. In order to avoid the UMTS system base station receiver desensitization, overload, and/or inter-modulations, a sufficient isolation between the PHS and UMTS base station antennas should be maintained.

Impact of the interference depends on the frequency separation between the PHS and UMTS systems in the same geographical area. The frequency separation dictates the magnitude of the required isolation. Obviously, the larger the separation the smaller the isolation required to alleviate the interference.

This analysis uses the equipment specifications defined by the standards. However, the performance of actual base station equipment may often exceed the specifications in the standards and hence may require significantly less isolation than the values obtained in this analysis. As such, this analysis can be considered as the worst-case scenario approach.

In general, sufficient isolation can be achieved by spatially separating the antennas of the PHS and UMTS base stations and/or by employing filter attenuation at the PHS base station transmitters and/or the UMTS base station receivers. This analysis will focus on the required antenna isolation.

15.3.1.1 Isolation Criteria Figure 15.4 shows conceptually the mutual interference between two coexisting base stations. Interferences from the interfering base station transmitter may degrade the performance of the interfered base station receiver. Therefore, isolation between the two base stations is required. Isolation in this case means the path loss between the interfering base station transmitter antenna connector and the affected base station receiver antenna connector. The path loss includes the propagation loss through the air, the antenna gain, and cable loss of both stations.

Intersystem Interferences

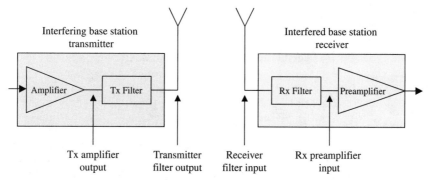

Figure 15.4 Interferences between two coexisting base stations.

The isolation requirements are calculated based on the following four criteria [4]:

- **Criterion 1:** The interfering transmitter's spurious emission power received by the affected system is 10 dB below the receiver noise floor.
- **Criterion 2:** The total interfering carrier power received by the affected system is 5 dB below the 1 dB compression point of the receiver amplifier.
- **Criterion 3:** Each of the 3rd order inter-modulation products (IMPs) generated by the affected receiver and caused by the interfering carriers is 10 dB below the receiver noise floor.
- **Criterion 4:** The total interfering carrier power attenuated by the affected system's receiver filters is 10 dB below the receiver noise floor to prevent receiver desensitization or blocking.

For UMTS base station receivers, Criterion 4 dominates Criterion 2. That is, if the isolation meets Criterion 4, it automatically meets Criterion 2 as well. Therefore, only Criteria 1, 3, and 4 need to be considered.

15.3.1.2 Isolation Requirement for Spurious Emissions
To calculate the isolation that meets Criterion 1, you will need to know:

1. The acceptable interference level at the affected base station antenna connector I_{aff}
2. The interference level transmitted at the antenna port of the interfering station falling into the affected station receiving frequency band I_{int}
3. The bandwidth adjustment factor (BWAF), defined as the ratio of the affected base station carrier bandwidth to the interfering base station spurious emissions measurement bandwidth

With the aforementioned information, the required isolation for meeting Criterion 1 is given by

$$L\_criterion\_1 = I_{int} - I_{aff} + \text{BWAF}, \quad (15\text{-}2)$$

where I_{int} is determined by

$$I_{int} = P_{Tx\_amp} + \text{ICR}_{Tx\_amp} - L_{Tx\_rej}, \quad (15\text{-}3)$$

where P_{Tx_amp} is the interfering base station transmitted power at the transmitter amplifier output, ICR_{Tx_amp} is the interference-to-carrier ratio at the transmitter amplifier output, and L_{Tx_rej} is the interfering base station transmitter filter attenuation in the affected base station's receiving frequency band. Assuming a noise figure of 3 dB for the UMTS FDD base station, the UMTS FDD base station receiver noise floor is –105.2 dBm (–174 dBm + 10 log (3840000) + 3 = –105.2 dBm). According to the PHS standards, the base station equipment must provide at least –93dBc/288 kHz attenuation of the emissions at the frequency offset between the PHS and UMTS FDD bands. That is, for a base station transmitted power of 27 dBm, I_{int} = 27 – 93 = –66 dBm. Therefore, for a base station transmitted power of 27 dBm, the required isolation based on (15-2) is given by

$$L\_criterion\_1 = (27-93) - (-105.2 - 10) + 10 \log (3840/288) = 60.4 \text{ dB}. \quad (15\text{-}4)$$

15.3.1.3 Isolation Requirement for Third-Order Inter-Modulation To calculate the isolation that meets Criterion 3, you need to know the third order intercept of the affected receiver amplifier and the affected system filter attenuation. The relationship between the third-order inter-modulation product and the third-order intercept of an amplifier is defined as

$$\text{IMP}_3 = 3\, P_{in\_Rx} - 2 \times \text{TOI}, \quad (15\text{-}5)$$

where IMP_3 is the third-order inter-modulation product, P_{in_Rx} is the input power of the affected system receiver amplifier, and TOI is the third-order intercept of the affected system receiver amplifier. The required isolation to meet Criterion 3 is given by

$$L\_criterion\_3 = P_{Tx\_ant} - P_{in\_Rx\_ant}, \quad (15\text{-}6)$$

where P_{Tx_ant} is the transmitted carrier power at the interfering base station antenna connector, and $P_{in\_Rx\_ant}$ is the received carrier power at the affected base station antenna connector (before the receiver filter). Assuming that the insertion loss of the receiver filter is 2 dB (i.e., $P_{in\_Rx\_ant} - P_{in\_Rx} = 2$ dB) and combining (15-5) and (15-6), you have

$$L\_\text{criterion}\_3 = P_{\text{Tx\_ant}} - [(\text{IMP}_3 + 2 \times \text{TOI})/3 + 2]. \quad (15\text{-}7)$$

The third order intercept (TOI) for the UMTS FDD is estimated to be −22 dBm based on the specifications in the standards [1]. Using (15-7) and assuming a PHS base station transmitted power of 27 dBm, the isolation required to meet Criterion 3 is

$$L\_\text{criterion}\_3 = 27 - [(-105.2 - 10 - 2 \times 22)/3 + 2] = 78.1 \text{ dB}.$$
$$(15\text{-}8)$$

15.3.1.4 Isolation Requirement for Avoiding Carrier Overload To calculate the isolation that meets Criterion 4, you need to know the total filter rejection provided for the radio frequency, intermediate frequency, and base band at the UMTS base station receiver. The required isolation is given by

$$L\_\text{criterion}\_4 = P_{\text{total\_Tx\_ant}} - P_{\text{total\_aff\_ant}}, \quad (15\text{-}9)$$

where $P_{\text{total\_Tx\_ant}}$ is the total interfering base station carrier power at the base station antenna port, and $P_{\text{total\_aff\_ant}}$ is the acceptable interference level at the antenna port of the affected station.

Blocking requirements in the UMTS standards do not specify tests for narrowband rejection in band I (that is, a 2100 MHz band). It is assumed that the narrow-band blocking requirement (at 2.7 MHz offset of the UMTS center frequency) specified for band II (i.e., 1900 MHz band) is met by the UMTS FDD equipment in consideration. The UMTS base station receiver filter should provide a 60-dB attenuation at the frequency offset between the UMTS FDD and PHS. Using (15-9) and assuming a 60-dB attenuation for the UMTS base station receiver filter ($P_{\text{total\_aff\_ant}}$ = acceptable interference + 60), you obtain

$$L\_\text{criterion}\_4 = 27 - (-105.2 - 10 + 60) = 82.2 \text{ dB}. \quad (15\text{-}10)$$

15.3.1.5 Antenna Isolation Calculations Based on the calculations shown in the previous subsections, the isolation required to meet all four interference criteria is 82.2 dB, as dictated by Criterion 4. However, it should be noted that this value is calculated based on the receiver rejection specification derived from UMTS standards [1]. If the actual UMTS FDD equipment characteristics are better than those listed in the standards, the base station isolation requirements may improve.

The isolation can be achieved by using spatial separation between the interfering system transmitting antenna and the affected system receiving antenna, and/or by adding filters at the transmitting and the receiving ends. The spatial antenna separation could be horizontal, vertical, or slant (horizontal plus vertical).

For horizontal separation, the isolation is given by

$$L_{h\_isolation} = 22 + 20 \log (D/\lambda) - G_{Tx} - G_{Rx}, \qquad (15\text{-}11)$$

where D is the antenna separation in meters, λ is the wavelength in meters, and G_{Tx} and G_{Rx} are the effective antenna gain (including the cable loss) of the transmitting and the receiving antennas, respectively. It should be noted that the antenna gain here refers to the antenna gain in the direction along the line connecting the center of the two antennas. When the two antennas are pointing to each other, it is the antenna's main lobe gain minus the cable loss. When the two antenna's are pointing in the same direction, the sum of the effective antenna gains normally is less than 0 dB for most commercial base station antennas.

For vertical separation, the isolation is given by

$$L_{v\_isolation} = 28 + 40 \log (D/\lambda). \qquad (15\text{-}12)$$

And, for slant separation, the isolation is given by

$$L_{s\_isolation} = (L_{v\_isolation} - L_{h\_isolation})(\varphi/90°) + L_{h\_isolation}, \qquad (15\text{-}13)$$

where φ is the angle between the horizontal line and the line connecting the transmitting and receiving antenna centers.

Consider this example: Assume the isolation requirement is 82.2 dB and the sum of the effective antenna gains is zero: if the isolation is to be achieved by using horizontal separation, the required separation calculated from (15-11) is 1023λ. If it is to be achieved by vertical separation, then the required separation calculated from (15-12) is 23λ. For PHS carriers at 1920 MHz, the wavelength is about 0.16 meters. As such, the antenna separation required is about 163 meters for horizontal separation and 3.7 meters for vertical separation.

15.3.2 Simulation of the Interferences Between UMTS and PHS Systems

The analytical approach discussed in Subsection 15.3.1 is based on worst-case scenario analysis. Its advantage is that it can easily be done without complicated calculations and gives the results that will guarantee that there will be no interference if the results are used. However, the analytical results are often too pessimistic and may not reflect the real situations. For instance, the 163-meter horizontal antenna separation is impossible to achieve if the PHS and UMTS base stations are collocated. In real engineering practice, to reflect the actual situations, simulations are often adopted to produce more realistic results.

In simulations, snapshots are taken to explore the real picture of the interference at independent moments. For each snapshot, the scenario is established by using the random variables (such as mobile locations) that characterize the interference phenomenon. With a large number of snapshots, the statistics of the desired parameters can be accurately obtained. From the simulations, you can obtain

1. The probability that the affected receiver receives a C/I that meets the requirements
2. The probability distribution of the mobile station transmitted power
3. The effect of system planning techniques
4. The effect of the performance of transmitter and receiver
5. System capacity performance
6. System soft handover performance

In most cases, system capacity performance is what people are looking for in a simulation.

The simulation assumptions and methodology are specified in 3GPP TR25.942 [5] and CWTS document [6]. Simulation results indicate that the UMTS system capacity loss due to a PHS mobile (PS) interfering with the UMTS base station is minimal, while the UMTS system capacity loss due to the PHS base station (CS) interfering with the UMTS base station is relatively large. The simulation results are in consistency with the analytical results.

15.4 Interferences Between UMTS and GSM systems

Intersystem interference between UMTS and GSM systems is most likely to happen between UMTS2100 FDD and GSM1800. The main concern is that the GSM1800 base station transmitter interferes with the UMTS base station receiver. Due to the large carrier frequency separation, other interference scenarios are less likely and will not be addressed here.

Assume GSM1800 is operating in the frequency range of 1840 to 1850 MHz for the downlink and UMTS is operating in the 1920 to 1935 MHz frequency band for the uplink. The transmitting and the receiving characteristics of GSM1800 and UMTS FDD are described in [7] and [1], respectively. The isolation criteria described in Subsection 15.3.1.1 also apply to the intersystem interference between UMTS and GSM1800 systems.

15.4.1 Isolation Requirement for Spurious Emissions

Again, assuming a noise figure of 3 dB for a UMTS FDD base station, the UMTS FDD base station receiver noise floor is −105.2 dBm. Based on Criterion 1, the spurious emission power that the UMTS base station receives from the interfering GSM1800 base station should be 10 dB below the UMTS FDD base station receiver noise floor. According to 3GPP specifications [7], the GSM base station transmitter must provide at least −30dBm/3MHz attenuation of the emissions at the frequency offset between the GSM1800 and UMTS FDD bands. Specifications also specify that for coexistence with 3G, the requirement must be changed to −96 dBm/100kHz. Therefore, for coexistence of GSM and UMTS, the isolation required to meet Criterion 1 is

$$L\_criterion\_1 = -96 - (-105.2 - 10) + 10 \log (3.84/0.1) = 35 \text{ dB}.$$
(15-14)

15.4.2 Isolation Requirement for Third-Order Inter-Modulation

As stated in Subsection 15.3.1.3, the TOI of the UMTS FDD base station receiver is about −22 dBm. Assume that the GSM1800 carriers are suppressed by 60 dB in the UMTS receiver front-end filter, the isolation required for meeting Criterion 3 is

$$L\_criterion\_3 = (43 - 60) - [(-105.2 - 10 - 2 \times 22)/3 + 2] = 35.4 \text{ dB}.$$
(5-15)

15.4.3 Isolation Requirement for Carrier Overload

Blocking requirements in the UMTS FDD standards indicate that for collocation with a GSM1800 base station, a UMTS FDD base station receiver (including RF, IF, and base band filters) should provide at least 114 dB of rejection in the GSM1800 base station transmitting band. Therefore, the isolation requirement for avoiding receiver overload is

$$L\_criterion\_4 = (43 - 114) - (-105.2 - 10) = 44.2 \text{ dB}. \quad (15\text{-}16)$$

15.4.4 Overall Isolation Requirement

The calculations in the previous subsections indicate that the isolation required to meet all four interference mitigation criteria is 86.3 dB, dictated by Criterion 1, if the GSM base station emission mask complies with the general GSM spurious emission limits. For coexistence with 3G systems, the GSM base station emission mask must meet the more stringent GSM spurious emission limits. In this case, the isolation required for meeting all four interference mitigation criteria becomes 44.2 dB, as dictated by Criterion 4.

15.5 Summary

For coexisting UMST2100 and CDMA2000 systems operating in adjacent bands, simulation results indicate that for collocation scenarios, the mutual interference impact on the UMTS and CDMA2000 downlink/uplink is negligible. Therefore, it is recommended that the adjacent-band system antennas be collocated to minimize mutual interference between base stations and mobile stations, and that sufficient base station antenna isolation be maintained at the same time to avoid base station to base station interference. For the worst-case non-collocation scenario (UMTS and CDMA2000 base stations are sitting at the each other's cell edge), simulation results indicate that mutual interference impact on the downlink is not an issue, but the impact on the uplink is not negligible for either system.

For coexisting UMTS and PHS systems, it is found that the UMTS system capacity loss due to a PHS mobile station (PS) interfering with the UMTS base station is minimal, while the UMTS system capacity loss due to a PHS base station (CS) interfering with the UMTS base station is relatively large.

For coexisting UMTS and GSM1800 systems, the main concern is that the GSM1800 base station transmitter will interfere with the UMTS base station receiver.

This chapter has provided the antenna isolation guidelines between UMTS and GSM1800 base stations, as well as the isolation guidelines between UMTS and PHS base stations. The estimated antenna isolation is decided by the UMTS base station receiver technical specifications and receiver filter rejection. If the actual UMTS base station receiver parameters are given, the recommended antenna isolation can be calculated accordingly. There are a variety of ways to achieve a sufficient intersystem isolation. Typical approaches include spatially separating the antennas of the coexisting base stations and employing filters at the interfering transmitters and/or affected receivers.

References

[1] 3GPP TS 25.104, v6.3.0, "Base Station (BASE STATION) radio transmission and reception (FDD)," (Release 6).

[2] 3GPP TS 25.101, v3.16.0, "User Equipment (UE) radio transmission and reception (FDD)," (Release 1999).

[3] Personal Handy Phone System Standard, RCR-STD28, Research and Development for Radio System, 1993.

[4] Handbook of CDMA system design, engineering, and optimization, edited by Kyoung Il Kim, Prentice Hall PTR, Upper Saddle River, NJ.

[5] 3GPP TR25.942, v6.3.0, "Radio Frequency (RF) system scenarios," (Release 6).

[6] CWTS PHS-cdma2000 Ad-hoc Group, "Analysis and simulation of intersystem interference between PHS and WCDMA systems."

[7] 3GPP TS 05.05, v8.17.0, "Technical Specification Group GSM/EDGE Radio Access Networks; Radio Transmission and Reception," (Release 1999).

Chapter 16

Comparison of WCDMA and CDMA2000

WCDMA and CDMA2000 technologies are the two most matured 3G technologies in the world. Wireless networks using these two technologies have been deployed in various markets around the globe. This chapter presents a brief comparison of WCDMA and CDMA2000. Similarities and differences between these two systems are addressed.

16.1 Similarities Between WCDMA and CDMA2000

WCDMA and CDMA2000 are similar in many aspects, including physical layer concepts, physical layer procedures, power control, function-based physical channels, terminology and so forth.

16.1.1 Physical Layer Concepts

Both WCDMA and CDMA2000 use the direct sequence CDMA technique. They use similar modulation and channel coding techniques on both forward and reverse links. In the forward link, both systems use the QPSK modulation technique. Both systems use convolutional and turbo coding techniques for coding. Both systems use soft, softer, hard, and inter-system handover procedures. Both systems use rake receivers for receiving.

16.1.2 Physical Layer Procedures

CDMA2000 and WCDMA have similar paging procedures. The quick paging channel in CDMA2000 is equivalent to PICH in WCDMA. Both channels are used to increase the standby time of the MS/UE. Access handoff in CDMA2000 is equivalent to cell reselection in WCDMA.

16.1.3 Channelization and Spreading Concepts

Both systems use orthogonal-code-based channelization on downlink for user separation. In CDMA2000, up to 128 Walsh codes are used on downlink in 1.25 MHz. In WCDMA, up to 512 OVSF codes are used on downlink in 5 MHz. For scrambling, pseudo random (PN) sequences are used in CDMA2000, while Gold codes are used in WCDMA.

16.1.4 Power Control

Both systems use fast power control schemes to ensure optimum power usage. In WCDMA, the DPCCH carries power control bits on both uplink and downlink. In CDMA 2000, the R-PICH carries power control bits on uplink and they are carried by FCH/DCCH/CPCCH on downlink.

16.1.5 Physical Channels with Similar Functions

Some WCDMA and CDMA2000 physical channels have similar functions. These physical channels are summarized in Table 16.1.

16.1.6 Different Terminologies for Similar Functions

In many places, CDMA2000 and WCDMA use different terminologies for similar functions, and use different names for the same concepts. A comparison of these terminologies is shown in Table 16.2.

TABLE 16.1 Physical Channels with Similar Functions

	WCDMA	CDMA2000
Downlink	Common pilot channel (CPICH)	Forward pilot channel (F-PICH)
	Paging indication channel (PICH)	Forward quick paging channel (F-QPCH)
	Synchronization channel (SCH)	Forward synchronization channel (F-SYNCH)
	Dedicated physical channel (DPCH)	Forward fundamental channel (F-FCH)
	Physical downlink shared channel (PDSCH)	Forward supplemental channel (F-SCH)
	Physical common packet channel (PCPCH)	Reverse enhanced access channel (R-EACH)
Uplink	Dedicated physical control channel (DPCCH)	Reverse dedicated control channel (R-DCCH)
	Dedicated physical data channel (DPDCH)	Reverse fundamental and supplemental channels (R-FCH and R-SCH)

TABLE 16.2 Terminology Comparison

WCDMA	CDMA2000
Node B	Base transceiver subsystem (BTS)
User equipment (UE)	Mobile station (MS)
UMTS terrestrial RAN (UTRAN)	BTS-BSC
Radio network controller (RNC)	Base station controller (BSC)
Scrambling codes	PN codes
OVSF codes	Walsh codes
Soft handover	Soft handoff
Softer handover	Softer handoff
Non-access stratum (NAS)	Network of MSC/VLR/HLR
Access stratum (AS)	Interface between MS-BSC-CN
Iu interface	A interface
Iub interface	Abis interface
Iur interface	BSC-BSC interface
Uu interface	Air interface

16.2. Differences Between WCDMA and CDMA2000

Although WCDMA and CDMA2000 have many similarities, they are different in terms of network synchronization, RF characteristics, channel structure, overhead, paging channel operation, interfrequency handoff, and intersystem handoff.

16.2.1 Network Synchronization

WCDMA uses an asynchronous mode of operation for FDD and a synchronous mode of operation for TDD. For FDD, each Node B has an independent time reference and does not depend on GPS satellites for network synchronization.

CDMA2000 uses a synchronous mode of operation and uses a common time reference for network synchronization. It depends on GPS satellites for network synchronization.

16.2.2 RF Characteristics

WCDMA was developed to work in the IMT-2000 band. A WCDMA carrier occupies a bandwidth of 5 MHz with a chip rate of 3.84 Mcps. Its spectral efficiency is 0.817 chips/sec/Hz.

CDMA 2000 was developed to work in NMT, cellular, PCS, and IMT-2000 frequency bands. A CDMA2000 carrier occupies a bandwidth of 1.23 MHz for the cellular frequency band and 1.25 MHz for the PCS frequency band. The chip rate for CDMA2000 is 1.2288 Mcps. Its spectral efficiency is 0.999 chips/sec/Hz.

16.2.3 Channel Structure

WCDMA has fewer physical channels but a more complex channel structure than CDMA2000. Logical channels are mapped to transport channels and then to physical channels. The UE acquires the synchronization channel first and then the pilot channel.

For CDMA2000, logical channels are directly mapped to physical channels. The mobile station acquires the pilot channel first.

16.2.4 Overhead

Compared with CDMA2000, WCDMA uses relatively more overhead bits. A comparison of the overhead is listed in Table 16.3.

16.2.5 Paging Operation

In WCDMA, the paging indication channel (PICH) is always associated with the SCCPCH. UE wakes up during the paging occasions to monitor the paging indicator on the PICH.

In CDMA 2000, a mobile wakes up to receive QPCH indicator bits and goes back to sleep and processes the information in the background. Neighbor cell search for idle handoff is also processed in the background. Therefore, the standby time of an MS is longer than that of UE.

16.2.6 Inter-Frequency and Inter-RAT Hard Handover

WCDMA uses compressed mode operations for both inter-frequency and inter-RAT hard handover if the UE has only one receiver. CDMA2000 does not use compressed mode operation. Instead, it uses candidate frequency search messages and universal and general handoff direction messages for hard handoffs.

TABLE 16.3 Comparison of WCDMA and CDMA2000 Overhead

WCDMA	CDMA2000
Control channel (DPCCH) bits always exist regardless of data channel activity	In-band signaling (signaling information is part of the voice or data stream)
AMR vocoder rate information bits are sent as overhead bits	Uses blind rate detection, saving overhead bits
Power control bits are sent as overhead bits at 1500 times per second	Power control bits are sent by puncturing information bits at 800 bps
Signaling messages are generally long	Signaling messages are relatively shorter

16.3 Summary

This chapter has presented a brief comparison of WCDMA and CDMA2000 from purely technical viewpoints. Both similarities and differences between them have been discussed. Similarities include physical layer concepts, physical layer procedures, power control, and function-based physical channels and terminologies. Differences lie in network synchronization, channel structure, overhead, paging channel operation, RF characteristics, and interfrequency and inter-system hard handovers.

Index

1x-EVDO, 254, 255
3GPP, 35
16-QAM, 248

A

absolute radio frequency
    channel number, 7, 8
abstract syntax notation one, 39
access class, 49, 105
access failure rate, 313
access service class, 105
access stratum, 8, 11–15
acknowledgement (ACK), 85
    positive, 85, 105, 252
    negative, 85, 105, 150, 252
acquisition indicator, 150, 151, 228
acquisition indicator channel (AICH),
    17, 150, 158
active set, 188
    size of, 188
    update of, 67, 75
active set update message, 69, 196
active set update procedure,
    69, 188, 192
adaptive modulation, 247, 248, 267
adaptive modulation and coding (AMC),
    247, 248, 267
adaptive multi-rate, 1, 22
additive white Gaussian noise (AWGN),
    261, 274
adjacent channel, 10, 292, 329–333
adjacent channel interference power ratio
    (ACIR), 329, 332
adjacent channel leakage power ratio
    (ACLR), 329, 330–332
adjacent channel selectivity (ACS),
    329, 331, 332
admission control, 5
Alliance for Telecommunications Industry
    Solution (ATIS), 35
antenna audit, 303
antenna isolation, 319, 336, 339
Association of Radio Industries and
    Business Solution (ARIB), 35

authentication center (AuC), 5
automatic retransmission request
    (ARQ), 251

B

band-pass filters, 325
bandwidth adjustment factor (BWAF),
    337, 338
base station antenna:
    down tilt of, 292
    front-to-back ratio of, 293, 296
    gain of, 291
    horizontal beam width of, 291, 296
    null fill of, 292
    side lobe suppression of, 292
    vertical beam width of, 291, 296
base station controller (BSC), 346
base station desensitization, 316–318, 327
base station identification code (BSIC),
    183, 206
bearer services, 1
bit error rate (BER), 172, 231
bit rate, 272
block error rate (BLER), 170, 310
blocking rate, 294
    access, 313
    resource allocation, 313
body loss, 277, 278
broadcast channel (BCH), 17
broadcast control channel (BCCH), 17, 19
broadcast/multicast control, 177

C

CAIT, 305, 306–308
call control, 2, 12, 55
call origination, 305, 308, 311
call origination success rate, 311, 320
call termination, 305, 308, 311
call termination success rate, 311
capacity, 271
    cell, 271, 272
    hard, 271
    pole, 274, 275
    soft, 272

352 Index

carrier frequency, 7
carrier overload, 339, 342
cell breathing, 301
cell broadcast service (CBS), 50
cell loading, 200–202, 275–277
cell radius, 272, 275, 304
cell reselection, 177
   intra-frequency, 177, 183, 184
   inter-frequency, 177, 183, 184
   inter-RAT, 177, 183, 184
   ranking process of, 181
cell throughput, 313
cell update, 27, 29
   procedure of, 68
cellular frequency band, 348
change control, 51
channel activity factor, 274, 275
channel allocation, 5
channel bandwidth, 248, 275
channel coding, 127
   convolutional, 127, 128
   turbo, 127–129
channel mapping, 18–23
channel quality indicator (CQI), 248, 251
   feedback cycle of, 264
chase combining, 252, 256
China Communications Standards Association (CCSA), 35
chip rate, 114, 138, 141
ciphering, 63
   MAC layer, 103
   RLC layer, 89
circuit-switched domain, 3, 4, 5, 13
cluster, 303–309, 313
cluster optimization, 306–309, 313
code allocation, 253, 262, 267
coded composite transport channel (CCTrCh), 78, 111, 121
common control channel (CCCH), 17
common control physical channel, 139
   primary, 139
   secondary, 139, 140
common packet channel (CPCH), 18
common pilot channel (CPICH), 17, 148, 158
common traffic channel (CTCH), 98
compressed mode, 206
connected mode, 25
connection frame number (CFN), 32, 64, 103
connection management, 3, 12, 58
Consultative Committee on International Telegraphy and Telephony (CCITT), 30
control plane, 13, 15, 79
   circuit-switched, 13, 14, 32
   packet-switched, 13, 14, 32
core network, 2, 4, 11

COST 231-Hata model, 285
coverage area:
   cell, 26, 139, 277
   donor cell, 320, 321, 323
   repeater, 318, 320–322
coverage area shrinkage, 323
coverage probability, 278–280, 287
   area, 279
   edge, 279, 280
CPICH Ec/Io, 168
CPICH received signal code power, 162, 169
cumulative distribution function, 280
cyclic redundancy check (CRC), 16, 121

D

dBd, 291, 292
dBi, 278, 284, 291, 292
dedicated channel (DCH), 17
dedicated control channel (DCCH), 17, 218
dedicated physical channel (DPCH), 25–27, 114
   downlink, 154
   uplink, 156
dedicated physical control channel (DPCCH), 17
   downlink, 123, 134, 139
   uplink, 139, 156
dedicated physical data channel (DPDCH), 17
   downlink, 134, 139
   uplink, 139, 156
dedicated traffic channel (DTCH), 17
detected set, 189
dimensioning tool, 287, 294–296
discontinuous reception (DRX), 163
discontinuous transmission (DTX), 16, 121, 129
donor cell, 315–324, 327
downlink shared channel (DSCH), 22
DPCH offset, 155, 156, 169
drive-test-based optimization, 303, 310
dropped call rate, 310, 311
DRX cycle coefficient, 48, 57
DTX insertion, 16, 78
   first, 130
   second, 132
dual polarization, 293
dynamic power allocation, 254
dynamic resource allocation control (DRAC), 51

E

EDGE, 2, 11, 36–38
effective base station receiver sensitivity, 326
effective noise figure, 283, 316, 325

Index

EIRP, 284, 326
enhanced full rate (EFR), 1
Erlang, 272, 294
European Telecommunications Standards Institute (ETSI), 35, 37
expiration timer, 43–45, 51
extension bit, 84, 86, 87

F

FACH measurement occasion, 52, 178
fade margin, 278–280, 326
fast fade margin, 272, 280–284, 326
fast power control, 2, 215, 248
feedback indicator bits, 22, 138
feeder cable loss, 277, 278, 326
fiber link repeater, 318, 319, 322
filter rejection, 339, 343
final assembly code, 31
forward access channel (FACH), 17
forward error correction, 115
frame error, 320
frame protocol, 263
frame synchronization, 164, 174, 290
frequency division duplex (FDD), 2, 111
Friendly viewer, 310

G

Global Mobile Suppliers Association (GSA), 35
global positioning system (GPS), 54, 347
global systems for mobile communications (GSM), 1–4
GMSC, 4
Gold code, 118, 119
GPRS, 2–4, 11–13
GPRS mobility management (GMM), 9, 12, 13
GPRS support node, 4
  gateway, 4
  serving, 4

H

handover, 185
  algorithm of, 188
  successful rate of, 313
  hard, 185, 188, 348
  inter-frequency, 198, 213
  inter-RAT, 204
  soft, 185, 191, 192
  softer, 185–192
hashing collision, 164
Hata model, 285
header extension, 86, 87
high-speed dedicated physical control channel (HS-DPCCH), 18, 156, 256
high-speed downlink packet access (HSDPA), 247–252

high-speed downlink shared channel (HS-DSCH), 18, 95, 256
high-speed physical downlink shared channel (HS-PDSCH), 18, 256
high-speed shared control channel (HS-SCCH), 18, 23, 256
home location register (HLR), 5, 29
horizontal separation, 319, 339, 340
hybrid automatic repeat request (HARQ), 247, 252, 263
hyper frame number indicator (HFNI), 88
hysteresis, 49, 180, 204

I

ICT Standards Advisory Council Canada (ISACC), 35
idle mode, 24
IMT-2000, 347
incremental redundancy, 253
information rate, 112, 272
initial acquisition, 159
  procedure of, 159
in-sequence delivery, 16, 79, 86
integrity protection, 63
international mobile equipment identity (IMEI), 31
international mobile subscriber identity (IMSI), 29
interface:
  Iu, 4
  Iub, 5
  Iur, 5
  Uu, 4
interference, 329, 331–337
interference margin, 276
interference-to-carrier ratio, 338
inter-frequency measurements, 71, 183, 187, 198
interleaving, 16, 121
  first, 131
  second, 132, 133
inter-modulation, 321, 336
  third order, 338, 342
International Telecommunications Union (ITU), 30, 37
inter-RAT measurements, 52, 71, 204
intersystem interferences, 329
  between UMTS and CDMA2000, 332
  between UMTS and GSM, 341
  between UMTS and PHS, 335
isolation requirement, 325, 337–342

L

Lee model, 286
line-of-sight, 266, 318
link budget, 282, 326
loaded cluster optimization, 306

loaded throughput, 312
location area:
  code of, 48
  identity of, 30
  update of, 13
log files, 307, 308
logical channels, 17
low chip rate, 2

M

MAC architecture, 95
MAC header, 97
  for dedicated logical channels, 98
  for common transport channels, 100
MAC-I, 64
MAC RACH procedure, 104
market representation partners, 35, 36
master information block (MIB), 47
matched filter, 136
maximum C/I, 251
measurement control message, 71
medium access control (MAC), 77
macro diversity, 5
micro diversity, 5
mobile country code (MCC), 29
mobile network code (MNC), 29
mobile subscriber identification number (MSIN), 30
mobile switching center (MSC), 4
mobility management, 3, 11–13
modulation, 136
monitored set, 188
multi-code, 21, 155, 248
multiple input multiple output (MIMO), 247

N

narrow-band interference, 320, 323
NBAP, 217
near-far problem, 301
neighbor list, 188, 189
  incomplete, 302
network synchronization, 347
nordic mobile telephone (NMT), 347
node B, 5
noise figure, 274
noise injection margin (NIM), 317
noise rise, 276
non-access stratum (NAS), 12–15

O

observed time difference:
  SFN—SCN, 169
  SFN—SFN, 169
Okumura model, 283
optical transceiver, 319

orthogonal variable spreading factor (OVSF), 5, 112
OVSF codes, 112–119

P

packet data convergence protocol (PDCP), 15, 77
packet-switched domain, 4
packet temporary mobile subscriber identity (P-TMSI), 6
paging, 9, 55
  indicator of, 57
  occasion of, 57
  procedure of, 32, 56
  record of, 55
  response of, 56
  type 1 of, 55
  type 2 of, 58
paging channel (PCH), 17, 48
paging control channel (PCCH), 17
paging indicator channel (PICH), 17, 149
path loss, 278–280, 282–285
penetration loss, 282
periodical reporting criteria, 72
persistency check, 104–107
personal communication systems (PCS), 285
personal handy-phone system (PHS), 329, 335
physical channels, 17
  timing of, 158
physical common packet channel (PCPCH), 18
physical downlink shared channel (PDSCH), 23, 346
physical random access channel (PRACH), 17, 151
physical random access procedure, 160–163
pilot discrimination, 319
pilot pollution, 301
point-to-point protocol (PPP), 15
polling bit, 86, 87
positive feedback, 319, 321
post processing tool, 305, 307
power control, 215
  closed loop, 217
  inner loop, 219, 232
  outer loop, 218, 230
  open loop, 216, 229
power ramp step, 160, 162
power rating, 291, 296
power weighting, 16, 111
PRACH preamble, 20, 151
  initial power of, 162, 174, 239
  length of, 158
  ramping cycle of, 104–108

Index

processing gain, 273, 275, 295
propagation environment, 272, 273, 295
proportional fair packet scheduling, 251
protocol data unit (PDU), 77
  length of, 81
  size of, 81, 100
  size index of, 263
protocol extensions, 40
  critical, 40
  non-critical, 40
protocol states:
  Cell_DCH, 25
  Cell_FACH, 26
  Cell_PCH, 28
  URA_PCH, 27
pseudorandom noise, 118, 172
public data supporting network (PDSN), 11
public land mobile network (PLMN), 30, 179
public switched telephone network (PSTN), 11

Q
QPSK, 136, 248–250
quality of service (QoS), 3, 273

R
R-criteria, 180
radio access network, 2, 4, 5
radio bearers, 15
  radio access, 80
  signaling radio, 80
  establishment of, 65
  reconfiguration of, 66
  release of, 66
radio frame equalization, 137
radio frame segmentation, 131
radio link:
  establishment of, 165
  failure of, 167
  restore of, 167
radio link control (RLC), 77
radio link repeater, 315, 318, 319
radio network controller (RNC), 5
  drift, 263
  serving, 30, 103, 224
radio network temporary identity (RNTI), 29, 30
  cell, 27, 30
  UTRAN, 27, 29–31
radio resource control (RRC), 39
radio resource management, 268
rake receiver, 301, 345
random access channel (RACH), 17, 138
rate matching, 129, 137

receiver sensitivity, 274–276, 319, 325
repeater, 315–325
  donor antenna of, 317, 319, 321–324
  donor link of, 317, 322
  forward link gain of, 321–325
  reverse link gain of, 321–323
  subscriber antenna of, 315, 319, 321
reset ACK PDU, 87, 88
reset PDU, 87, 88
reset sequence number, 88
retransmission control, 16, 251
RF optimization, 299–304, 309, 310
RLC mode, 90, 220
  acknowledged, 79, 85, 86
  transparent, 79–82
  unacknowledged, 83–86
RLC sequence number, 64, 90
RNSAP, 225, 262
root-raised cosine filter, 136
round robin, 250
round trip delay, 312
round trip time, 172
routing area, 25
  code of, 48
  identity of, 30
  update of, 25
RRC connection, 58
  release of, 63
  request of, 59
  setup of, 60
RRC sequence number, 46, 47
Rx-Tx time difference, 170

S
S-criteria, 180
scheduling information, 41–45, 48, 54
scheduling priority indicator, 263
scrambling code, 118, 119
  identification of, 160
  left alternative downlink, 211, 212
  planning of, 289–291
  primary, 119, 136
  right alternative downlink, 211, 212
  secondary, 119, 136
scrambling code group, 119, 140
  identification of, 159, 174, 290
SDU discard, 90
sector verification, 303–306, 313
segment count, 47, 48
segment index, 46, 47
service data unit (SDU), 77
session management, 3, 12
shadow fading, 278–281
short message service, 1, 3, 12
signal-to-interference ratio (SIR), 171, 215
site selection diversity transmission (SSDT), 156

slant separation, 340
slot synchronization, 159, 174
slow fading, 279
snapshots, 340, 341
soft handover gain, 281
spurious emissions, 321–323, 337, 341
standard deviation, 279–281, 311
static power allocation, 254
status PDU, 86–89, 91–93
stop and wait (SAW), 252
space time transmit diversity (STTD), 143
successive preamble power, 239
supplementary services, 3, 12
synchronization channels (SCH), 143
    primary, 143, 158, 173
    secondary, 143, 158, 173
synchronization codes, 115
    primary, 119, 120
    secondary, 119, 120
system frame number (SFN), 31, 46, 57
system information, 40
system information block (SIB), 41–55, 74

T

target channel type field (TCTF), 97
transmission control protocol (TCP), 77
Telecommunications Industries Association (TIA), 35
teleservices, 1
temporary mobile subscriber identity (TMSI), 30
TEMS, 310
terrain data, 295
thermal noise, 168, 273, 274
thermal noise density, 284
third-order intercept, 338, 339
time division duplex (TDD), 2, 111
time to trigger, 192, 196–198, 200–204
tower mounted amplifier (TMA), 315, 325–328
TPC bit pattern, 219, 220, 233
TPC commands, 219–224, 232–235, 241–243
transmit code power, 171
transmit power control (TPC), 155
transport block, 122
    set of, 122
    size of, 123
transport channels, 17
    fixed position, 130
    variable position, 130
transport format (TF), 122
    set of, 122
transport format combination, 123
    indicator of, 123, 134
    set of, 123

transport format resource indicator, 265
transmission gap, 171, 207–211, 214
transmission gap connection frame number, 208
transmission gap pattern, 208–210, 214
    length of, 208, 209
    sequence of, 208–210, 214
    sequence identifier of, 208, 210
transmission gap starting slot number, 209
transmission rate, 247, 267
transmission time interval (TTI), 78
type approval code, 31

U

UE capability, 255
UE internal measurements, 72
UE receive power, 307, 308
UE transmit power, 282
    maximum, 236
    minimum, 237
UMTS terrestrial radio access network (UTRAN), 2
universal mobile telecommunications system (UMTS), 1
universal subscriber identity module (USIM), 3
unloaded cluster optimization, 306
unloaded throughput, 310, 312, 314
uplink DPCCH power control preamble, 229
uplink inner loop timing, 233
URA update procedure, 68
user data protocol (UDP), 77
user equipment (UE), 2, 6
user plane, 13, 15, 79
    circuit-switched, 13
    packet-switched, 13, 15, 32
UTRA absolute radio frequency channel number (UARFCN), 7, 8

V

vertical separation, 319, 340
Viper, 306, 310
virtual active set, 198
visitor location register (VLR), 4, 30, 347
voice activity, 274
voice IP (VoIP), 6
voltage standing wave ratio (VSWR), 293

W

Walsh code, 112, 346, 347
wideband code division multiple access (WCDMA), 2